TO THE LAST SMOKE

STEPHEN J. PYNE

TO THE LAST
SMOKE

AN ANTHOLOGY

THE UNIVERSITY OF
ARIZONA PRESS

TUCSON

The University of Arizona Press
www.uapress.arizona.edu

© 2020 by The Arizona Board of Regents
All rights reserved. Published 2020

ISBN-13: 978-0-8165-4012-9 (paper)

Cover design by Leigh McDonald
Cover photo: *West Cinder prescribed burn, Twin Falls BLM District,*
August 2010 by Kari Greer/BLM
Interior design & typesetting by Sara Thaxton
Typeset in Adobe Caslon Pro, Avenir LT Std, and Bebas Neue

Library of Congress Cataloging-in-Publication Data
Names: Pyne, Stephen J., 1949– author.
Title: To the last smoke : an anthology / Stephen J. Pyne.
Description: Tucson : University of Arizona Press, 2020. | Includes bibliographical references and index.
Identifiers: LCCN 2019052046 | ISBN 9780816540129 (paperback)
Subjects: LCSH: Wildfires—United States—History. | Wildfires—United States—Prevention and
    control—History. | Forest fires—United States—Prevention and control—History.
Classification: LCC SD421.3 .P964 2020 | DDC 634.9/6180973—dc23
LC record available at https://lccn.loc.gov/2019052046

Printed in the United States of America
♾ This paper meets the requirements of ANSI/NISO Z39.48-1992 (Permanence of Paper).

CONTENTS

SLOPOVERS

HERE AND THERE

AUTHOR'S NOTE

WHEN I DETERMINED TO write the recent fire history of America, I conceived the project in two voices. One was the narrative voice of a play-by-play announcer. *Between Two Fires: A Fire History of Contemporary America* would relate what happened, when, where, and to and by whom. Because of its scope it pivoted around ideas and institutions, and its major characters were fires or fire seasons. It viewed the American fire scene from the perspective of a surveillance satellite.

The other voice was that of a color commentator. I called it *To the Last Smoke*, and it would poke around in the pixels and polygons of particular practices, places, and persons. My original belief was that it would assume the form of an anthology of essays and would match the narrative play-by-play in bulk. But that didn't happen. Instead the essays proliferated and began to self-organize by regions. The upshot was a nine-volume suite of regional reconnaissances.

The advantage of this approach was that I could include many more places and I was able to frame each regional volume with essays about the region's traits and how it sat within the whole. The opening essay would explain what was characteristic and unique about each region; the closing essay, how to place it within the national narrative. The internal organization of the essays followed a logic that seemed appropriate to the identity of the region at hand. The American fire scene is both a federation of agencies and a confederation of regions. *To the Last Smoke* sought to convey that reality.

But my original intuition remained. Nine short volumes still added up to some 1,800 pages, a daunting mound of text for even the most ardent fire enthusiast. The argument for a single-volume sampler was compelling. *To the Last Smoke: An Anthology* is the result. It includes essays I liked and that speak with special power for each region, but it does not include the thematic frames. Those belong with the regional volumes and were not intended to stand alone. The one exception is a new essay on California that examines the 2017 and 2018 seasons. These occurred years after I wrote my California survey, but reviewers felt that the anthology presented an opportunity to address what seems an extraordinary outburst. I agreed, and "This Time Is Different. Maybe" is the result.

The essays reflect a light edit from those that appeared originally in the separate volumes of the series. Mostly these involve minor glitches of grammar or wording. Occasionally, it means deleting passages or sentences that effectively repeat text that appeared earlier. In the series this didn't matter because some recapitulation was warranted to reset the theme, and the volumes were separated. Gathered into a single book such overlaps can be annoying. This volume, too, dispenses with a note on sources. Those brief bibliographical essays were intended to buttress the regional volumes. For individual essays the endnotes seem ample.

The series title? I began my career as a smokechaser on the North Rim of Grand Canyon in 1967. That was the last year the National Park Service hewed to the 10 a.m. policy and we rookies were enjoined to stay with every fire until "the last smoke" was out. By the time the series appears, more than 50 years will have passed since that inaugural summer. I no longer fight fire; I long ago traded in my pulaski for a pencil. But I have continued to engage it with mind and heart, and this unique survey of regional pyrogeography is my way of staying with it to the end.

Funding for the project came from the U.S. Forest Service, Department of the Interior, and Joint Fire Science Program. I'm grateful to them all for their support. The University of Arizona Press deserves praise as well as thanks for seeing the resulting texts into print. Thanks, too, to Kerry Smith, who copyedited the entire, now-10 volumes with care and good humor and spared me from my worst grammatical self.

TO THE LAST SMOKE

FLORIDA

A Tale of Two Landscapes

H IS CAREER WITH the Army Corps of Topographic Engineers took Lieutenant Joseph Christmas Ives to two of the more intractable regions of the United States. In 1857–58 he led an expedition up the Colorado River to a rumored Big Cañon. He believed the expedition would make him famous, and it did, but not least because he pronounced that his was the "first and would doubtless be the last party of whites to visit this profitless locality," what the world would come to know as the Grand Canyon. The army's concern lay with possible supply routes to the so-called Utah War. What is often forgotten is that the redoubtable lieutenant's previous assignment had taken him to the Big Prairie region that slashed across Florida north of Lake Okeechobee. It was "a comparatively unknown region" whose "natural features oppose great obstacles to the prosecution of surveys and explorations." In brief, the place flooded seasonally. In this case the army's interest lay in pursuing the Seminoles south.[1]

As transportation routes Ives's opinion of both sites was not far wrong. Their future, however, did not belong with passing armies and insurrectionists but with tourism. For the Canyon the scene was primarily geological. For the Big Prairie it was biological. Eventually the Grand Canyon became among the best-known units in the national park system. A patch of Big Prairie became the flagship unit of the Florida Park Service (FPS).

• • •

Historically, the region around the Myakka River boasted a typical Florida landscape, seasonally overflowing either with water or flame. The scene came with one enormous geographic anomaly, however, a large swathe of savanna pocked with wetlands, pine "islands," and oak-palm hammocks and seasonally washed by flame, what was known as the Big Prairie and later as the Florida dry prairie. The area seemed so barren that early surveyors skimped on their markers in the belief that no one other than migratory hunters and ranchers would find enough value to visit the place routinely, and no one was likely to homestead. Unsurprisingly, settlement was slow, as it was throughout interior Florida. A few low-grade railways supported high-grade longleaf logging and scabby naval stores; the woods were too scattered and sparse for serious milling. Staked ruts for wagon trails marked routes of travel. Mostly, cattlemen ran cows and kept the landscape low and open through annual burning. By 1930 fewer than 39,000 people occupied the region now comprising Manatee, Sarasota, and Charlotte Counties, and nearly all resided on the coast. Lightning, transients, and ranchers burned most of the land on an annual or biennial rhythm.[2]

The big change came in 1934 when the interests of the Sarasota Fish and Game Association merged with a New Deal that was busy scouting sites for Civilian Conservation Corps (CCC) camps. The land acquired to support Camp SP4 became the nucleus for Myakka River State Forest and Myakka River State Park, which in turn evolved into the flagship site for Florida's protected state lands. The enrollees promptly did what they did everywhere: they built infrastructure, planted trees, and fought fires. Since the biotic core was prairie, not forest, and the biological dynamics relied on routine burning and flooding, the CCC inaugurated an era of ecological irony because they broke both flood and flame. Among the infrastructure they bequeathed, along with roads, weirs, drainage canals, and dikes, was a policy of fire exclusion outfitted with hundreds of miles of firebreaks and hundreds of flapper-wielding enrollees. They stopped deliberate burning, and they tried to stop nature's fires as well. In this the fledgling Florida Park Service, initially overseen by the state Forest Service, embraced the collective wisdom of nature preservationists everywhere.

Outside the Myakka River State Park the land continued to burn—ranchers saw to that; and what they missed, lightning scavenged. Other conservation agencies acquired sites nearby. In 1941 some 19,200 acres were bought from rancher Fred Babcock to become the Charlotte County Preserve, which over the next 40 years grew into the 65,675-acre Babcock-Webb Wildlife Management Area (WMA) administered by the Florida Wildlife Commission. Interested mostly in fostering further hunting, the wildlife commission encouraged ranchers to burn under a regimen of grazing leases. In 1962 the site hired a biologist, who added his own reasons for burning. As money for land acquisition kicked in during the 1970s, and especially the 1980s, more parcels around the region transferred from the private to the public sector into what enthusiasts began (hopefully) referring to as the Myakka Island, the whole of the watershed linked by wildlife corridors as well as by streams. Except where converted to houses and malls, most of the new lands persisted in their old fire practices. Paradoxically, the most ostensibly protected places, the parks, were the exception, and Myakka River, which became a park in 1941, was, at enormous cost, protected from the flooding and firing that had made it flourish.[3]

The Florida Park Service divorced from the Forest Service in 1948. The split made sense since parks and forests had different purposes, but it left park management without the economies of scale—intellectual as well as material—that a larger institution offered, and it left park administration without a conduit for resource *management* rather than a simple program of strict protection. By 1969, as Florida was voting in a new constitution, the FPS readied itself for a major overhaul of its own. The reformation sent its new chief naturalist, Jim Stevenson, to Tall Timbers Research Station to see for himself what the brouhaha was about. At the time the FPS was the only public land agency in Florida that did not prescribe burn. Stevenson recalled, "We permitted no prescribed fires on state parks. Fire was just the worst possible thing that could happen, and we would risk life and limb to put out any fire that occurred."[4]

Ed Komarek gave the naturalist a field tutorial. ("I believe Ed Komarek could sell a forest fire to Smokey Bear," Stevenson later wrote.) He left convinced that the FPS had to burn. Shortly afterward Betty Komarek conducted the first prescribed fire on a state park (Falling Waters). By then 20 years had passed since the Park Service had achieved autonomy and spun off from the emerging fire program of Florida agencies. Over the next 20 years it tried to make up for lost time and sought to put fire in as vigorously as it had previously struggled to take it out. In 1989 Jim Stevenson informed a Tall Timbers fire ecology conference that over the past year the agency had burned 110,000 acres in 66 parks.[5]

The agency seemed to be holding its own. But by the time Stevenson had gone to Tall Timbers in 1969, 35 years of attempted fire exclusion had already passed. Myakka River's ledger of fire removed was, in reality, the record of a crushing fire debt.

• • •

From the vantage point of the present, the tale of two "natural areas"—Myakka River State Park and Babcock-Webb WMA—can make painful reading. The one compensation, perhaps, is the old adage that great heroes require great villains. The more menacing the challenge, the more admirable the achievement. The contrast in fire histories is stark.

The more rudely managed site, the WMA, had never abandoned burning. From the time it was gazetted, cattlemen had continued to burn under leases, and when an official biologist arrived in 1962 he tweaked the regimen but left it intact. Unlike the Florida Forest Service, the preserve did not burn primarily for fuel reduction as leverage to help control wildfire but for habitat suitable to cattle, deer, and quail.

For the first 20 years the fire regime was classic Floridian: it free-ranged as much as the cattle. The legendary Buck Mann described the routine on his ranch just south of the park. Every year he drove his Jeep over the dirt ruts that passed for a trail, and whenever he spied a patch that needed burning he struck and tossed out kitchen matches. The fire would "burn wherever the wind carried it. It would burn until it reached a wet swamp, marsh, or hammock. Sometimes the fire burned for days." There were no control lines, no burn plans, no prescriptions, no backup crews for suppression, no firing strategies to lessen smoke, no standards for burners, no radios, no Nomex clothing or hardhats. There was no accommodation for species—the cattle moved out of the way, the other creatures thrived on the postburn flush. The dry prairie revived from its seasonal fire drought. There was no particular attention to neighbors, who were all ranchers; flames blew past fence lines as easily as over gopher holes. If Buck hadn't burned his neighbors' lands, they would have burned his.[6]

For the next 20 years, as the WMA expanded, fires remained long burning, tacking and veering with storm and wind, which meant they were large. As hunters relied more on vehicles (which made roads) and as population began to swell along the coast, managers gridded the preserve into 160-to-240-acre blocks etched by plow lines, which made control easier and long-lingering smoke less annoying to neighbors and less hazardous to I-75, which cut through the site. For the next 15 years the fire rotation flickered around two to three years. For the last 15, up to 2010, it ratcheted down to one or two. Essentially the land burned as often as it could.[7]

The contrasts with the iconic Myakka River State Park are stunning. The blessings Myakka River enjoyed at its birth turned into banes. The Park Service was statutorily charged with preserving "representative portions" of "original natural Florida," which it interpreted as overthrowing the inherited rural regime. It excluded the cattle and the hunters, substituted tourists for traditional transients, and banned any burning. The CCC gave it the clout to physically attempt to replumb and replant the landscape and to fight fires. Enrollees platted the landscape with firelines. Not until Stevenson had his epiphany and the FPS saw that it could not restore cherished landscapes without restoring fire did managers again pick up the torch.

Meanwhile, the Big Prairie was mutating into the Big Palmetto Patch, and the Florida grasshopper sparrow, burrowing owl, Sherman's fox squirrel, and crested caracara were gone. A sun-loving flora was being literally overshadowed, and a fauna of spotted skunks and gopher tortoises that depended on them was becoming rare. The "original domain" was choking on an obese, rough, and thickening saw palmetto, and ecologically suffering as if from biotic diabetes. The dry prairie had become the fire analogue of a drained wetland that was rewatered only by damaging floods. So it was with fire: savage wildfires broke out, and suppression replaced nourishing ash with plow lines.

But in a place where, in the absence of cleansing burns, palmetto can shoot up to head-high thickets, shrubs can often overtop grasses and forbs in less than a dozen years, and pine islands congeal into backlot woodpiles, all capable of stoking conflagrations after as little as five years, bad fires promised to replace good ones. Bad fire was the original ecological invasive: these fires did not burn as they should, much as Norway rats did not behave as possums. Myakka learned that putting fire back in was more formidable than taking it out. After being excluded for more than three decades, probably fire acting alone was not enough to restore itself. Reinstating something like fire's old regime was bound to be tricky, often crude, and occasionally violent. The earliest efforts were all of the above.

The project began with efforts to understand what the conditions were at the time the park was established. Written accounts, original survey notes and maps, aerial photographs, interviews with old-timers, all pointed to an open savanna with few trees, most of them clustered into pine islands, and Myakka as a place that burned relentlessly and benignly. The first prescribed burns followed the formulas of the day: winter fires, backing fires, tightly leashed controlled fires held within hard lines. But fire is what its context makes it: fires in a beaten biota behave like trodden-down burns. They could barely hold against the continued pressure to biotically bury the prairie and overgrow the pines. They might maintain the existing scene but could not reverse the trends, much less restore the old. Managers would have to reverse drainage by filling in canals, fight off exotic plants that burned on different rhythms, and probably crush the rough. The Myakka fire staff realized, too, that the park no longer received fires from adjacent lands, that roads and boundaries had grown into unburned hedgerows. Fire's habitat was as fragmented as those of caracara and gopher tortoise.[8]

In 1985 the staff adopted more aggressive measures, and a feistier mix of fires. They experimented with brush cutting and roller chopping to bring some mechanical leverage against robustly rooted ligno-tubers, intrusive woods, and the ever-more-entrenched palmetto. They loosed heading fires that might bring some thermal leverage as well, hoping a blast of heat could shock the system in ways that slow-cooking backing fires could not. They kindled growing-season fires in an effort to reintroduce a more natural regime. They watched fires hop and skip across freckled landscapes. They accepted some harsh burns and scorch-killed trees as the cost of paying down the park's fire debt. Yet even as the program's tempo quickened, countervailing frictions threatened to drag it to a halt.

The 1985 season was a horror across the state. On May 16 a wildfire brought much of Flagler County to its knees and incinerated 131 homes at Palm Coast. At almost the same time, what Robert Dye called "a letter based on a second hand description of a successfully conducted burn, claimed that 'it was by the grace of God my horses were not killed and property destroyed,' was addressed to the Governor." The governor passed the buck to the Department of Natural Resources, which proclaimed a moratorium on all burning in state parks through the summer season. Later, Myakka River endured its first off-site escape fire (onto an adjacent ranch), which kindled another ban, this one specific to the park.[9]

Yet, even as it became harder to burn, the necessity for still more burning intensified. Without tame fire, feral fire would overrun the park like a pyric version of cogongrass. By now resident species were being federally listed and the Florida dry prairie itself was recognized globally as a habitat at risk. Myakka River State Park was in danger of reclaiming its iconic status but as an emblem of unintended consequences. In the short run, a careerist knew it was safer to do nothing than to do something. But Myakka River was blessed with managers like Dye and Paula Benshoff, who remained on the scene for their whole careers. They were willing to bet on the long run, and that meant getting control over the fire regime by getting fire back on the land on their terms. The park received more equipment, cut broader perimeter fire lanes, and reinvigorated its experimentation. No one was fully happy with the amount burned, but they were no longer falling behind. Myakka became an exemplar of fire management for the park system overall.[10]

By the 1990s the prairie was being slashed and burned—chopped by tractor-drawn rollers and burned with a diversity of line fires, spot fires, heading and backing fires, dormant and growing-season fires. Fire managers aimed for a two-year rotation on restored lands and continued conversion of lost lands to a condition that better approximated the "original natural" state. But nature doesn't operate on a nightly news cycle or bureaucratic budgets. That revanchist rough could not be stockpiled in caverns like inert nuclear waste: it grew, and it was ever ready to burn. It was seemingly impossible to dislodge palmetto once it sank roots and thickened into swathes. A missed fire or two might allow the rough to reclaim what exhaustive labor had restored, and a handful of years without the right kind of fire is enough to lose what decades of onerous labor had clawed back. While students of public administration might speak loosely of "institutional ecology" or some other kindred metaphor, the Florida Park Service ran on energy flows and nutrient cycles very different from those that operated on the Florida dry prairie. Probably, too, part of what was missing was the biological interactions that fire had catalyzed, and now no longer worked—a dynamic version of fragmentation. By the onset of the 21st century the park had devoted as much time to restoring fire as it had previously invested in excluding it, yet a backlog of burning still brooded over the scene.

More than nature pressed back. There were social pressures not to incinerate a favored recreational site or to immerse a booming Sarasota County with smoke. Particularly, granted the amount of migration into the region, public education required a continual program of messaging to keep the rough of public opinion at bay. As the Myakka Island concept of a regional consortium of linked lands devoted to conservation gathered momentum, it added to the stockpile of mandatory burning, but it was harder to get money for maintenance than for acquisition. No one's name went onto a plaque honoring their investment in biennial burning. No one would be memorialized for saving fire as they might be honored for saving black bears.

Nor did all the restrictions come from the outside. Bureaucracy imposed an increasing roster of restraints. Burning required more than just plans, prescriptions for flame and smoke, checklists for equipment and crew size, and

Florida Forest Service authorization. Crews had to pass training and fitness standards. The Keetch-Byram drought index established the statewide conditions for determining whether a particular site could burn or not. Individual parks had to submit to agencywide policies such that a failure in a panhandle park might affect how Myakka did business. The restraints were cumbersome, but necessary; they served as social buffer zones, the legal equivalent of perimeter fuelbreaks. Still, the program that had reinstated fire from the 1970s onward was a far cry from Buck Mann tossing kitchen matches from a Jeep as his father had from horseback. But the fire managers of the early years well into the late 1990s enjoyed a freedom of action that the next generation would never know.

The dilemma characterized the entire system, however. The need for fire exploded, while the capacity to burn did not keep pace. Over the Florida parks system as a whole, the anxiety deepened that fire managers would be able, with ever more strenuous effort, to hold the line, but little more. In 2008 the FPS burned 89,000 acres on 66 units, although critics noted that four parks accounted for 80 percent of the total and that it appeared that some places might be simply blackening acres, not burning for identifiable ecological goals. The fretters worried that, with respect to fire, they could make the monthly payment but might not be able to pay down the accumulated debt.

Still, 40 years after Jim Stevenson had made the pilgrimage to Tall Timbers and got religion, the Florida Park Service had successfully gone from a program of fire exclusion to one that burned its "fire-type" holdings on a roughly four-year cycle. A third of its units met their minimum targets. Wildfire accounted for only 2.4 percent of acres burned. The agency began to track acres burned for maintenance apart from those fired for restoration.[11]

• • •

The state parks' story is a cameo of Florida fire generally. Most land continued to burn routinely (not always wisely or well) until it fell under formal protection. The Florida Forest Service estimated that a third to a half of those lands not legally protected burned annually. Then came the great transition. The postwar boom transferred land from rural status to urban or public, and with that shift came fire exclusion or, out of the attempt at exclusion, a remaking of flame from tame to feral. In many places, the continuity of fire broke, as it did at Myakka River. In others, it persisted, though altered, as at Badcock-Webb WMA. The modern saga of Florida fire hinges on that transition.

The Florida Forest Service accepted prescribed burning at the same time legislation closed the open range. In principle it was possible to substitute forestry's hand on the torch for ranching's; but the free-range fires of the cattlemen far outstripped the more tightly controlled burns of state-sponsored forestry, which were intended not to spring-clean landscapes but to batten down fuels as an aid to controlling wildfires. By 1970 formal protection extended throughout the state. Between the closing of the range and the saturation of fire protection lay 20 years in which burning lagged, and during which, for the Florida Park Service, only wildfire was possible.

The state's astonishing land-acquisition programs quickened the move from rural landscapes to preserved ones. Again, burning lagged, as administrators waited to install infrastructure, train and equip crews, sponsor research on burn plans and prescriptions. Even when the torch passed, a step or two was lost in the handover. In the case of small units, the burning never happened or occurred only in token daubs. In others it came too late, and fire had to be restored. As with a lost species, lost fire first required a suitable habitat if it was to flourish. Nature, however, didn't wait for the bureaucratic planets to come into alignment. It sprouted fuels. A handful of years was enough to mutate the fire regime from annual surface burns that brushed and skipped over the landscape into spasms that incinerated drought-readied rough and slammed against suburban sprawl. The old regime could not wait 40 years, or 20, or even 10 for a new one to seamlessly replace it. Five or six years without fire meant that burning was no longer a casual task in an afternoon drive but a costly and potentially onerous bureaucratic exercise. The legacy of even a decade meant a debt of lost fire that might never be made up. Few places burn as often and widely as managers believe necessary.

Paradoxically, those pieces of protected nature that came later were often better disposed to fire than those that had come early. Places that had endured a fire-free phase would have to be biologically rebuilt in order for fire to do its ecological work, and it was not obvious that fire alone could remake the old setting. Flame remained, as ever, an interactive technology.

• • •

Myakka River State Park demonstrates the old adage that it is easier to tear down than to build up. The oddity is that, with Florida fire, doing nothing could be the same as tearing down. Not burning might do as much harm as clear-cutting or overhunting; reinstating fire was far trickier than replanting pines or creating nesting trees for osprey.

It was not enough to cease suppressing fires. They had to be deliberately restored, and since fire and land were entangled that meant that both fire and land would have to coevolve into whatever condition people sought. It wasn't enough to plant, slash, or uproot exotics without burning, but neither was it sufficient to dump fire on the land. It was a process in which each interplayed against the other, while people constantly relearned from each new outcome. During his conversion experience, Jim Stevenson recalled that Ed Komarek had noted that it had taken 30 years of mismanagement to trash the parks, and it would take 30 years of "proper management to restore them." That seemed a long time, even if the perspective offered a pleasing symmetry. It also underestimated the time required.[12]

How to restore fire has proved a far more vexing task than how to suppress it ever was. Yet a gamut of strategies has appeared that ranges between coaxing and coercing. A Florida advantage is that the sites mostly remain working landscapes, so there is no ideological block against manipulating them by using chemicals, chain saws, masticators, and tractors. The western parks, dominated by wilderness concerns, have fewer options. That's one distinction, east and west. The other is the mix of urgency and patience that fire managers bring to the task. In the Far West decades may pass before the land has so altered that putting fire back is complicated or ruinous. In Florida one has only a handful of years before returning fire means restoring it. Oddly, it is the westerners, however, who are keenest to make the revolution happen instantly. The southerners have a greater forbearance toward the long rhythms of history.

No theory dominates practice. Restoring fire—nature's agent of creative destruction—to nature's economy is not unlike restoring capitalism to an economy stagnated by decades of state communism. One strategy is shock therapy: dump it on the system and let the land sort out the mess. The informing belief is that fire is inevitable, the landscape can absorb the blow, and the sooner it happens the better, since it allows renewal quickly. To wait is to worsen. Especially in natural settings, the invisible hand of nature will guide a rapid recovery.

The Florida strategy favors coaxing. It takes a longer view, expressed in a more incremental approach; landscapes segue one into another. Ideally, restoration proceeds bit by bit, as the undesirable parts of the landscape are removed, and with fire as catalyst, the desired parts revive. Retaining canopy trees, for example, even if they are the "wrong" conifers, keeps fine fuels on the ground through needle cast, which allows fire to persist. Likewise, keeping old burning practices, although they are not precisely targeted, holds fire on the scene. Once broken, fire and fuel, unlike bones, do not rebond stronger than before. Throughout, the hand of management is visible.

As the Myakka story shows, coaxing does not mean dawdling. The park does not have the luxury of reassembling the landscape piece by piece in cautious sequence. It's rather an exercise in coprocessing. It embodies what Horace once expressed as *festina lente*—hasten slowly. The prepark generation burned under circumstances that seem today unspeakably lax; the generation that restored fire after three decades of abusive exclusion operated under tighter constraints, both socially and ecologically, and confronted a crushing backlog of bad ecological debts. The generation that succeeds them will face an environment more tightly coiled still, requiring additional effort simply to hold existing levels of burning. Instead of uprooting planted slash pine, filling in ditches, and loosing fires, they must cope with such ecological game changers as Brazilian pepper trees, West Indian marshgrass, Cuban tree frogs, tighter regulations on particulates in smoke, closer suburbs, landscapes broken up and spliced together without regard to fire. They will have to burn both more and less, both faster and hotter, yet all with finer precision. Using flame they will have to beat back the unwanted rough, while, using ideas, they will have to fight off the ceaseless challenges to burning that crowd the margins like invasive weeds. Throughout, they will need the will to burn, which may ultimately depend on what William James once termed "the will to believe."

Today the distinction between maintaining and restoring still has meaning. In the future the two concepts may merge. It may be that Florida's experience with fire restored is an artifact of a particular historic time in which a remarkable transition in land use demanded a set of fire practices different from those that had previously kept fire on the land. That experience seemed, at the time, a unique moment. But as pressures mount, it may become the new norm. It may be that the only way to maintain fire is to continually restore it.

One Foot in the Black

N THE AGE BEFORE Standard Orders and rosters of Watch Outs, safety training relied on a few bits of folk admonition. To raw crews: "Stay with your foreman. He has been to many fires and is still alive." Against a crown fire: "Pray for rain and run like hell." And for fast-paced hotline: "Keep one foot in the black," which recognized that the safest place to be if a fire roused to fury was in its already-burned patches.

It's a wonderful phrase, though, and with a bit of metaphoric tweaking it might stand equally for the admonition to a landowner in a fire-prone place that working the land means working with or against a line of fire and that controlled burning is preferable to wild burning. Land management is not planning: it is doing. If you are serious about running a place that holds or needs fire, you have to work with one foot always in the black.

That line—between the black and the green—might equally divide nature organizations. There are the greens that study and plan and exhort, and there are those that must also work the land, that straddle an advancing line of flame and must keep one foot in the black. In fact, that's not a bad way to characterize the Nature Conservancy (TNC).[1]

• • •

The institution evolved out of a desire by a group of ecologists for "direct action" to protect natural areas. In 1951 the ambition incorporated as a nonprofit organization. The Nature Conservancy created a network of chapters, beginning in New York, that eventually spread throughout the United States. Over the next decade it established its trademark strategies: land acquisition, conservation easements, partnerships with public agencies, and a reliance on local volunteers. What it bought it had to manage, although in some cases only until the land could be sold to a public agency. Most of its initial purchases were small—60 acres of the Mianus River Gorge, 6 acres of the Bantam River marsh. Not until 1965, with funds from the Ford Foundation, did TNC have a full-time president.

What drove the Conservancy into fire was its acquisition of prairie in the upper Midwest. The only way to sustain prairie was to burn it. In 1962 it conducted its first burn on the Helen Allison Savanna Preserve outside Minneapolis. It was done, like the rest of its land stewardship, by local volunteers. Four years later, however, Katharine Ordway, heir to the $3 million company fortune, began donating money to buy tallgrass prairie, first in Minnesota, and then throughout the Great Plains, on what seemed an industrial scale. As it burned more, the Conservancy, which had no internal fire culture, came into contact with places that did. In Wisconsin it met an academic clique interested in restoration and accustomed to work on small plots of oak prairie savanna. As it moved into the Flint Hills of the central plains it encountered a stubborn culture of burning that dated back to settlement, when farmers found the hills too rocky for plows and abandoned it to ranchers, who found they had to burn to boost forage and protect against wildfires. The Nature Conservancy became, by default, a fire agency. By 1985 its involvement had advanced sufficiently for Mark Heitlinger to write a general manual for fire operations.[2]

What made TNC different from most environmental advocacy groups was that it owned its lands. What made it different as a fire agency was that it emphasized burning for ecological goals. What put the program on steroids, however, was land acquisition in the Southeast, particularly Florida. No landscape burning is simple or inherently safe (people even lose campfires all the time). But tallgrass prairie is relatively homogeneous; burning the same sites over and over builds experience that leads to competence and confidence; and most larger prairie sites abut other lands of similar composition that can accommodate some spillover flames. Little of that applies to Florida. The Conservancy had to burn on a scale and under conditions that exceeded what it could expect from local volunteers, and Florida marinated the organization in a regional culture of prescribed fire very different from that of the Midwest. The Southeast office decided it needed a dedicated fire staffer. In 1986 it hired Ron Myers, a

University of Florida PhD with fire experience who was then working at the Archbold Biological Station.

If its Florida experience was the pivot point for TNC as a national fire organization, and eventually an international one as well, Ron Myers was the bearing on which that transition turned.

• • •

His career is in some way a cameo of the national story. He grew up in one fire culture and converted into another. A Californian, he spent a summer on a timber stand improvement crew on the Shasta-Trinity National Forest in 1965; he got on a few fires and acquired a yearning for more. The next year he enrolled in forestry at the University of Montana and worked that summer as a fire control aid at Glacier National Park. He was at Glacier in 1967 when the Flathead and Glacier Wall fires broke out and spent a month on firelines. The effort to suppress those fires, what appalled many critics as mechanized brutality, helped push the National Park Service into its policy of fire by prescription. The next year was empty of fires. In 1969 he joined the Missoula smokejumpers for a season full of them. He learned fire by fighting it in the settings that, since 1910, had most shaped national policy.

Then he commenced a long reeducation. In 1970 he enrolled in the Peace Corps and for two years taught at the National Forestry School at Siquatepeque, Honduras. That brought him into contact with a living culture of rural burning in many ways alien equally to smokejumping and to the formal learning he promulgated through lectures. He was astounded to find pine forests that burned routinely, deliberately fired to good effect, though sometimes to bad. When his tour ended, he signed on as a seasonal at Everglades National Park, then making the great transition to full-spectrum fire management under Larry Bancroft. He yearned to return to Latin America and decided he would need an advanced degree to do it. Education was a means to get into the field.

He enrolled at the University of Florida. Its forestry program targeted pine plantations and had little passion for international work. He took up botany, studied under Jack Ewel, did field research at what became Big Cypress National Preserve, which educated him into the techniques of field biology, and as a teaching assistant developed ecology courses for nonmajors, which taught him how to translate complicated ideas for a wide audience. For his doctoral research he returned to Mesoamerica under an Organization of American

States fellowship to study Tortuguero National Park in Costa Rica from 1977 to 1980, a sodden rainforest without fire. He returned to the university as a post-doc, during which he coedited with Ewel and contributed chapters to a book, *Ecosystems of Florida*—still the standard reference. From 1982 to 1986 he worked as an assistant research biologist at Archbold Biological Station on Florida's Lake Wales Ridge. There, he developed the station's first fire management plans. In 1986 he signed on as the Nature Conservancy's first fire staffer. In 1988, the year Yellowstone National Park scorched TV screens for a long summer, he became its National Director of Fire Management and Research. He anchored the program in Florida at Tall Timbers Research Station.

He hired Paula Seamon, and together—in Seamon's words—they "grew" a national program. Myers proved indefatigable. Since TNC did not have a resident fire ecologist, he had to evaluate all its sites, write fire plans, and train local workers, with extensive mentoring chapter by chapter. All those sites known to need fire. All those that needed fire but didn't know it. And not only TNC holdings but all those lands and partners with whom the Conservancy had agreements, had arranged easements, or undertook mutual operations. These became a serious commitment as TNC expanded rapidly. It had completed a Natural Heritage Network for all 50 states in 1974, launched an International Conservation Program in 1980, and, the same year Myers assumed the director-ship for fire management, signed an agreement to help manage the 25 million acres held by the Department of Defense (DOD). A year later TNC commenced its Parks in Peril program, targeting Central and South America and the Caribbean, and acquired the Barnard Ranch that became Tallgrass Prairie Preserve. In 1991 it inaugurated its Last Great Places initiative.

Fire management was integral to it all. At the onset of the new millennium, TNC had a dedicated fire staff of three—Myers, Seamon, and Jeff Hardesty—to oversee the lot.

• • •

The challenges were enormous. Its fire staff had to plan, train, and leverage. The tiny group had to transform conviction and determination into programs that could meet not only the astonishingly disparate needs of TNC but that would allow its stewards and volunteers to participate with partners. Each response rippled through the organization and, with surprising quickness, through the national and even international fire establishment.

First, there were the internal needs of the Nature Conservancy. It had to manage fire on prairies, scrub pine, barrens, and wetlands to promote endangered species and eradicate noxious invasives and do it all on a shoestring. To restore prairie and free-range bison, Tallgrass Prairie burned on a vast scale but through grasses, amid a ranching landscape that also burned, and within a legal environment of strict liability. To sustain habitat for the endangered Florida scrub jay at Tiger Creek Preserve along Florida's Lake Wales Ridge, a prescribed fire team needed to set high-intensity fires in tightly bounded patches, often with collaborating agencies, within a regulatory environment that required Florida Forest Service authorization. To fashion habitat favored by the Karner blue butterfly, Albany Pine Bush, which was managed by a consortium of six landowners, had to burn among discontinuous patches nestled awkwardly within Amtrak lines, two interstate highways, assorted state and county roads, the city of Albany, major electrical transmission lines, a landfill, three nursing homes, assorted private lands and houses, a police-fire state complex, and a trailer park with a cluster of propane tanks.

But the Conservancy also had to reconcile two distinctive fire cultures. One culture was a tradition of prairie burning maintained by ranchers. Its core lay in the Flint Hills, with flakes split off across the plains. While chapters undertook similar burning with the earnestness that seemed endemic to TNC enterprises, the practice had a quality of studied relaxation, not leisurely but measured, like the rolling hills and homogeneous grasses in which it happened. The Conservancy absorbed the other tradition from its full-immersion baptism into Florida's fire scene. Burning here was concentrated, more and more bounded, potentially violent and unforgiving. It was a domain that favored toughened cadres for whom burning was something done year-round, not part of an annual ritual of spring cleaning. No one would dispatch chapter volunteers from the plains for fireline duty off site. Florida's TNC crews could join any brigade anywhere. Between those polarities lay a scatter diagram of sites, through which it seemed impossible to draw any line of administrative correlation.

Ron Myers, however, argued that this gaggle had to be yoked to a national standard. The complexity of island ecologies made burning more technical, liability from a burn gone bad could ruin the Conservancy, and to share operations with state and national partners TNC needed to accept common norms for training and qualifications. Unlike government entities the Conservancy had to buy insurance on the open market, which meant meeting social norms of accepted practice to avoid charges of negligence when escapes happened. If

TNC was to burn in Florida and receive liability protection, it had to satisfy state certification for burners, and if it was to join programs on federal lands, it had to satisfy national criteria. That pushed the Conservancy to adopt National Wildfire Coordinating Group (NWCG) standards, much to the consternation of many chapters.

The move gave TNC a place at the table and the opportunity to influence the character of such training, particularly with regard to prescribed fire. It could inject objectives for ecological burning into manuals that otherwise boxed in burning to fuel reduction. It could validate a fire organization based on prescribed fire rather than suppression. In fact, the national courses that emerged under NWCG evolved out of TNC material. Public agencies even sent their staff for TNC training. The Conservancy was doing what it did best: it built capacity, leveraged small resources into wide results, and served as a catalyst for making ideas operational.

The Florida program sprang out of the scrub-pine sandhills of the Lake Wales Ridge, as tough a landscape to burn as any in the country, and one ratcheted more tightly as 11 agencies oversaw 63 protected sites even as urbanization spilled south from Orlando. In 1991 the Florida chapter helped organize a Lake Wales Ridge Ecosystem Working Group (Ron Myers had come to the Conservancy from Archbold Biological Station at the south of the Ridge). In 1992 Disney Corporation created the Disney Wilderness Preserve at Kissimmee and funded TNC handsomely to run it for 20 years. The 12,000-acre project brought modern equipment and the imperative for smartly trained staff, and in response the Conservancy hired mostly out of the Florida Park Service. Among the fire community it became a valued partner and a neutral enabler. When, that same year, the controversy over seasonality of burning was peaking, the Florida Game and Fresh Water Fish Commission sponsored Ron Myers and Louise Robbins, also with the Conservancy, to write a summary review and make recommendations. Full integration, however, came with the state's 1998 trial by fire. TNC stepped up, joined the suppression effort, and contributed several strike teams, which proved particularly valuable for nighttime burnout operations. Because it met standards, TNC could participate, and afterward the Florida Forest Service contracted with the Conservancy to train the National Guard to meet them also.

The next stage in its Florida evolution was to formalize those firing teams. In 1999, as the Conservancy's fire program was bidding for national standing, it conceived a Florida Scrub Jay Fire Strike Team (later renamed the Lake Wales

Ridge Prescribed Fire Team) as a dedicated body to assist that consortium. The team had six members, two of them rated as Burn Boss II, along with three Type 6 engines and two ATVs. The fire team proved particularly valuable in grappling with the burning backlog problem, as a vanguard to conduct those first, tricky burns in overgrown settings. It soon became a vehicle for training and exporting to other TNC chapters; the Florida Park Service replicated the idea with a Backlog Abatement Team. Similar consortia congealed for northeast and central Florida, and the experience was passed along to the Maine Forest Service and Army National Guard, the U.S. Forest Service (USFS) in Colorado, and the Bureau of Land Management in Alaska, and then, as the fire initiative went global, so did the expertise of the Lake Wales Ridge Prescribed Fire Team.[3]

What happened in Florida, in brief, happened nationally. In the late 1990s the U.S. Forest Service seconded a staffer to TNC for a year. That experience led to a Fire Roundtable held at Flagstaff, Arizona, in 2000 to explore collaborative interests, an idea that then became embedded in the National Fire Plan. A presentation at the National Interagency Fire Center led to a general cooperative agreement with all the federal land agencies. Amid an increasingly polarized (and paralyzed) national establishment the Conservancy was viewed as an honest broker, a facilitator. The outcome was the Fire Learning Network, a spectrum of national training, and active partnership with LANDFIRE (Landscape Fire and Resource Management Tools). In 2006 the concept was replaced by a five-year "Fire, Landscapes, and People" program that continued and expanded the agenda. TNC seemed to be everywhere, and everywhere a van der Waals force that helped hold factious alliances together, and everywhere welcome. All this was in addition to TNC's own swelling population of sites; in some years the Conservancy prescribe-burned more acres than the National Park Service. And it was complemented by a stunning Global Fire Initiative.

The Conservancy's international program picked up steam just when Ron Myers assumed the directorship of its national program. It was, again, an ideal collusion of institutional needs, CV, and personality; TNC's national fire director was the perfect point man for its international campaign. With 80 percent of TNC global efforts directed to Central and South America and the Caribbean, Myers could draw on his personal past in Honduras, Costa Rica, and Puerto Rico to craft practical plans and build local capacity. Over the next 15 years he designed and sparked to life fire management programs and the training to make them happen in Mexico, Belize, Guatemala, Honduras, Costa

Rica, Cuba, the Bahamas, Trinidad, the Dominican Republic, Peru, Paraguay, and Argentina, as well as short assignments for training and reconnaissance to China, Mongolia, South Africa, and Indonesia. In 2006 Myers codified the experience in a glossy-page publication, *Living with Fire—Sustaining Ecosystems and Livelihoods Through Integrated Fire Management*, a shockingly readable document in which the jargon of bureaucracy and science melted away in favor of direct language, categories that could hold a driptorch, a vision of fire ecology that included people, and a universal "framework" of fire management that quietly dumped the 10 Standard Fire Fighting Orders in favor of 10 "guiding approaches."[4]

The 28-page document was a personal testimony of a singular life, the synthesis of a Montana smokejumper turned world-class prescribe burner, a botany PhD who learned from campesinos in Honduras and Mexico, and a pragmatist who knew that ideas only had meaning as they were expressed on the ground. In many ways, too, it was the TNC model translated and exported: networks of small chapters trained to basic standards of safety and effectiveness while grounded in and sustained by local communities, to which outsiders might offer guidance but would not administer. The contrast between lumbering, costly public agencies in the United States struggling to meet prescribed fire targets and a Conservancy rapidly scaling up its programs on a shoestring with a permanent staff that could barely fill a Type 6 engine repeated itself overseas. In countries like Mexico it provided the leverage to allow local interests and institutions to take root.

Then the Great Recession took the wind out of TNC's sails. Like a body in shock that withdraws blood from the extremities, the Conservancy laid off a fifth of its staff, pulled back from international commitments, and shuttered its Global Fire Initiative. After 23 years of service Ron Myers was given notice.

• • •

The Nature Conservancy fire program was so small in its formal constitution—bureaucracy seems too strong a term—that it reflects to an unusual degree the places that nurtured it and the personalities of its officers. Its design favored the local over the large: that's how it was able to scale out rather than up.

Much of the TNC experience, in fact, had wider significance. The Conservancy was not tied to the eccentricities of particular landscapes, as the Tall Timbers model was. Nor was it pushed and pulled by legislative mandates, like the Florida Forest Service. Nor was every decision potentially a federal case, as with the national forests and parks. It demonstrated that one could burn for biodiversity as well as for fuel reduction. It showed that a vibrant fire organization did not have to evolve out of a suppression program. In places like Lake Wales Ridge it could make prescribed burning an extreme sport, as exhilarating as smokejumping. It made a good partner, ready to roll up its sleeves and do.

The key was land. The Conservancy was accepted not simply out of goodwill but because it was a landowner, with a landowner's rights, prerogatives, and responsibilities. It acquired lands by purchase just as other conservation lands in Florida did, and it managed lands for others (notably, DOD). It was not simply an environmental advocacy group like the Sierra Club, nor a research facility severed from operations like the University of Florida Botany Department or the Canadian Forest Service. It was a committed land manager who could not evade problems or wait upon research to suggest remedies or defer to legislatures and courts. But neither was it originally chartered to manage fire-type landscapes as Tall Timbers Research Station was. TNC burned to support its mission to nurture habitat and save species. It saw in fire a shared need and a collective threat. What remained unburned threatened not only its own holdings but its neighbors; what was burned badly by others could slop over onto its property. Whatever its founding expectations, it had to handle fire, and it had to learn how to do so.

What makes the Conservancy's Florida story so powerful is that the two models—Florida's and the Conservancy's—found ways to collude rather than collide. A fire strategy based on prescribed fire, a commitment to landowner rights, a determination for multiagency cooperation, a preference to be proactive— these were genetic properties of Florida fire agencies, yet they were also traits bred into TNC from its conception. Each emerged from that chrysalis stronger. TNC became a primary vehicle for translating the Florida model outside the pyric and political eccentricities of the state. Equally, its Florida phase gave the Nature Conservancy a second leg to stand on, even if its foot rested on a torch-equipped ATV. It toughened the institution's skills, kept it from being isolated into hobby burners, and helped carry its larger message about land, biodiversity, and burning to a wider world. After its Florida workouts, it punched well above its weight. It was the only private conservation organization anywhere that had in-house capabilities to manage fire.[5]

Even so, that alignment of tumblers could only fall into place and unlock the potential if there was someone skilled enough to work them. If, in those years, the Conservancy and the Florida model seemed to be everywhere, it was because Ron Myers was too.

• • •

It's seductive to anthropomorphize institutions. It simplifies analysis to treat as one entity what holds scores or thousands of refractory agents, and from a literary perspective, anthropomorphizing allows for a more direct narrative since institutions can serve as characters. Yet despite Supreme Court rulings that corporations are "persons," they aren't. What happens happens because people act or are acted upon; they ponder, fear, react, believe, stumble, leap, argue, fight. They choose. Often these acts get merged into a statistical composite that we call an institution. But sometimes the people involved are so few and they achieve so much that the institutional story and their own are virtually the same. That seems to be the case with TNC and fire. Jeff Hardesty, Paula Seamon, and especially Ron Myers took a Conservancy fire operation hardened in Florida and propagated it countrywide, and then overseas.

As the Nature Conservancy celebrated the 60th anniversary of its incorporation, the 25-year span during which it had stabilized its own fire needs and become a national firepower shrank in proportion to its longer history. The program that Seamon and Myers "grew" had grown up. The program had built capacity, it had adopted national standards, and it had internalized its lessons learned. The chapters could take care of themselves, a new directorate for the Conservancy decided. The Fire Learning Network replaced the national directorship as a means to link scattered holdings. The chapters could strengthen by bonding with their local cooperators. The national Coalition of Prescribed Fire Councils could institutionalize what had been held together by the strong nuclear force of personality. The Florida chapter assisted projects in Belize and the Caribbean. It appeared, too, that the new regime believed (hoped) that TNC could achieve its mission with less fire, that fire was not integral and universal, that Florida was an exception rather than a norm. What happened in Florida could stay in Florida.

It couldn't, and those expansive years in scrub pine and palmetto did not flare and expire like a flaming match head. They left behind a hard deposit that would endure, among which are lessons for all fire organizations about nimbleness, commitment, pragmatism, calculated risk taking, reconciling big ideas and big institutions with local knowledge and small communities. The experience bonded prescribed fire to ecological burning, not simply as a tool to reduce logging slash or as a sideline for suppression organizations. It showed that even a single person, without obscene wealth or inherited influence, could shape the national scene, the way a small switch can turn on a powerful dynamo.

When his tenure with TNC ended, Ron Myers stayed on the line. He had reached retirement age yet wished to continue doing what he had done for a lifetime. He opened a consultancy. He kept his personal ties with Mexico. The U.S. Fish and Wildlife Service hired him to devise fire management priorities for the National Key Deer Refuge at Big Pine Key, about 25 miles from Key West and 100 from Miami. One kind of temperament might see that as a virtual exile, fire management's version of Devil's Island. Another—Ron Myers's—could see Big Pine Key as the latest in a litany of protected sites, a tightly bounded place that needed fire and had little room for either error or dithering, a place where the Florida model met Latin America and the Caribbean. It was a place where a man with tenacity, conviction, and verve might continue to keep one foot in the black.

Fire 101 at Star Fleet Academy

T
HE VISITOR COMPLEX AT Kennedy Space Center is a theme-park paean
to propaganda and the human presence beyond Earth. Two astronaut
mannequins loom over the entrance. The rocket garden is a gleaming grove
of those missiles (and only those) used to loft capsules stuffed with people.
A robot explorer exhibit makes clear, through anthropomorphized mechanical "scouts," that those plucky spacecraft exist only as "trailblazers for human
explorers." What is not said, what doesn't need to be said, is that the voyage
beyond begins with a blast of engineered fire.

The visitor complex is where it is because it abuts NASA's major launch facilities. Those pads are at Cape Canaveral because being nearer the Earth's equa-
torial bulge grants a boost to rockets that more temperate zones can't deliver
and because flights arc over the Atlantic, which can absorb a lot of failures. The
Cape juts out so far because it is really two barrier islands, with the interior one,
Merritt Island, separated from Canaveral by the Banana River and from the
mainland by the Indian River, thus further isolating the installation.

These are the geophysical reasons. Had someone better versed in natural history pondered the site, he would have identified it as one of nature's
great installations for pyrotechnics. There is more lightning in central Florida
than anywhere else in North America; and more to the point, there is as much
lightning-kindled fire. Merritt Island is where lightning's ceaseless countdowns

launch fire heavenward in boiling clouds of smoke. It is where, in Fred Adrian's memorable phrase, fire gets managed along a "wildland-galactic interface."[1]

•   •   •

When NASA decided President Kennedy's call to go to the Moon required an enhanced launch facility, it expanded, by purchase and condemnation, from an Air Force installation on Canaveral into Merritt Island. That gave it a lot of buffer zone, but one that needed an administrative presence. In 1963 NASA signed a memorandum of understanding with the Interior Department to have the U.S. Fish and Wildlife Service manage those surrounding lands. The landscape proved to be more than an inert swamp of gators and skeeters. It burned. The FWS soon realized it could not simply attend to eagles and migratory waterfowl. It had to cope with all the species of fire.

The fires were relentless. Although heaviest from May through September, the main storm season, lightning historically started fires every month save October. Meanwhile, fires from human causes occurred constantly. The refuge found itself fighting fires, and before long, lighting them in an effort to beat down some of the scrub that fueled the wildfires. NASA realized that its launch facility was very far from a clean room: fires burned around the base like feisty raccoons and smoke swirled like herons. Unless contained, they could threaten the infrastructure, halt launches, jeopardize instruments and optics that had to be quarantined for months (like the Hubble Space Telescope), and interfere with shuttle landings. Moreover, although seemingly constant, wildfires were unpredictable in occurrence and spread. Highly politicized, billion-dollar programs were hostage to the whims of giant electrostatic matches foraging the launch facility for kindling.

•   •   •

Yet, in those years, the FWS was not the sophisticated fire operation it eventually became. It was itself overmatched. The agency knew a lot about ducks, egrets, and otters. What it knew about fire was local, personal, and institutionally unsystematic. Firefighting crews were "militias"—call-ups of local staff primarily hired as refuge managers, biologists, mechanics, or wildlife techs. They fought fires with surplus military vehicles outfitted with pumps, or with small tractors and plows, and with a lot of burning out. Prescribed fires were routine, even tedious affairs on small plots, a step above matches tossed into grass and subject to oversight by NASA's charged calendar. The agency stood apart from an evolving interagency fire community that pushed hard for common standards in equipment, safety gear, and training. In average years the staff got its fire chores done with no more fuss than it did the other varied tasks demanded of a refuge with too little money and too many calls on its time. Wildfires got knocked down, prescribed fires got lit. Yet year by year the amount that needed to be burned fell behind, a backlog that wildfire chewed on like turkey vultures on carrion. And then there were the exceptional years, the ones that accounted for most of the acres and crises.

On June 8, 1981, lightning set three fires, one of which shifted suddenly when prevailing southeast winds were overtaken by violent downdrafts from a thunderhead blowing from the west. The fire front's abrupt change surprised and overran a two-man tractor crew. Both men died from their burns. One was the son of a politically connected federal judge. Congressional hearings subsequently blasted over the Fish and Wildlife Service with the force of the fatal Ransom Road fire. The tragedy acted on the FWS much as the 1967 fire that consumed Apollo 1 did on NASA.[2]

The Merritt Island fire, following another fatality burn in 1979 at Okefenokee, brought money and a national fire program to the agency. Merritt Island National Wildlife Refuge scaled up, its personnel and hardware becoming interchangeable with fire crews, equipment, standards, and practices elsewhere. In particular, it committed to prescribed burning as the best available strategy for containing wildfire. It faced the same suite of challenges as everywhere else in the national matrix of exurban sprawl and abandoned wildlands that went under that lame label, wildland-urban interface (WUI). But Merritt Island also had to deal with NASA, which was mightily irked when wildfires upset missions but was even less tolerant of fires set by nominal partners. In addition to the metastasizing checklists that everyone else had to consider, Merritt Island also had to seek approval from NASA, which was not keen on having long-planned missions to other worlds delayed by smoke on this one or having shuttle-weary astronauts remain in orbit while refuge fire crews burned wetlands and flatwoods to assist wood storks and the Atlantic salt marsh snake.

But NASA did not hold all the cards. There was a compelling case for doing the burning more or less on a schedule rather than leaving the task to lightning, arsonists, and off-road catalytic converters. No less, the FWS had the Endangered Species Act on its side; in fact, it administered the act. The Kennedy Space

Center might sit next to the Astronaut Hall of Fame, but Merritt Island was a biotic chamber of threatened and endangered species (10 listed, 93 of state and federal concern), most of which depend on habitats shaped by fire. Especially noteworthy is the Florida scrub jay. Merritt Island holds one of the three viable populations in Florida, and the most vigorous one.

In 1993 when NASA wanted a new facility to support the International Space Station, it sought a site within prime scrub jay habitat. It got approval, but in return it had to support restoration of an equivalent patch of scrub jay habitat elsewhere on the island. The refuge staff built that landscape through industrial-strength slashing and burning. There was no alternative to fire: the fastidious jay not only demanded sand pine of a particular height but open land adjacent on which to forage. Lightning and scrub jays dealt the refuge two powerful trump cards.

This was not the classic wildland-urban interface, where fires start on one side of the fence and threaten the other and where one group wants fire (and accepts its smoke) and other group does not. At Merritt Island each side had to accommodate the other; each provides windows for the other's operations. With coordination NASA can launch and land without smoke and flame as checks; what stalls its launches are the shuttle's flaws, not the refuge's fires. For its part, the refuge burns an average of 20 percent of its burnable land annually. It fires off the marshes every 18–36 months; the flatwoods, every 3–5 years; the scrub, every 5–10 years. It leaves unburned many sites adjacent to restricted facilities. Convective columns cohabit with rocket plumes.

Still, the geographic setting replicates—even if it exaggerates almost to parody—the national fire scene. On the cover of the refuge's 2003 fire plan there is an aerial photo that shows, in one snapshot, a launch pad to the north, an ordnance dump to the west, an explosive gas storage tank farm to the east, buildings and contaminated soil to the south; just beyond are restricted zones where deadly force is authorized. But then, as nature showed in 1981, it too can call upon deadly force against intruders. In the euphemism of the day, all this makes quite a challenge.

• • •

Among the theme-park schlock that infests the visitor complex, there is a show called *Star Trek Live* in which actors, under the auspices of Star Fleet Academy, dress in costumes, do skits, and explain the rationale for the human colonization of space. The show runs next door to an exhibit about what resources Mars offers colonizers and what life within a habitat module on the flanks of Olympus Mons would be like. Why a putative superpower would use a TV-show-turned-movie-franchise to promote a major expression of its ambitions and status, how one fantasy might be acting out another, is a question best stepped around, like meeting an irritated alligator at a pond.

It suggests, however, that the real wildland-galactic interface is more implacable than the garden-variety WUI because it combines commerce and politics with utopianism. The spectacle at Merritt Island slams together two incommensurable visions of the human future. One seeks to use controlled fire to leave the planet. It takes as axiomatic that the ultimate security and perfectibility of humanity lie in leaving Earth. The other proposes to use fire to make the home planet, the only one we will always reside on, more habitable. It accepts as a practical and moral charge to enhance what we have and to preserve its interstellar uniqueness. It assumes that our true destiny is not to understand how to live like hamsters on Mars but how to live like human beings on Earth.

The Everburns

THE FLORIDA STORY flows south.

It ends when the land spreads into a wide delta of limestone and sedge rimmed by mangroves and sea. At the Everglades water, fire, and people converge to make, in Marjory Stoneman Douglas's famous dictum, "one of the unique regions of the earth," a place like "nothing anywhere else." Here, subterranean aquifer and surface overflows merge into a single stream; and here the fires set by lightning and those by people, the regimes hidden by evolutionary time and those visibly spilling out from the flow of settlement, converge. The pieces that make up this scene are common to Florida generally. It is how they come together and heighten that make the Everglades singular.[1]

Everything shrinks to a minimum, as though the Earth's landscapes were infinitesimals approaching a limit. Relief is measured in inches; peat domes and sinkholes replace ridges and swales. The dense biotic mosaic of the subtropics thins into patches of pine rockland and hardwood hammock on a sea of saw grass, like atolls in the Pacific. Rivers become sheet flows moving only hundreds of feet in a day. Ecological dynamics dissolve into a dialectic of water and fire, rising and falling, coming and going, flickering like Schrödinger's cat from the fabled river of grass into a river of fire. The Everglades could as easily have been named the Everburns.

Or it once could. Like Florida generally the Everglades is a broken landscape. A century of hydrologic engineering and ragtag plundering have shattered the rhythms of flooding that annually passed through the Glades. The waters that flooded hummocky landscapes and once spilled south from Lake Okeechobee to pass over the land like thick ooze have been diverted into canals for agricultural fields and cities; have been halted by ditches, roads, and developments; have been drained away or dammed or released at the wrong times or come laden with toxins. The regularity of fires broke down in the same way. The fireshed fragmented. Fires were too few, or too intense, or out of sync. Water and fire, each emulated the other. Each needed the other.

The Glades's depth relied on its breadth. The rhythmic flows of water and fire depended on a wide watershed and a broad biota that could absorb and buffer. Like a graded stream that adjusts its profile to accommodate water to debris, the Everglades's two flows adjusted their profiles to the amount of work required, the one reconciling water to sediment, and the other flame to fuel. But settlement ripped away those borders, flaked off chunks of the biota, and slashed its size in ways that left it starved for water, stripped of critical species, and exposed to fires that could burn down instead of out.

As the water fell, the fires burned deeper and spread wider. The canals and roads that water had dissolved previously now disintegrated through muck burns. In 1920 hammocks that had for most of a century provided sanctuary for Seminoles burned away. Ernest Coe, who began campaigning for park status in 1928, argued that "fires have always swept the Everglades and they are going to continue to do it." But these fires had become pyric mutants: they didn't burn in the old ways. In 1935 soil and water conservation authorities successfully urged the legislature to establish a fire control district. It had mixed results, swatting out small burns, though proving ineffective against frost and drought (or its artificial surrogate, drainage). When Daniel Beard conducted his wildlife survey in 1938, he estimated that over the past year fire had burned half the piney woods, 80 percent of the Everglades prairie, 30 percent of the coastal prairie, and 5 percent of the Ten Thousand Islands Coast and "about the same amount of the Cypress." In 1941 a fire "of incendiary origin" blasted over 250,000 acres and left crews scrambling to get out of its way.[2]

In 1945 the full magnitude of the tinkering became unblinkingly clear. Drought settled over the Glades, as it had for millennia, but this time the wetlands had lost their capacity to shrug off the drying, to sink into deeper holes like its gators and wait for the rains and spillage to return. Surface fires burned to the new water table, where they smoldered wretchedly through peat for days, for weeks, spewing an apocalyptic pall of smoke that gagged the region. The river of water shriveled to a trickle; the river of fire flooded, its flames scouring its historic channel down to bedrock. In Douglas's outraged words, "The Everglades were dying." The Indians, with "the stoic faces of fatalism," saw the "end of their world." The white man, at last, realized the consequences of his fecklessness. For the first time the "problem of the Everglades was seen whole."[3]

• • •

The Everglades is a landscape of ideas as much as of egrets and thunderstorms. Behind the hand that worked guns, axes, and dredges lay a mind that imagined what the place meant, what it might be good for. In this sense, again, the concepts, beliefs, and tropes of imagination that guided fire practices in Florida also flowed into the Everglades. Here philosophies of fire management converged with a special clarity made possible by the apparently simplified landscape and its intensified scrutiny by outsiders. How the Everglades coped with fire became a cameo of Florida fire, and of American fire generally.

The great burn of 1945 gave urgency to the movement to legally protect the Everglades from further abuse. Two years later the Everglades became a national park, the first established primarily to protect biological values. The wretched peat fires had acted on south Florida as cutover slash had in the Lake States 50 years before, kindling national outrage and leading to schemes to protect "from fire and axe," or in the Everglades, from fire and dredge. From the onset, fire protection was fundamental to the park's purpose: the issue was literally existential. If rangers could not prevent the return of fires like those of 1945, there would be no park.

But ideas changed park fire regimes as much as they affected its hydrologic regimes. Those ideas came from the outside and had to express themselves amid the unique setting that was the Everglades, where, no less than other settlement schemes, they sparked unintended consequences. The narrative of Everglades fire is thus a chronicle of national notions meeting local circumstances. Those notions involved fire control, fire management, and fire restoration; the circumstances were a large patch of pine upland (Long Pine Key), expansive sloughs and wetlands, and a broad fringe of marshes and mangroves. Meanwhile, the park's borders changed as it absorbed inholdings, acquired additional lands, and confronted intense urbanization to the east.

There was no avoiding fire: it was endemic, frequent, and indomitable. It would come even if people vacated the place. Saw grass burned over standing water, hardwood hammocks burned in drought, mangroves burned after frosts, and slash pine rough could burn anytime. A subtropical place in which something was always blooming meant that something was always burning. Equally, there was no avoiding scrutiny. The park resided next door to that urban conurbation known as Miasma (aka Miami), while its iconic status made it visible across the country and vital to national conservation organizations. Inevitable fire, implacable attention, a fire park in an agency undergoing reforms in its fire policy—the Everglades became an anomalous leader in the national story.

Initially this meant fire control. Ending the burning was a reason for the park's reservation in the first place. As its first superintendent, Daniel Beard put it, "Under present circumstances" fires of "all kinds" must be "prevented, or extinguished if they start." In fact, aggressive firefighting, perhaps leading to fire exclusion, had long been a policy of the National Park Service (NPS), stiffened by the Forest Service's adoption of the 10 a.m. policy in the spring of 1935. Both the policy's target time for control and its purpose had little meaning in south Florida. In the Northern Rockies, where the policy was birthed, 10 a.m. marked the breakup of the evening inversion, which led to quickened burning; this had little pertinence in a tropical landscape as flat as a bureaucrat's desktop. Nor could fire be excluded. Lightning and people had burned the place as often as it could carry fire. Daniel Beard, whose 1938 report for the Interior Department provided the scientific basis behind park legislation, and who became the park's first superintendent, began his observations on fire by repeating the old saw that "Florida burns off twice a year." It was, he concluded, "hardly an exaggeration."[4]

Still, the NPS commenced with a control policy that, as the park's chief scientist noted, consumed much of the staff's energy but did little to alter the fundamentals. Meaningful fire control was impossible without water control, and until the park was reflooded, its staff could only swat fires as they might mosquitoes. Newcomer rangers were appalled at local fire brigades who simply burned out from the nearest road; it was better, they thought, to attack the fires directly. But bulldozers, plows, pumpers, tracked Bombardiers and Thiokols, even glade buggies, too often tore up the landscape (and got themselves torn up or mired in muck). Fires burned across deep-pitted pinnacle limestone, through decadent saw grass overtopping knee-high standing water, amid coarse subtropical rough that could carry fire any month. Tactics used elsewhere proved as helpless as pulaskis; and ideas floundered as much as tactics. Not least, change happened quickly. The effects of fire control in Oregon or Minnesota might take three or four decades to become apparent. In south Florida the outcome was visible in three or four years. Even casual staff, cycling through the park for a careerist tour of duty, could see the consequences.

In 1951 park biologist William Robertson began an investigation into what specifically fire did in the Everglades. Completed in 1953, "A Survey of the Effects of Fire in Everglades National Park" became an instant classic, one of half a dozen seminal fire studies in the century. While he concentrated on the slash pine uplands, whose dynamics most resembled Florida fire ecology generally, every part of the Everglades system had its fire regime, and a nationally based Park Service policy was wildly out of sync with all of them. Removing fire was damaging the biota the park was intended to preserve, while actual firefighting with tracked vehicles and plows left lasting scars. A 1956 Fire Control Plan sought to contain fire but also to limit the damages mechanized firefighting did by restricting their use and to replace ad hoc firelines with a system of permanent roads.[5]

Behind the appreciation for the special difficulties of fire control lay the prospect that prescribed fire might be necessary as a surrogate for wildfire. In 1958 Everglades received authorization to conduct controlled burning in its pinelands, the first and sole exception within the national fire program. In 1962 wildfires swept much of the park; the next year, 1963, the Leopold Report catalyzed a shift in NPS fire policy, proposing that parks generally should be "vignettes of primitive America" and that any restoration would have to include fire. The report culminated in new guidelines for natural area management published in 1968 that sought to reintroduce fire where it had been lost and where ecosystems had suffered in consequence. Everglades was again at the forefront. In 1969 it expanded its prescribed fire program into saw grass prairies. Soon afterward it sponsored research by University of Miami professor R. H. Hofstetter to update Robertson's classic.

Twenty years after Robertson's report, Hofstetter issued his own, complete with a long litany of recommendations. These were largely incorporated into the revised Everglades Fire Management Plan of 1973, further amended in 1974. As "fire management" replaced "fire control" in the title, so the intention of the plan was to "restore" fire to something approximating its "natural" or presettlement state. To this end, it parsed the Everglades into three biomes, each with its own prescription for burning: a coastal belt of marsh, prairie, and mangrove; the sloughs of saw grass and wet prairie; and the pine rocklands. Some fires would be left to roam, whether kindled from lightning or people. Many would be set deliberately. And some fires would be suppressed, particularly those when drought pushed the water table too low or when fires threatened inhabited borders; but even here "control" could mean a range of responses from direct attack to loose herding or backing off to barriers from which to burn out. Other parks were adopting similar strategies, but none could approach the Everglades for actually getting fire on the ground or through the sedges. The plan was approved the same year massive fires again descended.

While the program seemed radical on the national arena—and received prominent attention—it actually nudged the park into practices that had long characterized south Florida. A few problems were pragmatic, like keeping smoke out of Miami and fire out of deep-peat hammocks. But many were metaphysical. What was the "natural" fire regime? Should it include the anthropogenic fires that had coexisted with the emergence of the Glades over the past 5,000 years? And what did "natural" mean when the hydrology had been utterly replumbed, when the rhythm of fires had become atonal, when flammable exotics like melaleuca and Australian pine threatened to overrun burned areas, when there was always a Cape Sable seaside sparrow, a Florida panther, or an indigo snake that was endangered depending on how a place burned? It was simple to imagine untethered lightning fires reinstating a regime in the High Sierras or the remote Gila Wilderness. It verged on scholasticism to argue the merits in the vast marshy terminus of a peninsula that had had its indigenous character erased, its flanks crowded with avid invaders and a giant megalopolis, and its fundamental processes rewired, polluted, and deranged. The one axiom was fires would happen. They would come whether or not people could reconcile the thesis of the past with the antithesis of the present into some kind of synthesis for the future. The place would burn regardless.

So the debate continued, one of many environmental themes at Everglades that captured national attention but had a very particular meaning in this very peculiar place. After all, how many universals could play out in a unique setting? The Everglades had an unblemished record of frustrating grand schemes with perversely unexpected consequences. Still, the Sierra parks and the Everglades made the odd couple of NPS fire management, and they attracted some of the premier fire officers in the service, with a few like Larry Bancroft working in both. For the NPS it was an era of exhilarating experimentation that only slowed, not ceased, in 1979 with efforts to standardize and nationalize practices. The park fire program flourished.

Then came the Yellowstone conflagrations of 1988. Even as the episode forced "prescribed natural fire" programs to cold start, which meant parks and forests had to redo and resubmit their fire plans, Yellowstone's potlatch brought huge sums of money and attention to fire management in the Park Service. What had been for most parks a tangential program, not easily put into visitor services, muscled into prominence. The outcome for Everglades was a sophisticated rewrite of its fire plan that modernized terminology, accommodated the boost in resources, and absorbed another management unit—a block of newly acquired land to the northeast (with actual purchase scheduled for 1995) that would regulate the water flow through Taylor Slough but enormously complicated the fire program because it included agricultural (and Hispanic) squatters, along with fireweeds like melaleuca and because it would press against the ravenous sprawl of the Gold Coast. The plan sustained the ambition to restore "natural fire regimes," although it craftily sought to finesse what this meant in practice. The essence was it sought to get more of the right kind of fire on the ground.

Until Big Cypress National Preserve arrived to make a tag team, no one in the NPS did that with more routine success. Even so, no one was content with how much got burned, and all worried about the increasing constraints imposed by outside interests and ideas. It was as though the invasive Burmese python had the park fire plan in its coils. With every operational hiccup or review-inspired gasp for breath, with every call by researchers for another study or reconsideration, or with every new advocacy group for a cherished species, the coil tightened. The process only went one way. Fire officers could hold, not expand, operations. Still, as fire management commanded more national attention through the 1990s, the program got additional money and personnel. The debates raged incessantly about how, when, and with what restraints to do the burning; but the burning went on. Compared to Everglades and Big Cypress, the rest of the National Park Service looked like a gaggle of fire hobby farms.

The arrival of the National Fire Plan and the Comprehensive Everglades Restoration Plan both bolstered and hobbled the Everglades fire program. The park enlarged and enriched its fire staff, but that effort dimmed in comparison with the $8.6 billion, 30-year project to reinstate more of the historic hydrological regime. Once again, water and wildlife dominated the scene, reducing fire to an epiphenomenon, a looming disaster that apparently could only be set right by a restored hydrology, or a threat to a swelling population of endangered species, each of which had sharp-eyed (and sharp-tongued) citizen advocacy groups peering across the park boundary. Fire had no such constituency; it had no Endangered Process Act to give it legal standing; no park had been specifically chartered to showcase fire. The park staff included one fire ecologist but, as a fire crewman shrugged, seemingly "hundreds" of hydrologists. Although a revised fire plan, accompanied by a formal environmental assessment, was headed for approval in 2011, in many ways fire's status remained where Marjorie Stoneman Douglas had left it 60 years previously. It was seen as a potential predator on the park, worsened and made prominent by the ruinous heritage of mismanaged water that threatened to drain the river out of the River of Grass. At best, to many laymen burning seemed ameliorative; it did not seem essential.

The larger fire community saw it differently. However hobbled matters appeared within the park, when viewed nationally the fire program at Everglades had been, from its origin, a powerhouse that helped move Florida fire from exceptionalism to exemplar. Everglades had sponsored the first comprehensive research on fire ecology, had conducted the first authorized controlled burns, had created the first prescribed fire program, had executed some of the earliest prescribed natural fires, had constantly challenged the National Park Service, and through it the national fire community, to rethink its premises. The Everglades idea pressed against its surroundings, and against all reasonable odds it had helped convert adjacent land into national preserves, wildlife refuges, state forests, and water conservation zones and had helped share their fire programs. No one doubted the significance of fire to Everglades. What was uncertain was the value of the Everglades experience elsewhere.

Within the national park system Everglades remained an anomaly. But then that nominal system was itself anomalous. Unlike other countries, the United States did not have a national park act that established new parks under a common policy. Instead, Congress or presidential proclamation established each park or monument with a separate act and purpose; what united the parts was a common agency to administer them. Even so, Everglades sat awkwardly in the national economy of fire. Ideas imported from elsewhere failed to take or like invasive plants they inflicted damage; and practices invented at Everglades struggled to survive when exported. Its fire program, that is, was much like the park.

There were two ways to characterize this outcome. In one, the place was sui generis, a fire autarky. Only in the most abstract and academic senses did it relate to fire elsewhere in the country. In the other, Everglades represented a valid if unquestionably unique synthesis of national ideas and local circumstances. It was as though the entire country had flowed southward into Florida and then stopped and congealed at the end of the peninsula. In this configuration Everglades was synecdoche for fire in Florida, and maybe more.

• • •

Long narratives can distill into people as well as places. At Everglades it gathers with uncanny appropriateness into the career of Rick Anderson.

His forbearers were among the wave of Scots-Irish that washed ashore at Oglethorpe's Georgia colony where they hoped to work off their debts and return to Ulster. Neither happened. In 1732 the Andersons and the McClellands first married. Then the clans began a slow, long trek south through the advancing marchlands of the Creek and Seminole wars. By the 1830s they were around Gainesville. They continued to probe south down the Lake Wales Ridge. The Andersons settled mostly onto farms, the citrus industry helping to root them. The McClellands yielded to frontier wanderlust and their herding heritage. Southwest Florida was still raw in the 1940s when Rick's grandfather logged and hunted in the Big Cypress Swamp.

In 1957 Fort Myers offered the closest hospital, so Rick was born there. Mostly he grew up amid citrus and cattle near Dade City farther north, spending much of his time around Green Swamp, and always he heard the old family stories about life on the land. There were fires everywhere: they were just there, like the clouds and the summer thunderstorms and the piney woods. He grew up around burners and with stories of burning humming in the background like the summer's mosquitoes.

But the boom was on. Old Florida was vanishing: the longleaf were cut out and the big cypress gone from Big Cypress, the tick-ridden and free-roaming cattle were dipped and fenced, and the free-ranging fires increasingly prescribed by law and plow lines. In 1976 he joined the Florida Forest Service, right out of high school, eager, as he puts it, to find a job that would pay him to work with fire. The monster Turner #10 fire at Big Cypress found him among the futile suppression crews and brought him into contact with the National Park Service. He quickly realized that the NPS paid far more than FFS and transferred. Then he learned that being a part of a national agency meant he could go elsewhere in the country.

He went to Yellowstone in 1985 and was there during the great firefest of 1988. The experience left him full of questions: he'd started his career with a megafire, but the Yellowstone plateau and the Absarokas were vastly different than the Big Cypress and Lake Wales Ridge. Park researchers tired of answering his endless queries. Go to college, they told him. He was too old, he replied; he'd be in his mid-30s when he graduated. He would be in his mid-30s anyway, they answered. So in 1989 he enrolled at the University of Montana, which had a sterling fire curriculum and roosted among one of the country's great fire cultures. For two summers he went to Belize for research, and he used that experience to segue into a master's degree in ecosystem management. In 1994 he returned to a National Park Service still flush with its post-Yellowstone funding.

He transferred to Olympic National Park, a cold swamp that, unlike Big Cypress, still had most of its giant trees but its only fires were in campfire rings. Within a year he moved south to Saguaro National Park, outside Tucson. The park ranged farther over the Rincon Mountains than the saguaros on its mid-slopes. It looked oddly familiar: this was a drier version of what he had grown up with, the western yellow pine replacing the southern, the muhly and fescue substituting for wire grass, summer lightning kindling fires like a Fourth of July picnic. What the Rincons had that central Florida did not was real relief: it had slope and a crenulated texture that created a biotic mosaic that in Florida had depended on differences in soils and water tables. He quickly built up a prescribed fire program that pioneered landscape burning, that used terrain to direct and contain long-burning fires, igniting at the summits and letting the flames work their way down. It was a top-down system built from the bottom up. Where, nationally, prescribed fire modules were still in the garage or spun wheels in bureaucratic mud, he found ways to get them on the road and give them traction. He worked around concerns with the Mexican spotted owl and a wary Forest Service—nothing here was as complex as it was in Florida.

Then came a prescribed fire brouhaha as great as Yellowstone's natural fire season. On May 10, 2000, crews at Grand Canyon kindled a prescribed fire that blew out of control and forced an evacuation of the North Rim. The next day crews at Bandelier National Monument faced the same passing cold front, lost their fire, and watched it burn into Los Alamos. Rick reckoned that burning would become much more difficult for the agency. He resigned and returned to Florida to conduct research burns for the Archbold Biological Station at Lake Placid. Within two years he moved to the Nature Conservancy, where he worked with Ron Myers and became fire management officer for the southeast region, which also included the Caribbean and Central America, bringing him back to Belize. In 2005 he re-upped in the Park Service and what he calls his "briar patch" to become Everglades's fire ecologist. Then he took the job he was "born to do." In 2008 he became the park's fire management officer.

In reflective moments he muses about restoration and redemption and what faith and good works might do for a place that, as his Florida ancestors might have put it, was "rode hard and put away wet." As he sees it, in the early years the park burned too much. Then it burned too little. It's extraordinarily resilient in the face of fire, and not resilient in fire's absence. The agency should not seek to "restore" so much as sustain the "native" ecology—a deliberately vague goal—for which it must regain control over fire, not so much to reduce fire's damages as to exploit its benefits. Too often fire was seen by park administrators as an afterthought, as something that might control itself if the water regime got properly reestablished. Rick Anderson saw it as a positive force. Fire is what will jolt the reassembled parts to life. In the River of Grass fire must do what flooding does elsewhere: it flushes the channel clear of the debris that would otherwise dam it into senescence.

Yet he knows how often and easily the Everglades have frustrated the ideals, schemes, ambitions, and yearnings of those who have come to it to remake the land according to conceptions of agricultural reclamation, tourism, national parks, or wilderness. The national saga of fire unleashed, of fire suppressed, of fire tolerated, and of fire restored have each cycled through the Everglades, which received them and then rejected them in its time-honored way, by redirecting their consequences. It had defeated simple ideas of fire exclusion, of natural fire, of prescribed surrogates for restored fire, and might well do the same for ideas of redemptive fire. All were noble visions—ideals by which to enhance the Everglades, not chop it up or drain it off into sugarcane or cattle or commercial hunting. But the Glades do not distinguish among intentions. They only know what does or doesn't happen.

The outcome is uncertain. Too little is known and too much has been done to assert we can predict what will happen next, assuming it can even be done according to plan. But it seems wholly appropriate that someone of Rick Anderson's pedigree should oversee the next phase. Fire management in Florida has worked poorly when ideas get imposed from the outside; it has thrived best when old Florida hybridizes with newcomers or when indigenes cycle to the outside for an education before returning. Rick says he has no single vision. He wants to make fire serve the land, which means being nimble, being opportunistic, mixing fire types and timing, making up for what is lost in one place with gains in another. It means seizing control over a critical process for a landscape in rehab, or at least reasserting the significance of fire to the biota for whatever rehabilitation or restoration might come to it.

To most visitors the wild Everglades seem immutable, or they evolve with the sluggish tempo of its creeping waters. To those familiar with its biota, the place, as Rick puts it, "moves on fast forward." It moves with the speed of fire.

• • •

At the start of the new millennium two national endeavors were authorized that would affect restoration at Everglades. One was the Comprehensive Everglades Restoration Plan, a 30-year federal-state undertaking intended to reinstate more of the historic hydrology. The other was the National Fire Plan (NFP), catalyzed into political existence by the horrific 2000 fire season. The NFP soon segued into the Healthy Forests Restoration Act of 2003; together they directed considerable attention and funding to fuels treatments, to presuppression projects to contain large wildfires, and, implicitly, into projects to return fire to fire-starved landscapes. That was also the year wildfires ripped unchecked through the Northern Rockies and an escaped prescribed fire blew through Los Alamos, and it was the year Rick Anderson showed you could come home again and returned to Florida.

Everyone could understand, or thought they understood, what restoring water meant. Ever since Marjory Stoneman Douglas's classic, the Everglades had been characterized as a River of Grass. (Douglas had written the book for the Rivers of America series, hence the title.) Rivers are about water. You restore a river by restoring its water. The idea that you might have to restore it by burning, that it needed fire to maintain its organic sediments at grade, was, at best, counterintuitive. It was axiomatic that burning could not save Everglades: rehydration had to come first. Rehydrating might even be sufficient since everything else would literally flow from it.

That was unlikely, and it was a misreading of the Everglades biota. The prevailing sense endured that fire mattered because it was a problem or could become one; that fires had gone feral because the hydrology had failed; that prescribed fires were something you did because, if you didn't, even worse fires might result. There was little appreciation that fast fires complemented the slow waters. Moreover, restoring water did not threaten surrounding communities, endangered species, air quality, visitor safety, or staff. No one would, even in principle, wish its waters away from the Glades. But it was possible for many to imagine its fires gone, like the removal of an infestation of nasty exotics.

Yet the reality was that the Everglades was a dialectic of water and fire. Without fire the system would choke on its peat. Species would flee to find habitats that suited them. The saw grass—the very emblem of the Everglades— grew in ways that held oxygen in stem cavities such that it accelerated combustion. Saw grass grew to burn. Fire would not only come, as the rains would; it would overflow, as the Glades routinely did. It was not simply inevitable: it was essential. Rehydration might hold the biotic pieces of the Everglades mosaic together, like a watery grout. But it would be fire that kindled them to life.

When Rick Anderson returned, then, the park was abuzz with plans for restoration. He thought in humbler terms—redemption rather than restoration. He knew that Everglades would not, even in principle, reflood with fire as it sought to do with water. There would be no $8.6 billion program to widen the borders for fire, to broaden the room for burning. The burning would fall behind. All-out restoration was a fool's errand. What made sense was to accept the broken land as it was, and as it might be after it was once again watered, and to think about what might be possible for renewal.

Renewal doesn't have the cultural cachet of restoration, particularly in the sense of rewilding. It doesn't come with the moral radiance surrounding redemption, and its aura of attrition and promise of atonement. But it is a way to keep fire on the land in ways that will allow the future to recover and to draft from flame in new and unanticipated ways. The land does far worse unburned than burned poorly. If it holds fire, it will remain malleable enough to be resilient and allow future fire practices to evolve. The goal of the draft 2010 fire plan would assure that fire and fire management capabilities were not lost.

That's not the fanfare of a slogan that will rally the public or rouse politicians. It's the hard-won wisdom of someone who grew up in the fading light of Old Florida and returned to a neon-lit New Florida. It's the distilled lore of a Floridian pedigree operating in an abused wetland that lacks the water to quench its ecological thirst, a vast oft-trashed marsh that fronts a ravenous megalopolis and two American Indian reservations full of toxic history, a landscape rife with threatened and endangered species, poised to be overrun with exotic pyrophytes and fauna, lusted after by developers of super jetports and the Turkey Point nuclear power plant that claims a right-of-way through Taylor Slough, and blessed and cursed by environmental groups with designations as a World Heritage Site, an International Biosphere Reserve, a Wetland of National Importance, and the Marjory Stoneman Douglas Wilderness. It's a singular place without analogues elsewhere in the world. It will require a fire program of equal singularity.

All in all, not a bad place for the flow of Florida fire history to end.

CALIFORNIA

State of Emergency

BIG, BURNING, BOISTEROUS—that California should have formidable firepower was always more or less a given. The state is too large and its fires too prominent to ignore. But the shape of state-sponsored fire management was far from foreordained. Local authorities have fire responsibility for roughly a third of California; the federal agencies, for another third; and the rest falls to the state. But it was never obvious how these entities might unite and certainly not inevitable that the State of California would establish an in-house firefighting force and extend urban-style fire services throughout some 31 million acres on a scale that has made it the biggest of its kind in the world, the third-largest fire agency in the United States, and a gravitational disturbance to the national commonwealth.

All states have foresters, and all assist with fire protection beyond cities and the federal estate. Many, like New York's, were created to staff state parks or forests. Texas established a forestry bureau, which came to assume rural fire protection responsibilities (or oversight for volunteer brigades), but by 2010 (and keeping with that state's anti-institutional instincts) its permanent staff numbered a scant 375. Alaska's Division of Forestry manages 20 million acres of state forests and furnishes fire protection for 150 million but has little presence outside those lands. Probably the closest cognate is Florida's Forest Service, which maintains general rural fire protection for most counties and became a national pioneer in institutionalizing prescribed fire. But none compare in scale and heft to CalFire.[1]

Some quirk of California geography and life allowed a puny Board of Forestry to bulk up into a behemoth and have its fire obligations dominate the others. The simplest explanation is the spasmodic tempo of California history, specifically its proneness to cataclysms. No single institution could by itself cope with such catastrophes; each new crisis forced the state to fill in the gaps between other fire agencies and to seek alliances among them; each cataclysm boosted overall capacity a quantum level. So effective did the response become that the system not only reacted massively to cataclysms but paradoxically discovered it could not survive without them. It found new ways to live beyond its means because, in California, the mean has meant little. It is the extreme event that drives history. California became a permanent state of emergency.

• • •

In 1885 the state created a Board of Forestry. It was timely if toothless, part of a false dawn across North America to halt untrammeled slashing and promiscuous burning. The board was able to hire a few "agents" and enroll citizens to help enforce laws about starting and fighting fires, with little effect. It withered away by 1893.[2]

In 1905, the same year the national forests were transferred to the Bureau of Forestry, renamed the U.S. Forest Service, the California legislature passed the Forest Protection Act. The act reconstituted the Board of Forestry, appointed a state forester, and granted him authority to designate volunteer fire wardens who could, in turn, enforce forest and fire laws and impress citizens to aid during emergencies. The act further allowed counties to organize "fire districts" (at their expense). It granted to the forester responsibility for the state's park, Big Basin. And it permitted the state forester, in times of "particular fire danger," to staff fire patrols with the cost borne by counties. In effect, California did on a state level what was happening on the national scene. Appropriately, the state's first hire was E. T. Allen, from the U.S. Forest Service.

The state, in practice, contributed little. The counties paid for patrols, the fire wardens were volunteers, and the only effective fire protection force was that furnished by the newly endowed USFS, supplemented in Southern California by organizations committed to protecting watersheds. In 1911 the Weeks Act allowed for formal cooperation and grants in aid between the federal government and qualifying states ("qualifying" meant the state had to contribute funds to the common cause of fire protection). Mostly, on-the-ground firefighting was the work of private range, timber, and watershed associations, which contributed labor in kind and occasionally funds for trails and fuelbreaks.

The breakthrough came in 1919 with two new laws. One reconstituted (again) the Board of Forestry, which eventually came to be known informally as the Forest Department. The other permitted the state forester to create administrative units and appoint state fire rangers to supervise them and granted some funds. The upshot was, in principle, an integration of government fire services. California could now join the Weeks program and claim federal subsidies; and it could operate on rural lands in ways jointly financed with the counties. In 1923 a public outcry halted an 80 percent cut of the forestry budget. Instead, the legislature enacted statutes under which counties could create fire protection districts and forest landowners could be charged for fire protection by the state if they did not maintain a "fire patrol" on their own. Few counties outside of Southern California took advantage of the act. The upshot was to increase state responsibility.

Still, an actual presence was meager. The 1919 program consisted of four patrolmen hired seasonally. In 1922 the Forest Department erected its first fire lookout. By 1923 the state had 16 rangers, 4 inspectors, and 2 lookouts. The next year Congress upgraded cooperative fire programs with the Clarke-McNary Act, which quadrupled the federal contribution. Cooperative agreements with national forests allowed for mutual aid along shared boundaries. Twenty counties contracted with the state to provide some level of protection. But field results were still lean. In 1927 the Forest Department had a staff of 28 rangers, 6 patrolmen, 7 inspectors, and 9 lookouts.

The reality was, the state never appropriated enough to do the task it set for itself. Instead, it looked up to the Feds for grants and down to the counties, fire districts, and landowner fees to staff for protection—and to a state emergency fund, whose expenditures fluctuated from $50,000 to $300,000 annually, "a huge sum compared with the regular State Forester's budget of those days," as Ray Clar observed dryly. The state had effectively miniaturized the national system, with itself assuming the role of the federal government disbursing to the counties as the USFS did to the states and relying on emergency supplements to cover shortfalls. Increasingly, the cost of big fires forced it to consider a quasi-permanent staff of patrolmen instead of pickup labor. The apparent logic of fire argued, to many minds, for the state to create its own fire service for rural lands rather than outsource the job to others. The evident logic of politics and finances argued, however, that such a conception was a chimera.[3]

The Great Depression changed the calculus. Again, as though an echo of the federal Forest Service, California's sponsored a comprehensive survey of the issue (the Sanford Plan as surrogate for the Copeland Report), found additional funds and staffing, and built out an infrastructure in short order. Emergency monies and the Civilian Conservation Corps helped match means with ends. California became the largest and most audacious arena for CCC presuppression projects, of which the 650-mile-long fuelbreak known as the Ponderosa Way may claim special honors for hubris. Between the New Deal's Works Progress Administration and the CCC, emergency programs erected over 300 lookout towers, 9,000 miles of telephone line, 1.16 million miles of roads and trails, and numberless fire stations, tool sheds for smokechasers, and office and storage buildings.

With local options flattened by the crisis, the Department of Forestry stepped in to make the case for a California-wide "master fire plan." Its essence was the belief that the various jurisdictions of governance should each be responsible for their own lands, which would leave Forestry to assume fire protection on the remaining landscapes as designated by the state. The plan would do for fire what parallel schemes would do for water. No longer would Forestry concentrate only on sites of high-value timberlands and watersheds. As the only fire department capable of reaching much of rural California, it would extend its mantle over the countryside. It could provide consistent, measured protection—the only governmental entity equipped for the task. The legislature was ambivalent, appreciating the value of the service but alarmed about funding it. The general fund was in deficit. The CCC was being decommissioned. Repeal of the Compulsory Patrol Act of 1923 was imminent. For two years floods not fires had submerged requests for more revenue.

What spared collapse was catastrophe. The threat of war led to a scheme for civil defense, prompted by the War Department and the State Council of Defense—what became the California Fire Disaster Plan. As an existing network already responding to emergencies, the California Department of Forestry (CDF) was enrolled and designated as a statewide dispatch system. The scheme was entirely rational, and almost wholly hypothetical. For over two decades California had only begrudgingly built up its rural fire capabilities, preferring to decentralize and cooperate in ways that left the state more a broker than a player. A lot of shoving went on during legislative scrimmages, but neither side could effectively move the other.

Then came Pearl Harbor, and written plans doffed hardhats and staffed engines and lookouts in the expectation of attacks or sabotage on the mainland. At this time Forestry was under the administrative aegis of the Department of Natural Resources. Its director, Kenneth Fulton, proposed a dramatic escalation in state expenditures in anticipation of the extraordinary effort the war would demand. It was a sum calculated to jar the most jaded observer—"considerably more than the total war-caused needs requested by 22 other State departments." In effect, CDF would become California's department of defense. The money came through. The master fire plan gave Forestry its marching orders.

Over the course of World War II, California completed the infrastructure— "all the essential features of a full blown ideal plan," as Clar remembered— that the CCC had begun. The state would furnish fire protection wherever its resources allowed. It would step in, if reimbursed, to augment county forces. It would finance emergency firefighting. The funds bestowed—money almost coerced—by the threat of war would become the new budgetary floor. The military added manpower to firelines. Conservation camps of both juvenile and adult inmates replaced the CCC, and then persisted as a permanent source of labor. In place of coordinating volunteers, CDF was on its way to becoming one of the dominant institutions for fire protection in the nation.

Of course problems persisted. Squabbling was incessant over who should pay how much and for what. What exactly were the expectations on state responsibility areas? How much should the state pay counties for contract services if they so elected? (There were five, four of them in the south.) How much should CDF evolve from strictly forestry issues into an all-hazard fire service? Always, too, even in flush times like the 1960s, the state failed to appropriate funds sufficient to what standards demanded. Instead, the funding gaps were made up through expenditures from an emergency fund, and if the state's account was in overdraft, through federal aid. A wartime emergency had created, almost overnight, a comprehensive system. A continual cold war on fire kept it running.

• • •

What makes the California scene distinctive, however, is not that a state agency swelled so large but that wildland and urban fire melded. Even as the state's landmass was being pulled toward one or the other pole at the expense of a rural middle, something caused those extremes to fuse, and to hold together for a common cause. Some factor had to act as a flywheel to keep the pistons that powered fire protection, whether in cities, parks, woodlands, or pastures, in sync. That is the historic role of the Office of Emergency Services (OES), and as its name suggests, what forced the disparate parts into a single engine of response was crisis.

No entity could cope with the scale of California calamity on its own; no one could keep constantly on hand all the materiel and personnel that an emergency might demand, and even planning for an "average worst" event ignored the nonlinear—very unaverage way—in which California cataclysms collided with California society every decade or so. When the winds shrieked and the flames poured over the ridges, neighbor had to help neighbor, and when that failed, the region reached further, as it did with water. But as much as stockpiled apparatus, an agency charged with responding needed communications adequate to ensure that requested help could talk to the requester, and that once on the scene the differently uniformed and equipped responders could speak to one another. It needed incentives adequate to break down tribal allegiances and protocols. It needed a profound external shock.

In 1941 the California legislature enacted a war powers act that bestowed on the governor authority over all civilian protection agencies, notably fire departments, in the event of attack or a declaration of war. The governor assigned that particular responsibility to the attorney general, who established a State Fire Advisory Committee to oversee fire protection across 10 civil defense regions. The group included representatives from the U.S. Forest Service, National Park Service, California Department of Forestry, state fire marshal, and the chiefs of the three largest municipal departments (Los Angeles, San Francisco, and San Diego). The state forester chaired the committee. The provisions moved from theory to practice shortly after Pearl Harbor. War against Imperial Japan and Nazi Germany made possible a collective fight against fire.

When the war ended, the military crisis was discharged into civilian life. In 1945 the California Disaster Act replaced the Advisory Committee with a similarly membered California State Disaster Council. Probably earthquakes could have replaced the threat of invasion, but major tremblors came too rarely and randomly. A state of continual emergency demanded a cataclysm that would recur in place and time with some regularity. California fire, particularly Southern California conflagrations, was ideal. The United States, too, commenced what might seem a permanent state of war, first with Korea, then the Cold War, and a succession of regional hot wars. Each external threat—the onset of the Korean conflict coincided with the Soviet Union's first atomic bomb—boosted the capacity for internal reaction to hazards of all kinds. The state established an Office of Civil Defense directly responsible to the governor, which included a Fire and Rescue and Emergency Services Branch and resulted in a Fire Disaster Plan.

Over the next decade the program underwent almost annual upgrades, and no less significantly, it found pots of honey to offset the dose of threatened vinegar. In 1951, in a civil-defense version of the Weeks Act, the federal government announced a program of matching grants for state and local authorities to acquire fire and rescue materiel. The projected windfall was significant—a hundred triple-combination engines, 29 heavy-duty rescue trucks, 100,000 feet of quick-couple pipe, and the basics of a statewide radio system. Gaining access to this cache was a powerful incentive for fire services to sign on since the engines, when not called out for emergencies, would be housed in local departments. To ensure equity as well as efficiency, the hardware needed software, however, so the Fire Advisory Board adopted a protocol for distributing the Office of Civil Defense's largesse to its members. All would be available to all. The old practice of cooperative fire and mutual aid, previously restricted to shared boundaries of cities and forests, swelled to cover 100 million acres. As in other matters, California assumed the role and scale of a nation in itself.

The pressures mounted while the Los Angeles Basin filled out and tract homes shouldered against the ridge spines, debris dams, and interior hills even as conflagrations seemed to come as often as the summer drought: 1953, the Monrovia Peak fire; 1954, the Panorama Point fire; 1955, the Refugio fire; 1956, the Malibu and Inaja fires. Southern California resembled a pyric fault zone, with each stressed patch rupturing in sequence. Then came 1961 with the Basin and the Harlow fires along the Sierra foothills, and, most notoriously, the Bel Air-Brentwood disaster, which burned the backlots of Los Angeles city itself. Each appealed to OES for support and each in turn stimulated the demand for more.

The sprawl of fires appeared to outpace the blistering urban growth. Even the proudest, most autonomous fire department could not keep pace. The hits kept on coming. The Decker fire, the Loop fire, the Canyon fire—the blowups were killing firefighters as much as they were leveling houses and unsettling watersheds. But it was not enough to send more hose: the flames were crossing the borders that separated not only incompatible land uses but fire services. The Forest Service trained to fight free-burning wildfire; counties and cities trained to protect structures and evacuate civilians. It was not easy to reconcile those tasks, and for all its impressive dimensions, the institutional edifice was, up close, full of cracks and loose boards.

Still, this being California, only a truly Big One could rattle the still-lingering complacency and confidence. The jolt came in 1970.

• • •

That year big fires blew a thunderous rain of sparks through the gaps. From September 22 to October 4, 773 fires broke out, of which 32 escaped initial attack, blackened 580,000 acres, burned 722 houses and some 200 additional structures, and inspired the then-largest mobilization in California history. Of the 32 big fires, all but 3 were in Southern California, including the monstrous Laguna fire (160,000 acres) and a complex that swarmed over the celebrity Santa Monica Mountains. Under OES direction firefighting forces converged from across the state, and then from around the West. They came from the Feds: the U.S. Forest Service, the Bureau of Land Management, and the National Park Service. They came from the state: the California Division of Forestry, the National Guard, Conservation Corps camps, and OES's own reserves of engines and support. They came from local authorities: cities from San Diego to Los Angeles to Oakland; counties from San Bernardino to Humboldt, and especially the "contract counties" of the South Coast—Los Angeles, Ventura, Kern, and Santa Barbara. Outside fire crews poured in, from Forest Service hotshots to the Southwest Forest Firefighters, Snake River Valley laborers, and local Hispanic field workers. Fire engines by the hundreds funneled south. Some 28 air tankers flew missions, and a flotilla of helicopters dumped water, retardant, and burnout flares. At a time when critics of the endless Vietnam War were arguing to "bring the war home," the fall of 1970 seemed to realize that ambition.[4]

The 1956 fires at Malibu and Inaja had advertised the impending crisis, the 1961 Bel Air-Brentwood fire had broadcast the message widely, and the lethal 1966 Loop fire had confirmed the high costs in money and lives. Each had yielded targeted reforms. But the 1970 fires sought to catalyze the whole, to imagine a collective response on a par with the state water plan or the reorganization of its university system. That charge fell to the Task Force on California's Wildland Fire Problem, which promptly and efficiently identified the usual suspects and prescribed the traditional cures. It recognized that the 1970 explosion had plenty of antecedents and would foreshadow many offspring if nothing substantial changed.[5]

The breakthrough came when Congress ordered another approach under the direction of the Riverside Forest Fire Lab. In 1971 a group headed by Richard Chase reincarnated the California fascination with systems engineering that dated back to Coert duBois (this time updated by experiences from the aerospace industry) and sought to make the *process* of firefighting work better. The tangle of jurisdictions and jumbled hardware could be—had to be—made to function much more smoothly. While it seemed unlikely that planners could shake tract homes free of wooden shingles, zone out construction in the wind equivalent of mountain debris fans, or even agree on the ultimate purposes of fire management, it should be possible to improve firefighting. The outcome was Firescope.

Firescope (FIrefighting REsources of Southern California Organized for Potential Emergencies) had its research charter approved in 1973, moved into field trials in 1975, and went operational in a graduated series of expansions from 1977 to 1979. Ideally, it sought a systems approach by which data would flow in, predictions about fire behavior and countermeasures would be generated, and a collective response from resources pooled by many agencies would follow. In practice, the elaborate gathering of information and modeling of fire behavior fell by the wayside, and what remained was the essence of the firefight itself, a means by which to coordinate personnel and equipment from scores of agencies to a common crisis. By means of sophisticated software, Firescope sought to reconcile a jumble of hardware platforms and the organizational cultures that operated them. It would do for individual incidents what OES did for statewide crises.[6]

Almost immediately, however, the chasm between wildland and urban fire services threatened to undo the enterprise. They had evolved in utterly different ways: all they had in common were those moments when flame put them both at a common border. Researchers were astonished that the two cultures not only had different terms for apparatus and operators but struggled to find terminology both could agree to. The example all cited was what to call a machine that squirted water. Urban fire departments called it an "engine." Wildland fire departments called it a "pumper." The terms reflected more than different classification schemas; they were markers of different occupational cultures, such that discussions quickly pivoted on the heritage and relative strength of their experiences. Yet each lexicographical difference was sand in the gears of common operations. Each difference was multiplied by scores of jealous jurisdictions.

All politics being local, and border exurbs being the shared focus, the provincial city and county fire departments came to dominate. They insisted that all vehicles that pumped water be called "engines," that fire officers be chiefs, battalion chiefs, captains, or engineers, not fire bosses, line bosses, or fire management officers, that standards for apparatus and performance come from the urban rather than the wildland side. Firefighters would wear turnout gear; fire officers would display bugles of rank on their collars. In the end, cooperation meant co-option. And because what happened in one part of California had to reconcile with the rest if emergency callouts were to succeed, the triumph of the urban model in the south meant its propagation everywhere.

Still, it took another 15 years and more catastrophic fires to replace the centrifugal forces of the assorted institutions with the centripetal power of OES. Firescope and the OES coevolved. In 1971 OES updated the California Emergency Plan, which included its Fire and Rescue Mutual Aid Plan. The 1977 fire siege, during which Firescope was vigorously tested, was followed by a further upgrade. In 1980, another big Southern California fire year, which included the ravenous Panorama fire, OES assumed full management for the program, and the National Wildfire Coordinating Group examined the program for possible national use. In 1982 the incident command system was rewritten into the National Interagency Incident Management System, and the federal government ended its contributions. In keeping with tradition, some 60 percent of the original system design remained unfunded.

The participating agencies appreciated that if they wished to realize the full opportunities proposed by cooperation, reforms would have to encompass the state and to embrace an all-hazard model. OES coordinated the effort to spread Firescope lessons northward under the auspices of the California Fire Information Resources Management System. The integration was completed in 1986, not only between north and south but between Firescope and OES Fire and Rescue Service Advisory Committee as well. The outcome was the California Emergency Management Agency and a commissioned needs assessment for the future. The reforms arrived just after the fire siege of 1985 and before the siege of 1987. Further stress tests on the system followed in 1991 with the East Bay Hills (Tunnel) fire and the 1992 Los Angeles riots. By allowing for instant cooperation when requested, the biggest fire departments in the country in effect got bigger.[7]

Firescope succeeded as few fire research projects ever have. Because the logic of cooperation and the catalyst of catastrophe demanded ever more, its operational core, the incident command system, went national and then international to underwrite a universal protocol for all-hazard emergency systems.

Interagency Management Teams went to Yellowstone in 1988. They went to the Twin Towers in 2001. They assisted in the recovery of the space shuttle *Columbia* in 2003. They joined the Hurricane Katrina response in 2005. Significantly, the National Incident Management System as a research and development program migrated from the federal land agencies that first sponsored it to the Federal Emergency Management Agency (FEMA), which now oversees its further development. That shift, however, had already been anticipated in California through OES. It was a familiar California story of an intrastate solution that became a national norm.

• • •

In 1945, the same year it passed the California Disaster Act, the legislature enacted a Forest Practice Act that established a permit system for range burning, and appropriated funds to purchase lands for a state demonstration forest. Forestry had to respond to controlled burning, wildfire, and timber harvesting. In its conception the agency remained a predominantly rural presence. But big fires, urban sprawl, and an institutional conscription for cataclysms all pushed the California Department of Forestry away from its origins.

The postwar economy, after a boom decade, shifted from commodities to amenities. Cattle moved from ranches to farms and feedlots, forestry meant recreation and biodiversity not timber harvesting, and wildfire bolted out of wildlands and the rural countryside for an urban fringe. Even with a third of its responsibility lands lightly populated, the counties (or local fire protection districts) that contracted with CDF for emergency services were either filling with houses and malls or had their institutional geography deformed by such developments. What happened everywhere in California happened with CDF: the urban model dominated. Meanwhile, even as demand increased, CDF's funding from traditional sources shrank, quickened by the tax revolt that culminated in Proposition 13 in 1978. Though it sought to update traditional practices with such measures as the Chaparral (later, Vegetation) Management Program, its change in context from managing land to servicing sprawl inexorably changed CDF.

The one constant was fire control. Even rabid tax protesters demanded protection, and big fires, as emergencies, stood outside routine budgets. More and more its fire mission defined the agency. Its original uniform patches featured a circle with a green conifer in the center. In 1979 its patches balanced two parts, one with a green tree and the other with a red flame in a triangle. In 1987 the agency changed its title to the California Department of Forestry and Fire Protection. In 1999 it eliminated the title "ranger" in favor of "chief." Collar brass now identified ranking. The next year CDF abandoned its classic green and khaki uniform, long the trademarks of a forester, for the navy blue favored by urban fire services, particularly Los Angeles County Fire Department. In 2006 the agency completed its transformation by relabeling itself CalFire. It had become an urban fire service in the woods.[8]

Paradoxically, by narrowing its land management mission into emergency response, CalFire had grown large. By the time of their shared centennials in 2005, it was second only to the U.S. Forest Service nationally as a fire and emergency agency. It exercised primary responsibility for 31 million acres and provided a degree of emergency services for 36 counties. It had a permanent staff of 3,800, a seasonal boost of 1,400, and a conservation (inmate) corps of 4,300 arrayed into 39 camps. It owned 23 air tankers and 11 helicopters. It had 58 bull-dozers and 38 aerial ladder trucks. It responded to over 5,700 fires annually— and more than 300,000 incidents. It had an operating budget of $775 million. How much it actually spent depended, as always, on big fires, busts, and sieges.[9]

• • •

The saying that "fire does not respect borders" is, like many truisms, a half-truth. Flames certainly ignore boundaries that do not, in fact, bound anything other than names, and the phrase comes with a tinge of scorn where borders signify political entities that attempt to divide on a map what nature holds in common. Yet borders can also join together what nature has sundered. Both trends characterize California.

The need to respond to overwhelming crises, particularly wildfire, forced agencies to cross lines. If an agency stayed only within its jurisdictional boundaries, it would fail. The scope of cataclysm would overwhelm it. Yet in devising ways to cross boundaries, California the state unified what otherwise did not have common cause. It cajoled, coerced, cooperated with, co-opted, and otherwise cojoined what nature had sundered. It bonded Sierra redwood with coastal sage, Los Angeles with the San Gabriels, San Francisco with the Ventana Wilderness, and that most fungible and intangible of all intrastate divides, Northern and Southern California. Northern California found it difficult to schedule prescribed fires since its crews could be sent south; fire seasons, north and south, were out of sync; but they were merged under the master fire plan dedicated to suppression. The impact of these reforms could transcend California altogether: an "incident" became a category dissociated from any particular land and its history. The incident command system could consider Maine in the same breath as Texas.

Those forced fusions demanded a powerful jolt of energy, and one that could repeat itself. California found that essential catalyst in recurring catastrophes, genuine or imagined. In the case of fire, the cataclysms were all too real, all too frequent, and all-too-often prone to border crossings. In the end, even California could not contain them. As both its critics and partisans have long believed (and feared), California, it would seem, is unbounded.

Cajon Pass

SAN GORGONIO MOUNTAIN, where the Peninsular Range pivots into the Transverse Range, stands as a sentinel to the Los Angeles Basin. To each side lies a major pass into Southern California. Between it and Mount San Jacinto, to the south, lies San Gorgonio Pass. Between it and the San Gabriel Mountains, westward, lies Cajon Pass. Here the Old Spanish Trail emptied into the basin. Cajon Pass was the historic portal that joined Southern California to the rest of the nation.

It remains so today. Each successor transport system has built on that early trail. The California Southern Railroad punched through to connect the Santa Fe Railroad to the basin and San Diego. Other utilities followed—17 in all. There are two more railways. State Highway 138 got a big sibling in 1969 with Interstate 15. Southern California Edison runs three high-voltage (500kV) power lines and two 237kV lines. There are four oil and natural gas pipelines and five fiber-optic cables. And for backpackers the Pacific Crest Trail runs over the summit. The I-zone refers here to infrastructure. Cajon Pass is a femoral artery to Greater Los Angeles. If something shut it down, the effects would ripple not only throughout the South Coast but the nation.

Yet Cajon Pass is also a natural corridor for wind. Cajon and San Gorgonio Passes are among the most routinely windy sites in the state. More significantly, they are prime portals for the seasonal Santa Anas that drive the most volatile conflagrations. This means fire will flow as much as rail traffic, natural gas, and digital bits. In this extreme half-built environment, however, fire cannot be tolerated. To close down the corridor for even a few hours would have effects that could cascade throughout the country for days. Cajon Pass is a corridor: it is equally a chokepoint. It must be kept open.

So it mandates fire protection, but of a peculiar sort. It requires a fire service that possesses the intensity of urban firefighting but can operate within a setting that most people would characterize as wild. Fire engine meets fire wind. High tech meets high geology. The Mormon Rocks station halfway through the pass has the second-highest call volume in the national forest system. Sycamore Pass station at the bottom has the third highest. (The highest, Oak Flats, lies on the west side of the San Gabriels.) Cajon Pass not only symbolizes the fire challenge of Southern California but is itself a forcer of that system.

• • •

The San Bernardino National Forest has long prided itself on a fire organization that can match its mountains. Fire is to the San Bernardino what timber is to the Olympic or recreation is to the Tahoe. Its on-forest resources—a fleet of engines, four hotshot and two Type II crews, a rappelling crew, helicopters, including two Type 1 helitankers, a dozer module, an air tanker base—are commensurate to that charge. It originated the idea of a Forest Service honor guard.

But the San Bernardino provides an infrastructure, and has a reach, well beyond its own borders. It hosts the Regional Interagency Wildland Fire Training Center. It oversees the San Bernardino Air Base, a regional facility converted from the decommissioned Norton Air Force Base—the largest and most modern in the country, and usable by air tankers and heavy helicopters (two of which are on contract), which it can fill with water, foam, or retardant. It operates a Federal Interagency Communications Center (FICC) that handles emergencies of all kinds, including law enforcement, across nearly 30 million acres. The center runs 24–7, 365 days a year—the busiest FICC in the nation. It is an urban all-hazard model adapted for fire in the Mojave Desert, the Transverse Range, and the edge city. All of this occurs within sight, literally, of the country's largest metropolis and amid the highest concentration of media in the world. It is an amphitheater in reverse: the fires burn as though projected onto a giant IMAX screen. The air base even has bleachers so the public can watch.

What drives the system is not just that the San Bernardino Mountains have fire, but that they have big fires that threaten massive assets and act out in full public display.

• • •

The firefight is the great set piece of the American way of fire. But Southern California has bulked it up with performance enhancers until it stands to the rest of the country as San Jacinto Mountain to the Salton Sea. It has slammed the big fire against a big built environment. Wildland fire has to deal with structures, and urban fire has to cope with brush and hills. As the space between flame and city shrank, the decision space for fire's management went up in smoke, and the distance between the fire traditions that had emerged for each realm disappeared until they fused into something distinctive to make one of America's informing fire cultures. Unlike Florida, it did not result from a new graft onto an old rootstock as fire practices devised for hunting and ranching adapted to new purposes. Unlike the Northern Rockies, it did not evolve by reincarnating old practices into new avatars, a novel way of living in expansive mountain wilds. Unlike the Great Plains it was not a seasonal ritual of working landscapes.

In Southern California fire management means fire suppression. It means pushing an urban fire service into the frontcountry of a mountainous backcountry. For firefighting, this is the big time, animated by an if-you-can-fight-fire-here-you-can-fight-it-anywhere attitude. Everything is magnified; the flames, the costs, the aftershocks. Between 1990 and 2010 some 85 percent of structures burned nationally in the intermix have been in California. Today, some 50 percent of the Forest Service's national fire budget goes to California, and over half of that to the four urban-flanking forests of Southern California. Public scrutiny gets broadcast through a media bullhorn. There is scant margin for error. There is little tolerance for failure. In Southern California the Big Ones not only leap over mountains. They can, through institutions, cross continents.

Cajon Pass is a portal to another world. Through it Southern California has drafted wealth, energy, and ideas from outside, even as it projects itself outward to the rest of the country. Stand at Cajon Pass and see what this means and how it happened.

Imperium in imperio

N 2011 TWO CENTENNIALS commemorated the origins of America's modern era of fire protection. The U.S. Forest Service celebrated the Weeks Act, which created the basis for cooperative fire protection between the federal government and the states and established a national infrastructure that still allows for common practices and mutual assistance. And Los Angeles County honored the creation, after several stutter steps, of its Forestry and Fire Warden Department, which evolved into a full-spectrum fire service.[1]

When they began the two institutions had much in common. They shared a common birth parent in forestry. Gifford Pinchot, the founding chief of the USFS, was the country's first native-born forester; and he had persuaded his family to endow the School of Forestry at Yale, whose first class graduated in 1904, the year before the Transfer Act gave the Forest Service responsibility for administering the forest reserves. Stuart J. Flintham, LA County's first forester, had graduated with forestry degrees from Cornell and Yale and then worked for the USFS before moving to California. The two agencies accepted common ambitions: to protect the land from fire and to reforest what was damaged. (The California Board of Forestry completed the triumvirate.) But they also differed in ways subtle and profound.

Over time these divergences widened into a chasm. The Forest Service worried over woods and an imminent "timber famine" and saw its mission as managing for what was termed "forest influences" on the public domain. Los Angeles County fretted over brushland watershed, attempted afforestation, and saw its mission as the protection of life and property in a metastasizing metropolis. At a national level, the difference is the bifurcation in how Americans live on the land that has split the countryside into wildland and city, each with its separate fire institutions. The Forest Service had begun with wildlands and over time had to cope with an urbanizing periphery. LA County began with an urbanized core and had to absorb a fractal wildland fringe.

Accordingly, they evolved very differently, despite a common pedigree in academic forestry and a shared conundrum in urban sprawl. They viewed the scene from opposite sides of the I-zone. The USFS protects houses reluctantly since those structures reside outside its jurisdiction and its mission, which it sees as overseeing the uninhabited public estate. The Los Angeles County Fire Department (LACFD) protects houses because that is its primary charge, and it manages yet-uninhabited lands so far as necessary to serve that goal. The competition, and cooperation, between those two visions define much of the contemporary California fire scene.

What happened in Los Angeles County is a cameo of California fire history. Whatever their hopes and ambitions, county foresters watched their fire mission overwhelm their forestry mission, and this scenario, pioneered by LACFD, was repeated at the state and federal levels. The metamorphosis of the California Board of Forestry into CalFire recapitulates the LACFD scenario; so does the U.S. Forest Service's evolution as Region Five; and, as critics darkly worry, so might wildland fire management in the country overall. Although history is fine-grained, full of idiographic events and the quirks of personalities and places, the visible outcome can resemble a kind of remote sensing in which any heat source strong enough to be recorded saturates an entire pixel. The effect is exaggerated but can be defining. Its peculiar fire scene filled the Los Angeles County pixel; the LA County pixel filled the Southern California pixel; the Southern California pixel filled the California pixel; and many elsewhere in the country fear the California pixel may fill the American pixel.

In this way LACFD is not, in brief, simply a convenient emblem of such changes. It has been a major driver of them. It pioneered a unique hybrid of wildland and urban fire services, and thanks to an interlocking system of fire suppression it projected that invention beyond its borders. Much of the character of California fire is an outcome of LACFD's vision, tenacity, wealth, and genius for publicity. For a century in California it has been the strange attractor, the not-always-seen but ever-present disturbance that tugs and yanks and pushes the quotidian world of fire protection.

• • •

Los Angeles County sprawls across 4,000 square miles, reaching from the South Coast beaches over the Santa Monica and into the San Gabriel Mountains. It embraces the major biotas, industrial ecologies, and urban landscapes of Southern California. The ontology of its fire protection system recapitulates the phylogeny of fire agencies in the state.

The county created a board of forestry that echoed the state's. Its director was titled Forester and Fire Warden. It thought in terms of forests and "forest influences," and since trees were sparse, the agency planted them, for which it quickly established a nursery. But the critical duty was fire control: fire protection for rural and small urban landscapes, for watersheds that regulated the vital water flow for town and agriculture, for the newly planted woodlands. The fire-flood cycle in the mountains was the leverage for public funding. Watershed councils contributed funds to build trails and fuelbreaks. The county distributed monies to the Angeles National Forest to do likewise. For a decade the county experienced an annual roster of 100 fires that burned 20,000 acres.[2]

It was the big fires and the bad years that mattered. The Big Ones burn off the most vegetation, set the most debris in motion, and most mobilize political will. The first crisis year was 1919 when two fires on the San Gabriels, one at 40,000 acres and the other at 75,000 acres, blew away the combined efforts of the county and federal foresters. Two years later the USFS convened the first national fire officer meeting at Mather Field, and two years subsequently, the state legislature allowed for the creation of fire districts, which set off a frenzy of constituted districts. Amid these rising expectations came the revolution that followed.

Everything happened in 1924. In California the fire protection districts began to gel; the country fire warden was deputized a state warden; the state contracted, with those counties that wished to participate, for fire protection on state responsibility lands. These funds became a regular budget item for the county. Essentially overnight, Los Angeles County acquired the third-largest fire department in the state. Meanwhile, Congress passed the Clarke-McNary Act, which greatly expanded the scope of the 1911 Weeks Act to allow for forested watersheds that might not contribute to navigable rivers; a trailing amendment specifically included the "brushland" watersheds of Southern California. The spigot opened for federal funds to the state. Since the state paid Los Angeles County to provide protection, that money flowed to county coffers. In fact, LA County claimed the lion's share of state monies. At the same time, the county and the Angeles National Forest signed a mutual aid agreement that allowed each agency to make initial attack as needed across their shared border and for reimbursement after the fire was out. The arrangement sparked one of the great tag teams in fire protection as the two agencies found ways to complement one another and compete for innovations. Within two weeks a wave of fires swept over California—the first fire siege, really, of the modern era. Over 50,000 acres burned on the San Gabriels. The Forest Service was so alarmed that it convened a national board to review the season, the initial installment in what would become a California tradition.

The outcome confirmed the basic institutional infrastructure for fire protection. Within weeks the largest county fire department in America—one of the biggest of any jurisdiction—sprang into existence. That it appeared to come from nowhere and crystallized, under crisis, almost instantly seemed peculiarly and suitably Californian.

• • •

As it emerged from this forced chrysalis, fire protection in Los Angeles County found itself split into two distinct operations. One was anchored in the fire districts, which matured into urban fire models as their landscapes ripened into cities. They were funded by property taxes; they used regular engines and painted them red; fire crews dressed in black uniforms. The other operation was the forestry or mountain division. This was a wildland fire brigade, more or less interchangeable with U.S. Forest Service or California Division of Forestry crews. It relied on county appropriations supplemented by state and federal funds. It ran foot crews and pack strings, and where vehicles were possible it invented hybrid pumpers, which it painted green. Crews dressed in light green work clothes. There was no way to confuse one group with the other. The county, in fact, exactly reflected the state overall. All that joined them was the informal sobriquet, the Los Angeles County Fire Department. There was no other easy way to fuse the two systems.

As the decades passed, the red overtook the green. The land was paved and pocked with houses; the nominal wildlands were crowded to the margins or left over the hills; industrial accidents challenged wildfires for the headlines. Highway accidents rose. Lookout towers came down. The money, the political interest, the future—all slowly turned red. Crises helped. When the San

Francisquito Dam broke, when an earthquake shook Long Beach, when a natural gas pump erupted into a fountain of flame at the Santa Fe Spring oil fields,
when the Pickens fire was followed by the LaCrescenta-Montrose flood, the
county turned to its fire department as the only agency capable of tackling the
required logistics. During the Depression, when labor became cheaper, the forestry division continued to innovate by using the Civilian Conservation Corps
and then creating comparable camps for juvenile inmates who were trained to
fight fire, a program subsequently expanded to include adult inmates. During
World War II, as labor dried up and war-industry growth overflowed, LACFD
turned, as it always had, to technology and cooperative programs to pick up the
slack. It made the case that it was appropriate to externalize funding—to insist
on outside contributions—since the war was a national undertaking. In 1940, as
war clouds blackened, the fire chiefs of the Greater Los Angeles area gathered
together to find ways to share resources during emergencies. The upshot was a
Mutual Aid Act, which was subsequently expanded throughout California by
the state legislature in 1941.

The postwar boom brought further reorganizations as urban sprawl outpaced
fire protection, as it did most other services. Each consolidation brought the
forestry division into greater harmony in structure and style with the general
fire department. By 1954 the two divisions became, for administrative purposes,
interchangeable. That year bad fires and floods led to an interagency Southern
California Watershed Council that sought to balance funding for the mountains. Usefully, the burns coincided with Operation Firestop, a multiagency
project aimed at transferring wartime science and technology to firefighting
(LACFD was a major sponsor). By now LACFD had responsibility for half of
a county urbanizing at breakneck speed. (In July Disneyland officially opened
amid former orange groves.) The next two months saw breakout fires that
exceeded those of 1954 and began killing crews. In 1956 it was Malibu's turn
(again). In 1959 the Woodwardia fire, started by a disgruntled firefighter, so
alarmed county officials that the board of supervisors declared LACFD Chief
Keith Klinger a "fire czar," free to order whatever he thought necessary. The fire
claimed the first death by an air tanker drop and the first air tanker fatality. The
two realms of fire, open flame and internal combustion, were converging with
thoughtless, lethal power.

But much as the landscape was rapidly urbanizing, so was the fire service.
Between 1954 and 1969 personnel had doubled to 2,500, and stations had multiplied from 80 to 113. All divisions shared this explosive growth. So fast had the

region sprawled, however, that LACFD had to rely on technology rather than staffing to meet the demand, particularly for wildland fire for which engines could not substitute, as they could in urban settings, for handcrews. The other strategy, also traditional, was to find cooperators who could fill needs during emergencies. The state's master mutual aid agreement, overseen by the Office of Emergency Services, was one solution; and after the 1970 season, so was Firescope. LACFD was among the Big Five fire agencies who signed on to the Firescope prototype. When the program went operational in 1976 to 1977, it was headquartered in LA County.

Behind the move to consolidate—an old theme for LACFD—was a worsening fiscal crisis. Even as development blew explosively outward, funding was imploding. Legislation in 1972, 1976, and especially 1978 with Proposition 13 gutted the property tax that financed fire districts and contributed to the county general fund. The federal Cooperative Fire Program also underwent a contraction that in principle (though not always in practice) deleted the generous grants to state forestry bureaus. In response the county consolidated further, hugely in 1990, and found ways to draft some lost monies from the state treasury. What had emerged as a clever practice—mutual aid—became a necessary one.

The big fires kept coming. Flames broke out; crews fought them and sometimes fell. Malibu Canyon was a fire bellows. Fire sieges gripped the South Coast in 1993, 2003, and 2007. The first two came on the heels of economic busts that followed the end of the Cold War and the dot-com bubble; the last coincided with the subprime real estate bubble that burst to become the Great Recession. Yet while property values may have tanked, they were still substantial: LA County embraced a wealthy part of the largest metropolis in a wealthy country. The assets under protection were immense; and the future promised more development, which would demand yet more intensive shielding. LACFD rose and fell with the regional economy generally. Or it did until the state and federal governments found they could no longer afford to subsidize California in the manner to which it had become accustomed. Only the war on fire, a conflict without a horizon, endured.

When the smoke lifted, LACFD was an urban fire service—an all-risk first-response outfit—with some peculiar landscaping in its jurisdiction. Fewer than 20 percent of its alarms involve fires of any kind. The forestry division shrank to a relict appendage. Although some 42 to 43 "badged foresters" remain on staff, their number is dwarfed by the scale of the agency overall, and they serve an urban constituency. Today, LACFD supplies fire services for 58 of the county's

88 cities. It has responsibility for the county's unincorporated areas, which it protects under contract from CalFire. It is the regional coordinator for California's Emergency Management Agency. It has mutual aid agreements with Santa Monica Mountains National Recreation Area and the Angeles National Forest. It is one of two departments qualified for international deployment for urban search and rescue. It is one of the five largest fire departments in the nation.

• • •

Its story, however, is more than big money protecting big-money assets. On the national scene its landscape and mission made LACFD an outlier, but an outlier that paradoxically claimed a nuclear core of American fire. The Los Angeles scene was not simply an index of national trends but an instigator of them.

Both LACFD and the Angeles National Forest pride themselves as agencies of "firsts." The Angeles was the first forest established for urban watershed, the first to house an experimental forest devoted to brushland, the first to devise preattack plans, among the first to create hotshot crews, exploit helicopters, helitanks, and helijumping, the first to implement the incident command system. It hosted the first national-level postseason review. It pioneered fuelbreaks systems. It was an urban-edge forest before the term was invented. But LACFD has been no less innovative; it is as proactive in fire suppression as Florida counterparts are in prescribed fire. From its origins it has turned to technology as the only means to cover expansive lands with explosive fires. The department used horse patrols and mule strings, invented a mountain fire truck, sent out fire guards on motorcycles, early turned to field telephones and radio, experimented with portable pumps and flamethrowers, adapted helicopters for night flying, flew pony blimps for reconnaissance, and tested air tankers of all sizes. It was an active sponsor of Operation Firestop. It fielded bulldozers big enough to swallow Abrams tanks and outfitted them with protective cabs. When fire broke out on Santa Catalina Island, it requisitioned hovercraft from the Marine Corps base at Camp Pendleton and sent them across the strait bulging with engines. It was a first mover for inmate crews. It field-tested ideas about defensible space in the form of brush clearance regulations. It flung a net of remote automated weather stations across the county. It had more firepower than many state forestry bureaus.

It was no less innovative with institutional technics. The latent disasters that lurked in its elemental matrix of air, water, earth, and fire meant that LACFD had to be self-contained. It needed the capacity to rebuild its own capabilities if it was to assist in a catastrophe, so it has its own corps of mechanics, plumbers, electricians, carpenters, and emergency generators; it can repair bulldozers, engines, and a fleet of helicopters. But its real reach extends much further because of its willingness to seek out mutual aid and use it for leverage. Its ancient alliance with the Angeles boosted both agencies; together they dominate the region. Its contract relationship to CalFire made it a state power. That triumvirate transformed Los Angeles County into a national presence. Paradoxically, its wildlands kept it a *fire* department, even as its urban disaster capabilities have sent crews to earthquake-blasted sites in Haiti, New Zealand, and Japan.

• • •

Its story is, by and large, the modern California story in miniature. Its de novo origins, begun rapidly from scratch; its "instant" creation as a fire department by a fiat of reorganization; its blistering postwar growth; its appeal to technological solutions amid a polyglot population that both swelled and turned over seemingly as often as El Niños returned; in short, an institution that constantly reinvented itself even as it metastasized—all this looks a lot like California overall. But then one reason why California fire looks the way it does is the irrefutable presence of Los Angeles County.

In 1995 LACFD had a budget of $500 million. After two booms and two busts, it operated with a "strapped" but still daunting 2011 annual budget of $900 million. Its Forest Service sibling, the Angeles, has the largest fire budget in the national forest system. Its state partner, CalFire, has largest state fire program in the country. Together they define an institutional matrix for the Southern California fire scene and one of the three prevailing national cultures of wildland fire. What makes the complex distinctive, however, is that Los Angeles County has not been merely the beneficiary of state and federal attention (and largesse), but a stimulant to them. Los Angeles County is the fire that won't die.

Mending Firewalls

I N SAN Diego County the landscape softens. The mountains are less towering, the valleys less yawning, the contrast between urban and wild less shrill.
Ownership is dappled; ranches and farms still spread over the higher plateau.
Oak savannas and soft chaparral mingle with hard brush and granite boulders
that appear to graze on the land like sheep. Orchards and parks daub the peri-
urban scene. But as the urban pockets gel and add layers like a concreting
nodule, the borders become more rigid, and a creeping suburban pastoralism
hardens into a fractal frontier with little to buffer the once rural from the fast
urbanizing.[1]

This kind of geography frustrates management, if you think of management as the application of explicit principles to distinctly bounded lands. Or as
Robert Frost famously put it, good fences make good neighbors. As landscape
geometry, however, San Diego belongs more with Schrödinger's cat than with
Euclid's *Elements*; administratively, it requires a conceptual leap, like moving
from Newton's laws to quantum mechanics. The world tends to seep between
those borders, to replace the exactitude of a drawn digital world with the sloppy
analogue of the real one. Moreover, borderlands are notorious as places in themselves, not simply a swirl of what fronts them. As Frost also noted, Something
there is that doesn't love a wall. That wisdom—both sides of it—applies to fire.
And it manifests itself, with paroxysms of violence, in Southern California.[2]

Certainly fire both respects and ignores walls. The essence of fire suppression is to erect one, called a fireline; and the goal of much presuppression, to create a fireline in advance, called a fuelbreak. Yet vigorous fires frequently leap over them. Some walls are both necessary and, when most needed, frequently useless. In this regard they reflect the most fundamental of conundrums in fire management, how to cope with the big fire, the rare but catastrophic event. But they also exhibit, in miniature, the dilemma of fire in the Southern California I-zone, in which society tries to mix what doesn't want to be together and to segregate what wants to join. That tension focuses on the line in the dirt intended to separate what, increasingly, has no separation. The fuelbreak, or its most recent avatar, defensible space, is where the battle has joined.

• • •

The fuelbreak has a long history in the region. Local authorities were advocating (and financing) fuelbreaks as dual-purpose firelines and trails as soon as the mountains were gazetted as forest reserves. There were fuelbreaks in the San Gabriels before the Forest Service took over their administration. They provided, in principle, a means of access, a method to break up continuous fuels, a rudimentary fireline ready to activate when needed, and a visible display of administrative resolve. They inscribed their message across the great screen of the mountains. Unsurprisingly, the nation's grandest experiment was of course California's Ponderosa Way.

Over the years fuelbreaks have displayed a cycle of senescence and regeneration much like that of their surrounding vegetation. After each disastrous fire season, existing fuelbreaks are scraped clean and widened and the system expanded. Then, they decay. They are expensive to maintain; other needs clamor for the money; critics scorn the ridgeline scratchings, which they regard as ugly and useless. The secondary system overgrows. Only the primary roads and those deemed most essential receive maintenance. Then the flames rush over the landscape, the public demands protection, and the fuelbreaks return. The life cycle of fuelbreaks, in brief, shows the same rhythm as the chaparral in which they are embedded.

In classic theory the fuelbreak assumes two forms. One breaks up the interior of the reserve in order to assist fire control. The principle is identical to that used in the built environment: create firewalls that retard spread and give firefighters time to wrestle the blaze into control. If sited in a forest plantation, the fuelbreaks would be incorporated into the design of planting. The most successful outside planted landscapes act as levees rather than dams; they flank the terrain-and-wind-directed flow of the flames rather than try to stand against them. The other variety of fuelbreak is intended to guard a reserve's perimeter; it serves as a biotic or fiery fence. This is rarely an integral feature, but rather something imposed onto and across a landscape that wants to behave differently. The reserve's borders might cross firesheds as they do watersheds, which means fighting the order of nature rather than working with it.

While the arrangement of the fuelbreak system has remained more or less permanent, its purpose has changed. The reason for persistence is easy to explain: this is where fire behavior and fire control—and politics—argue for breaks, and such considerations don't alter. But the nature of the fire threat does change. Originally, foresters recognized that the surest way to prevent big (and costly) fires was to contain small ones, and for this a network of access and buffers was necessary. Originally, too, they constructed a system to prevent the promiscuous (and sometimes malicious) fires set all around the forest reserve from entering.

But as the belief took hold that some fires were inevitable and good and that fire control could itself be damaging, the pressure for those interior fuelbreaks lessened. They became more an act of political posturing, a signal to the public that fire officers were wary and prepared. The bigger change involved the perimeter fuelbreaks. Increasingly, these became less critical for keeping fire out than for keeping it in. The catastrophic fires were those that kindled in the mountain reserves and then spilled out onto the surrounding countryside or, more properly, cityscape. The fuelbreak evolved away from a matter of protecting the reserve from the community and toward protecting the community from the reserve. It became a question of public safety, like debris dams and flood control channels.

It was one matter, however, to sculpt a fuelbreak along a continuous border across the flank of a mountain. It was quite another when the "border" was a speckled landscape of inholdings, fractal suburbs, and infrastructure nodes, or when, amid developed landscapes, it assumed quasi-natural features in the form of parks, greenbelts, and zoning for protected species that did the speckling. Critics bridled that by the time each enclave had its belt of clearing, there would be nothing left. The issue is particularly acute for San Diego because the city has 900 linear miles of ravines and some 55,000 houses along a fire-exposed fringe. If each entity put in the 100-foot or more clearances recommended by fire authorities, the practice would effectively wipe out relict landscapes. It would convert nature to city by stealth. It would, so the argument goes, create an urban desert visible from space.

Instead, critics want to focus on the structure itself, the home ignition zone, which is to say, the structure and its immediate environs. This is where surface fires and ember attacks actually take out houses, and it is often the sheer congestion of houses that carries the fire as each involved structure ignites those around it. The critics want fire protection to concentrate its hardening here, which would shrink the penumbral zone of clearing. If development is dispersed, flames would wash around the structures, and if concentrated, they would dissipate as they ran out of fuel. A wildland fire, or an urban conflagration, would shatter into manageable house fires and then dissolve.

Behind this conception are related disputes about what fire management means. It begins with an argument about the nature of protection, or to state the issue slightly sideways, do you manage fire at its source or its sink? The fire-as-source group sees the problem as managing the landscape where fire originates. The fire-as-sink group targets the places where fire strikes. The hydrologic equivalent is whether to control flooding by improving watersheds or by erecting levees and dams. The fire-as-sink group envisions the task as focusing on the assets at risk, both hardening individual structures and making the I-zone overall more resilient. It reasons that if the goal is to protect houses and the citizens who reside within them, then protect those houses directly, not remake the amorphous landscape from which the fires emanate in the vague hope of eliminating risk. In this view nature needs protection from overzealous and misdirected fire control. Each group defines mitigation differently. The source group focuses on the need to mitigate against fire; the sink group, to mitigate against unwonted fire prevention measures.

Each sees with different eyes the border between the fire and house, and accordingly each assesses fuelbreaks differently. The source group, intent on managing fire within its fireshed—not only to prevent escapes but for ecological purposes—wants interior fuelbreaks to help and perimeter fuelbreaks to halt as much as possible. It can cite an honor roll of successes. The Harris fire, where burning out along the International Fuelbreak spared Tecate, Mexico. The Border 16 fire, where flames crossed north over the border and took only a solitary house, the sole structure without defensible space. The Shockey fire, where a treated neighborhood in Campo shook off even Santa Ana–driven flames. The Banner and Angel fires, where burnouts along the Sunrise Fuelbreak shielded the town of Julian.

The sink group poses a counternarrative. It sees fuelbreaks as ineffective when they are most required. The region's vaunted defenses did nothing against the monster fires such as the Cedar and Witch that define the contemporary scene. Worse, they aggravate the abuses lavished on an already damaged landscape by creating grassy fuses to carry fire and channel invasive pyrophytes into unscathed chaparral. Giving every homeowner a personal fuelbreak in the form of overly expansive clearances called defensible space guts any hope of preserving a vestige of native plants and habitat. It means subversive urbanization by other means.

So the debate continues, each side more effective at criticizing the other than promoting its own agenda. Both groups, however, design with a particular fire and environmental risk in mind. But the Southern California scene is nothing if not dynamic. The representative fires keep morphing. The assets at risk keep moving. The only constant is the argument of neighbors across the wall, one rebuilding it and the other willing to let it decay.

Then the next conflagration comes.

•   •   •

Each side can point to failures and successes. Yet in the San Diego region there are two examples of enduring fuelbreaks that seem to perform exactly as designed. It's worth pausing to examine how and why they work and what their costs might be.

Fuelbreaks succeed best when they are integral to a built landscape, or when they are part of a planted agricultural complex. They have worked in pine plantations on the shores of the Baltic, the sand dunes of the Landes, and the Sand Hills of Nebraska. They have contained fires along railways lines, often combined with grazing, planting, and controlled burning. As greenbelts, they have shielded new communities, even cities. They have worked in India and Ghana to define and defend the boundaries of gazetted forests. In most instances the threatening fires do not scale up to conflagrations, and in no cases do fuelbreaks succeed by themselves, any more than firewalls will keep a building from burning down; but they buy time and assist firefighting.

They struggle when retrofitted or imposed over landscapes in defiance of terrain, wind, and fuels. When local conditions favor large fires, only very large fuelbreaks can help check them, and that effectively means type conversion, transforming whole landscapes, which in Southern California means housing tracts. Still, under less than extreme conditions, they can leverage fire suppression, helping channel a fire; and in miniature forms, as defensible space, they can encourage engine crews to stay with a house that they might otherwise yield to the flames. If fuelbreaks fail during extreme conditions, so do all the other strategies and maneuvers of fire management. The Big One continues to haunt wildland fire management.

There are, however, two exceptions, although they may prove the rule because they show what a strategy of fuelbreaks can cost. One is Camp Pendleton. On a map of regional fuelbreaks Pendleton's dense network sits like the textured surface of a grenade. The camp is laced with roads and fuelbreaks, all mandatory to contain the burns kindled by endless live-fire exercises. But the camp is also surrounded by a defensive belt, like a demilitarized zone, that is annually cleared and burned. For what it is designed to do, the fuelbreak system works. It is integral, dense, comprehensive. Very few fires leave Pendleton, very few enter.

The other example is the International Fuelbreak that spans some 40 miles of the border with Mexico. It originated, as so many large fuelbreak projects did, in the 1930s when the Civilian Conservation Corps program demanded public works commensurate with vast pools of unskilled labor. When the CCC left, so did the fuelbreak. It was revived in the 1950s by the California Division of Forestry, this time using California Department of Corrections conservation camp labor. It decayed again. It revived forcibly during the 1990s when Operation Gatekeeper sought to control illegal immigration from Mexico around Tijuana. The flow of migrants moved east. Abandoned campfires pocked the chaparral, escaped fires swarmed over the mountains.

The International Fuelbreak then joined other efforts to secure America's southern frontier. In 2002 an interagency alliance, using conservation camp labor, reconstituted the project. It began with a stretch two miles long and 300 feet wide, "with some vegetative islands for wildlife and aesthetics." Plans called for extending the reach through the mountains for 30 miles. In the 2003 fire siege, with the Otay fire driving south and west under Santa Anas, the initial attack incident commander successfully burned out along the break and controlled the larger burn "with very limited resources."[3]

Limited on the fireline, but not limited socially. Such fuelbreaks are a significant cultural investment; they have to tap national funds and purposes, for fire alone is an insufficient justification. A conflagration could rekindle their clearing but only some larger ambition could sustain them. And beyond apathy, or a distracted public, they carry environmental costs. Even with bottomless

CCC labor on hand, the Forest Service experimented with chemicals to keep the breaks clear. It had enrollees spread arsenic, and later, when CDF oversaw the network, agencies sprayed an herbicide that achieved notoriety in Vietnam as Agent Orange. The breaks became corridors for weeds and invasives.

Even a dense network still requires an active firefighting force; engines and fuelbreaks have to act in concert since protection is a social dynamic, not an inert piece of infrastructure. To commit to a model of fire control by fuelbreaks and engines, the threat from fire has to be high, for such landscapes can resemble the fire equivalent of a police state.

• • •

In 1989, four years before the cycle of Southern California conflagrations renewed, John McPhee published an essay that captured exactly the absurdist outrage that much of the country felt toward the region. "Los Angeles Against the Mountains" described the Sisyphean task of holding back the debris the San Gabriels continually shed, like a snake sloughing old skin. As with most Southern California stories this one involves fire.

The ecological narrative is simple. A big fire is followed by heavy rain that sends unstable regolith downstream. In extreme events the runoff gushes as a debris flow, as though a ridge liquefied and roiled down canyon full of mud and boulders. The social narrative is equally simple and predictable. It begins with debris dams constructed along the alluvial fans to stop the nuisance flows. The major debris events overrun them and fill them, which means the old ones must be cleaned out and new ones built farther upstream. The process escalates. The defining consideration, however, is what to do with the debris. With the landscape downstream built out—the alluvial fan has become a cone of tract homes and mansions—there is nowhere to dump the fresh debris, which means the dams must fail. The solution is an "elegant absurdity" by which the San Gabriel Dam, first erected in the 1930s, must be continually cleaned out and the only place to haul the fill is farther up canyon. "They take the debris and carry it back into the mountains," as a spokesman for the Department of Public Works explained to McPhee, "where they create a potential debris flow."[4]

Yet this is much the same formula for fire. Substitute fuelbreaks for debris dams, fuel for debris, conflagration for debris flow, all for the protection of private houses so they can survive in harm's way only because the cost of protection is absorbed by the public. Rather than fix behavior, in this case the real estate market, the region invests public monies (and national insurance funds), much of it from outside California, into infrastructure. The debris—fuel—regrows, and either it is mechanically cleared or it burns.

Its fuelbreaks thus do for Southern California what the state master plan for water does. It allows a city sited for economic reasons to hold out against prevailing natural forces. When its economy or its social resolve falters, the sea, for the one, and fire, for the other, rush in. Over and over, the problem repeats with a slight increment of intensity added to each cycle.

• • •

So it all continues. The fuelbreaks reappear in the same places. They cause the same disturbances. They boost firefighting capabilities against the lesser burns. They fail during the conflagrations. They reappear under new names or are modified to accommodate new realities—housing tracts rather than ranches, McMansions rather than barns. While in San Diego the contrast between mountain and plain is less intense, its discourse is just as fierce.

Apologists explain that anything built can only meet reasonable standards, not everything imaginable. Engineers design for a 50- or 100-year flood, not a millennial one, or for a 5.8 or 6.7 earthquake, not for a Richter 8. Similarly, fire agencies traditionally plan for an average worst event. But Southern California operates on extremes, not means; there is no "average worst." When it breaks, it tends to break completely. You might be able to stop the flow of flames but not the shower of sparks. For every reason advanced to keep fuelbreaks, there is a reasonable objection, and a reasonable objection to the objection, like debris dams backing up debris dams. The pragmatic solution would be to shift the argument about whether, abstractly, fuelbreaks are right or wrong, to whether, practically, in particular places, they are useful or not.

The real reason the discourse endures is because those wildland firewalls are tangible emblems of philosophical differences. They trace out separate conceptions of fire and its purpose, they define the border between competing ideas about how to live on the land. So even as they are scraped down and grow back, the old lines of debate persist, like buried fault lines that from time to time rupture under the stresses of deeper forces. Those walls are, in the end, the lines of negotiation between a nature that doesn't care and a society that doesn't want to worry.

The Big Ones

UNTIL MODERN TIMES city and country had always shared fires. American cities were typically rebuilt forests and burned with the panache and patterns of their surroundings. Countrysides were shaped by the demands of cities, both near and far, with slashed landscapes more prone to explosive fires. The same winds blew over both. The same logic of protection led each to prevent errant sparks, build firebreaks, and quench flames quickly.

During the frontier era, the two realms blurred. Then each matured and the fire scene calmed, as conflagrations disappeared from metropolises and were harnessed into the tamed rhythms of rural burning. More recently, in what might be termed a pyric postmodern phase, an out-migration of urban folk has begun recolonizing the countryside. This new frontier broke down the firewalls that had separated urban from wildland fire. The intermixing stirred by sprawl occurred everywhere, but it happened with particular vehemence in California, and most spectacularly, it burned. Urban fire returned, like an ancient plague once thought extinct and now revived in a more virulent mutation. City and countryside had to cope with fire along their shared fringe.

That was true north as well as south. The San Andreas Fault is less a geologic cut, neatly cleaving the crust into two sides, than a swarm of breaks as deep stresses release their strain and ripple through heterogenous rock. Even so, two regions along its complex trace stand out. To the south the swarm flexes into the Transverse Range. To the north a parallel fault swarm, highlighted by the massive Hayward fault, doubles the zone. The San Andreas proper runs to the west, and its echo, the Hayward, to the east. San Francisco Bay lies between them. The two dominant urban fires that frame California's fire century face each other with one of those faults at their back.

In the three centuries that span from Jamestown to San Francisco's immolation, a westering population had hastily erected towns and then watched them burn. The process of squeezing flame out of cityscapes came slowly and fitfully. Boston and New York continued to burn well into the 19th century. Only as buildings became less combustible (more brick and stone than wood), as flame became less abundant in daily life, and as firefighting became better organized did conflagrations reluctantly leave the metropolises and join footloose folk on the frontier. The last major urban fire on the East Coast was Baltimore's in 1904. San Francisco's fire two years later was the final conflagration of the settlement era.

Historians have come to think of the 20th century as a "short century," defined by the onset of the Great War in 1914 and the end of the Cold War in 1991. For fire history the corresponding events are the San Francisco conflagration of 1906 and the Oakland holocaust of 1991. The two cities pair off across the bay; so, too, they bookend a century. They are the Big Ones that haunt the imagination of California fire. They were big not because of geographic size but because they slammed into cultural centers that could register the shock. They killed. They incinerated, for the one, the major entrepôt of the western United States and, for the other, an elite community. They were fires that burst through nightmare into fact.[1]

• • •

In many ways San Francisco's fire history echoed that of the nation. The embryonic city was incinerated on Christmas Eve of 1849. The next year three fires swept through what was a high-order mining camp masquerading as a town. In 1851 the Great Fire destroyed three-fourths of the city, the same proportion as the 1906 fire. What the fire missed a second fire claimed later in 1851. Then, as the city passed through its accelerated adolescence, its fire scene settled down. Outbreaks more resembled episodes of domestic violence or saloon brawls than out-in-the-streets rioting. The big fires moved out to prospect the countryside.[2]

The mature city erected fewer combustible buildings and disciplined fires through a vigorous fire department. Paradoxically, without major burns to cleanse the cityscape—a kind of creative destruction—the city housed more and more relic structures from its wooden age. In October 1905 the National Board of Fire Underwriters investigated San Francisco and reported that 90 percent of the city's structures were wooden framed. No other major metropolis approached that figure, but then no other had progressed so suddenly from canvas tents to an edifice complex with monuments such as the Ferry Building and new city hall. Only 54 years before, the city had consisted of surveyor stakes in windswept grass.

When, at 5:12 a.m. on the morning of April 18, 1906, the San Andreas ruptured and sent earth waves rolling across the bay area, like a hurricane driving high seas before it, the city was vulnerable to fire; but it was not living on a knife-edge of conflagration. Rather than slow down settlement or compel rebuilding to national codes, the city had relied on its fire department to halt fires before they became large. For decades that strategy had worked because the number of fires was small and access relatively easy. An earthquake, however, changed that calculus. It kindled many ignitions, its rubble and ruptures checked movement, and it broke water mains. Firefighting alone might not succeed, and in fact, in the hands of the military and vigilantes who eventually took over the process, it proved fatally flawed as dynamited buildings and clumsy backburns almost certainly encouraged fire spread.

Given the density of the built landscape, it did not take much for stubbornly established fires to take out city blocks and then most of the city. A former chief geologist of the U.S. Geological Survey, G. K. Gilbert, was in Berkeley when the tremors struck. On Friday he took the ferry across the bay and recorded the fire's movement. "The westward progress of the fire north of Market Street has been checked chiefly by backfiring at Van Ness Avenue." Houses across the street were "blistered and had glass broken." The fire had blasted across at one place before being checked at Franklin Street. "Backfiring is in progress N. of Pac. Avenue, and apparently being carried to the waterfront." From that point "the fire rushes up the slope of Russian Hill, consuming block after block of houses—chiefly of wood. The flames work with wonderful speed. While I lingered, whole squares were consumed. An hour is probably enough to raze a square of wooden houses." Ever the scientist—when the tremors had awakened him, while still in bed he began timing the shocks and analyzing their direction by the swing of the chandelier—he measured the burning time for a two-story house. "Roof gone in 7'; first falling of wall in 9'; flaming ruins in 12'."[3]

This was primarily a fuel-and-ignition fire. The April weather was warm and fog free, but California's parching summer had not started, the city was not gripped by drought, and the winds were mild. Synoptic systems moved through the region, slightly out of phase with the fires. There was an episodic easterly breeze, and the first day, Wednesday, the winds shifted confusingly between damp northwest and dry east flows. The second day saw the winds blow from the east, drying the scene and driving fire away from the bay. On Friday the northwesterlies returned. On Saturday it rained. There were terrain effects, notably as fires moved up San Francisco's fabled hills. But mostly what propelled the burn was the density of available combustibles and the abundance of ignitions—those that started the initial outbreaks and those set deliberately or ineptly in attempts to stop it. The critical winds were apparently those created by the fire itself.

The 1906 fire was a plume-dominated conflagration, as Jack London reported from the scene. "It was dead calm. Not a flicker of wind stirred. Yet from every side wind was pouring in upon the city. East, west, north, and south, strong winds were blowing upon the doomed city. The heated air rising made an enormous suck. Thus did the fire of itself build its own colossal chimney through the atmosphere. Day and night this dead calm continued, and yet, near to the flames, the wind was often half a gale, so mighty was the suck." When it ended, all that survived was "the fringe of dwelling-houses on the outskirts of what was once San Francisco."[4]

The defining event was neither fuel nor wind but a 7.9 earthquake that overwhelmed the fire protection system of the day. Too many fires, too little water, too much social chaos—that was the fire triangle that shook San Francisco to its foundations. For the rest of the 20th century urban fires in developed countries followed a similar scenario. It took earthquakes, wars, or riots to break down the built landscape and its social infrastructure sufficiently to where it could carry fire. As the century progressed, the San Francisco scene was turned inside out as urban fire moved to the fringe and left an incombustible core. Cities were subject to occasional catastrophes in high-density structures but no longer to free-burning conflagrations.

The final tally reckoned 4.7 square miles burned—508 city blocks, or 2,832 acres. Some 28,188 structures were lost (88 percent of them wooden). The official roster of fatalities was 700, but observers believe the likelier total ran between

2,000 and 3,000 lives. The official narrative deliberately turned away from the earthquake as cause, since it was a geologic precondition that residents could do nothing about, to the fire, which they could. For the next century the city that had begun in a de facto fire rush ceased to burn.

• • •

Refugees poured across the bay into Oakland. The saga of Oakland's fire history thus began as an aftershock of San Francisco's. The newcomers met a fire scene not unlike San Francisco's before it urbanized. The landscape consisted of grassy hillsides dappled with copses of trees, often redwoods, nestled near springs. The golden hillsides burned regularly and harmlessly. As long as the city lined the wharf or served as an outpost to San Francisco and Berkeley, its urban fires were those typical of wooden towns everywhere. What changed the dynamic was when houses moved up the hills.

The East Bay was a miniature Transverse Range. The Hayward fault shouldered up an embankment of hills. A seasonal wind, known as the Diablo since it flowed past Mount Diablo, a prominent monadnock, was a northern Santa Ana ready to spill over that height. Settlement evolved sentiments to preserve the framing hills and East Bay backcountry as natural preserve. All this replicated the basics that made the Southern California fire scene so explosive, and as the city crept from the bay to the hills, it promised to re-create the south's cataclysmic fires. What checked this scenario, however, was the absence of the high-octane chaparral that made blowing-and-going fires along the Transverse unstoppable. The Oakland Hills had annual grasses, savanna oaks, and riparian redwoods. Fires might be frequent and annoying; they would not be ruinous.

While San Francisco hastily rebuilt, in Oakland refugees became residents, and the city grew as an alternative. By 1910 it had ballooned to a population of 150,000. Developers—the new forty-niners of California prosperity—knew "good sign" when they saw it. On the hills, arrayed like an amphitheater that looked out on the bay and San Francisco, they erected the Claremont Hotel, platted some 13,000 acres of land for suburbs, and began softening the windy grasslands with trees, shrubs, and ornamentals including broom that would add color, texture, and privacy, and (so they argued) might serve as fire windbreaks. They planted Monterey pine, a native from the Coast Range to the south. They planted eucalypts, an exotic from Australia. And, as easy money ebbed and flowed, they planted houses.

Settlement encouraged two trends. One filled up wildland with city, the other, cityscape with trees. Before settlement some 2 percent of Oakland is estimated to have been in woods. Between 1910 and 1913 the primary developer, Frank Havens, afforested perhaps several million eucalypts, mostly around the out- skirts. The City Beautiful movement encouraged internal plantings. By the late 1950s roughly 21 percent of Oakland was treed. As the city spread outward, it absorbed more of its adjacent wildlands or, more accurately, open space. In the mid-1930s probably half of the nominal city was vacant when citizens organized a tax district to support a system of parks. By 1988 only 20 percent remained open, and that was secured under the auspices of the East Bay Regional Park District. These "wildlands" claimed the hills and their backsides. The city spread below. The two trajectories—houses and trees—crossed in 1988.[5]

What joined the two realms was the Diablo wind. What did not join them was the kind of common administrative fire service developed by the contract counties of the south, the California Department of Forestry, or the U.S. Forest Service. The East Bay thus evolved a diminutive echo of the Southern Califor- nia scene, with a regional park system taking the place of national forests, but it did so without comparable fire institutions. Alameda County did not evolve along the model of the Los Angeles County Fire Department in which a sin- gle agency had to fuse fires from both wildlands and cities. Oakland remained, proudly, an urban fire department. It had the only apparatus capable to attacking a high-rise fire in the East Bay. It did not have air tankers.

• • •

As its landscape changed, so did the character of its fires. When Oakland was a wharf with back streets of shops and saloons, "recurring fires . . . almost every year swept over the hills," according to the *Oakland Tribune*. They did to the grasslands what sailors did to their ships each year when they careened and cleaned them of barnacles. As settlement pushed farther up the hills, flames and city clashed, and fires ceased to be seasonal nuisances and became historic milestones.

On September 27, 1923, a Diablo wind drove flames through the Berkeley Hills and into the campus of the University of California. Some 3,100 acres and 584 houses burned. Ten years later a Diablo-driven fire scorched 1,000 acres and 5 homes. In 1946 another 1,000 acres burned. While the early Diablos identified "smokers" as a cause, the newer ones charged arsonists. Meanwhile smaller fires under the influence of westerly winds burned 10, 30, and in one exceptional instance, 700 acres. These were big numbers for a city but negligible for grassy wildlands. Then in 1970 after a hard frost had killed many blue gum eucalypts and rained litter over the landscape, a Diablo wind powered fire through 204 acres and 37 houses. It was the first outbreak in what became the statewide 1970 fire siege.

Alarm among the wildland fire community was acute. The nearby University of California-Berkeley, after all, hosted one of the premier forestry and fire science programs in the country. When drought and frost returned in 1977, the prospects for East Bay fire found its way into congressional testimony. What boosted concern was the character of the settlement along and under the summit of the hills. These were the residencies of the East Bay elite, the Northern California equivalent to Malibu or the Hollywood Hills, even if they were more likely to house Nobel laureates than movie stars. They made conflagrations the celebrity fires of the Bay Area. A small but damaging fire in 1980 kept the pot boiling. A scheme to establish a demilitarized zone between parklands and city was deemed both damaging and ineffective, but led to a Blue Ribbon Fire Prevention Committee, chaired by William Penn Mott (later director of the National Park Service), which issued its report in 1982 and recommended a fuelbreak, although tempered by aesthetic considerations. Still, much as with San Francisco prior to 1906, fire's threat remained more vivid than its reality.

Then came the 100-year burn. On October 20 the Oakland fire department knocked down a fire that started as a warming or cooking campfire amid a spot of pine near Marlborough Terrace. Crews mopped up by soaking the perimeter lines. The next day, while crews were on site and rolling up hoses, the still-smoldering hotspots disgorged embers, the flames got into patches of Monterey pine litter untouched by hose lines. The rekindled fire raced up a largely grassy slope to the ridgeline. There the Diablo wind caught it, and the fire blew up. With stunning speed it burned out the basin below Grizzly Peak. It burned through the Parkwood Apartments. It burned out the Hiller Highlands. It burned out Grandview Canyon. It burned over Highway 24. In the first hour it consumed 790 structures, each of which scattered new sources of ignition. What became known as the Tunnel fire spotted over Lake Temescal. It burned through the Rockridge District. When the Diablo winds finally slackened and northwesterly winds returned, the main front—a swarm of new ignitions, building after building—headed southeastward into Forest Park. A new index of fire spread, homes burned per hour, made its appearance. Before the orgy of burning ended, 3,354 houses and 456 apartments were ash, and 25 people had died. Total area burned amounted to 1,600 acres. It was America's worst urban fire disaster since 1906.[6]

• • •

A cataclysm this horrific scatters reports and after-action reviews like spot fires. This one sparked reviews at all levels of government, from citizen groups to the National Fire Protection Association, from a mayoral task force to California Office of Emergency Services to the National Fire Protection Association and the Federal Emergency Management Agency. As with all major disasters the surveys identified many causes, most of which had to happen together to produce results so far off the scale. Those factors that governed fire behavior fell into two general categories. One pertained to the fire environment; the other, to fire suppression capabilities.

There could be little dissent from the observation that the East Bay Hills were a prime natural setting for fire. A mediterranean climate, seasonal foehn winds, terrain that could channel fire like a coal chute, and pyrophytic vegetation that encrusted the hillside—such conditions would argue for fire anywhere they appeared. That the fire occurred amid a drought, after frosts that killed eucalypts, then a record hot spell, worsened the circumstances; yet the values were "extreme, not exceptional," and even the exotic flora only acted as an accelerant by allowing embers to kindle surface fuels and fling sparks from torching eucalypts. The "wildland" fire, however, had burned upslope toward the summit, not through the structures, and in the end, the canyon's flora survived better than its structures. The fuels that mattered were the houses and especially their wood-shingle roofs. So close were the houses that they burned one to another, and so combustible were the roofs that they both received sparks and recast them into the wind. The character of the quasi-natural setting allowed the fire to start. The character of the city allowed it to spread.[7]

The capacity to fight the fire was badly compromised. After decades of boom, Oakland went bust in the 1970s. City services decayed, among them the capacity to maintain the kind of varied fire protection demanded by the mix of landscapes within the city. Over and again, the urban fire service failed to integrate with wildland counterparts. It did not know of the red-flag warning posted by CDF for the day of the fire. It did not understand how mopup in wildland fuels differs from overhaul in buildings. It did not appreciate how a city, full of internal firewalls, might be breached from the perimeter and find itself assaulted not from the streets but from the air. It had not reckoned with fire-induced power failures that neutralized pumping stations. It had not adopted the incident command system and could not function seamlessly with assisting agencies. It could not communicate on common radios (CDF officers resorted to telephoning dispatchers). It had three-inch hydrants, while adjacent cities used national standard two and a half. But even if compatibility had been perfect, the fire would have likely bolted away because it was moving faster than a fire department ever could. Something else intervened to break down the response.

That something was the wind. It did for Oakland what the earthquake had done for San Francisco. It simply overwhelmed the capacity to respond. The OES report noted haplessly that "a fire burning 400 or more homes per hour does not allow for normal firefighting tactics—either urban or wildland." Even mutual aid requires time to muster engines, planes, and personnel. In the first hour nearly 800 homes burned. Within two hours the conflagration had reached perhaps 80 percent of its final size. The narrow streets soon clogged with traffic and fleeing residents. It was not possible to move people and cars out as fast as the fire moved in. Converging fire engines met outgoing civilian autos. There was no Maxwell's demon in the box canyon to sort them out. There was no single flaming perimeter or high-rise to focus the action, only hundreds of individual fires—the firefight as mêlée.[8]

The subsequent committees, panels, boards, and task forces published hundreds of recommendations, ranked by priorities. Some involved simple changes in protocol (for example, getting daily fire weather). Many, however, required costly retrofitting, either by the city or residents in the hills (such as refitting hydrants and burying power lines). Given the parlous state of Oakland's finances, only a fraction could be enacted. But perhaps the most critical need was simply institutional: the East Bay needed the fire equivalent of its municipal utility district, or what Southern California had found with its county-CalFire-Forest Service triumvirate. The South Coast, however, had a few big, wealthy entities; the East Bay had many smaller and poorer ones. Even the 100,000-acre regional parklands distributed that largesse among 65 units.

Still, the reconstruction went forward. The neighborhoods rose from the ash, with better fire protection built in. After several stumbles, the East Bay Regional Park District was voted bonding authority in 2010 to expand. A Hills Emergency Forum gathered the various constituencies into a common conversation. A memorial—shaped like a gutted house with a missing roof—was erected at the intersection of Highways 13 and 23. The scars, both environmental and social, slowly healed. Ten years later the Hills Emergency Forum sponsored a review, and another 10 years later.

The threat remained dormant, not dead. A million people crowded against the hills in polyglot patches, perhaps 70 percent of them newly arrived since 1991. Parks claimed 10 percent of the landscape, a number that would rise. The municipal economy remained feeble. The Diablo still blew. Someday another Big One would shake the hills.

• • •

It seems the worst expression of pedantry to fuss over labels and classifications in the face of such calamity. Yet nine days shy of 10 years after the Tunnel fire the Twin Towers burned, identifying a new urban fire threat. How that problem was defined led to a decadal war on terror that bled the country white and may have diminished rather than increased its security. Definitions matter.

What kind of fire burned the East Bay Hills in 1991? Was it an urban fire, a wildland fire, or an intermixing of the two? Did it follow a Northern or a Southern California scenario? Did it result from breakdowns in fire departments or from a social fracturing that made the ability to control fire—long considered the very essence of civilization—too difficult? Were the parklands along the hills a threat to the city or the city to the parklands? Did the fire's narrative pivot on blue gums prone to torching, or on shake-shingle roofs receptive to firebrands? What sorcerer's stew of poisons and social incantation actually stirred in the canyons' cauldron? How the problem was defined would decide what solutions were suitable. People would argue over fixes because they disagreed over causes.

Most early commentators, including myself, saw the Tunnel fire as an example of the emerging intermix fire scene. They looked at the gusty summit where wind and housing, stacked like cordwood, met, and they shuddered. They likened the catastrophe to the 1990 Painted Cave fire that swept disastrously into Santa Barbara. To their credit many East Bay parks people, often staffed with early-retired fire officers from the Forest Service, recognized the potential for fires to bolt out of open lands into the city and exercised leadership in creating something like the consortia that had evolved to the south. Many, too, had long experienced the arduous and eccentric politics of public involvement. Patiently, yet with a sense of urgency, they applied those lessons to the hills.

Yet closer inspection reveals the Tunnel fire as an urban fire. The wildlands were adjacent, but their only contribution, after a fashion, was the Diablo wind, which would blow with or without dedicated parklands. The fundamental problem was that the city had planted structures, as earlier developers had eucalypts, where they didn't belong, and then did so in ways that violated even urban fire standards. If the outcome didn't look like an urban fire, it's because the character of urban settlement had changed. San Francisco filled up its hills until the entire tip of the peninsula was built over. The Pyne clan arrived during the gold rush, and a family story holds that the patriarch once won Nob Hill in a poker game before declaring that "nothing will ever live up there but the goats" and traded the deed away for something usable (a case of whiskey). In reality, the goats were driven off, and the hills populated by flocks of wooden houses. Oakland (and sister cities like Berkeley) kept the goats. They mixed pyrophytic landscaping with the wooden houses and had a backcountry from which fire and wind could come. San Francisco was how cities developed in the early 20th century. Oakland Hills was how they developed near the end of the century.

The prime mover is the push for high-end urban development, of which proximity to quasi-natural settings with expansive views are valued amenities. That is why the frontier between city and park exists and why quarreling is interminable about trade-offs regarding trees and houses. The story demonstrates, however, why California has the fire management system it does. Whatever the starting point, if the site is south of the San Andreas, or its East Bay offset, the Hayward, the pressures will drive the outcome to the same responses. If those measures fail, the fires will follow.

This Time is Different. Maybe.

CALIFORNIA—THE GREAT EXCEPTION—has long acted like a state-nation. By area it is America's third-largest state. Its population holds one American in nine. It boasts the country's biggest state economy, outranking its nearest competitor by 50 percent; if it were a country, its economy would rank between fifth and ninth in the world. Its media presence, from Hollywood to Silicon Valley, is even greater. It has produced model master plans for infrastructure, water, and education. Unusually, it balances public and land, each at 50 percent. And, spectacularly, it has a world-class fire scene.

California is built to burn. It has ideal rhythms of wetting and drying, its north has dry lightning busts that can kindle hundreds of fires with a single storm, it has autumn winds that drive flame and hurl embers miles ahead of a firefront, and it has monumental terrain that textures regimes and complicates control. If people vanished overnight, California would still burn vigorously, and frequently explosively, and fires would rush unbroken from mountain to sea. Onto this complex people have added ignitions, suppressed natural starts, introduced new fuels, rearranged old fuels, pushed settlements into fire floodplains, and generally challenged the natural order for fire supremacy. The earliest European sighting of California, Juan Cabrillo's expedition near San Diego, reported smokes; 450 years later, to the day, smoke still rose across the horizon. Unsurprisingly, California dominates national fire statistics as it does so many other metrics.

It has more fires, more big fires, and more telegenic fires than anywhere else in the country. It has the most damaging fires. It has the most firefighter fatalities (327)—practically a world unto itself. It has the five largest fire departments in the country. It spends the most money—no other state comes close. It holds one of three hearths for American wildland fire—a complex of research, equipment development, and institutions rivaled only at Tallahassee and Missoula. It absorbs a quarter or more of national fire investment and suppression funds. Historically, it hosted the light-burning controversy, devised systematic fire protection, arranged for the first aerial reconnaissances, catalyzed the conversion of military equipment to fireline use, developed inmate crews, held the first national conference of fire officers, promoted the concept of a wildland-urban interface, pioneered the use of natural fire, and generally acts like a gravitational body that distorts the entire space-time geometry of the national fire establishment. If it were a country, it would rank fifth on fire's league tables, lagging only such behemoth landmasses as Russia, Canada, Australia, and the rest of the United States. Unsurprisingly, what happens in California tends not to stay in California.

• • •

For a century California has been in a kind of pyric arms race with wildfire. Since 1919, when it devised its modern arrangement for fire protection on federal, state, and country lands, a predictable cycle has replayed. A big fire threatens, damages, or at least frightens settlements—population roughly doubles every decade, so the pressure to develop is relentless. A commission reviews the outbreak and makes recommendations. The only part of that package that gets funded is more suppression resources—more crews, more fuelbreaks, more engines, and more aircraft. The rhythm of drought and big burns quiets. Then it returns, and another commission reports, and more firefighting resources are ordered. Big fires become campaign fires and by the 1980s fire sieges. Meanwhile, what had been a California quirk, a pathology of the West Coast, moved east, and California-style fire services have begun to follow.

From time to time the state has tried to move from suppression to some fire restoration and a more nuanced and comprehensive control over threatened developments. Each effort faltered. Some prescribed fire occurs, and Yosemite and Sequoia-Kings Canyon National Parks have nurtured a natural fire program in their backcountry. But burning generally has been squeezed out of the landscape, even pastoral and agricultural fires; the Central Valley has the worst air quality in the nation and can tolerate little smoke. As population intensifies and as urban sprawl reclaims more and more of the rural countryside, the distance between the wild and urban shrink, the likelihood of damaging wildfires swells, and the political pressure to protect at all costs becomes unbearable. There is less space in which to maneuver—less geographic space, less bureaucratic space, less political space. The default setting to suppress strengthens.

The rhythm of bad fires has a parallel rhythm to forget, helped by the constant churn of population. For a while the fires are national news while they blow and go around and through communities. Then they become state and local news as cleanup continues and, months later, reports are issued. Meanwhile, the media has moved on to the next crisis. As with slash-and-burn agriculture, the window for planting is short. You can seed in the ash. Wait a year, however, and the weeds of quotidian life reclaim the scene and the opportunity is lost. What persists are the optics of fire protection—the visible emblem that the authorities are doing everything possible to fight the flames. The deeper substrate in which real reform must occur, actions that restructure the conditions, that alter the odds for conflagrations or damages, is overlooked.

It is hard to see what might break this pattern. Which is why the fire busts of 2017 and 2018 matter more than most. California burned. Then it burned again. Then it burned once more.

•   •   •

Each outbreak set new records.

In October 2017, with east winds—North winds, Diablo winds, Santa Ana winds—to drive them, big fires broke out both north and south. They kindled from human-caused electrical storms, mostly power lines. The Tubbs fire crashed into Santa Rosa, plunging from exurban fringes into the urban core, where it burned over 6,000 structures and killed 22 people. It was the worst urban fire since Oakland in 1991 and nearly doubled the number of incinerated houses—in fact, the worst wildfire in California history to that time and part of a complex that blackened 245,000 acres and took 44 lives. In December fires broke out in Southern California, culminating in the Thomas fire that, at

282,000 acres, was the state's largest on record. Two lives were lost to the fire, and 21 to subsequent debris flows. North and south, the damages were estimated at $8–10 billion.

In August 2018 a fire bust struck Northern California. The Carr fire burned into Redding and acquired notoriety for a fire tornado captured on video. The Ferguson fire burned around and into Yosemite National Park. The Mendocino Complex merged two fires to set yet another record for size, 460,000 acres.

Then came November, more ignitions, more east winds, more savage fires. Northern California claimed the most attention as the Camp fire blew through the town of Paradise, killing 85 people, while the Woolsey fire burned the Santa Monica Mountains and into Malibu. More houses incinerated, three people killed, and more media attention—that a dozen celebrities were among those who lost houses boosted public interest. The Paradise fire attracted international attention, and the way the disaster lingered, day after day while new deaths were discovered and hundreds of possible victims remained unaccounted for, assured that this fire would not blow through the media as it did the town.

• • •

The damages and stakes were higher—climate change, fuel buildup, and sprawl assured that. Each of the outbreaks on its own warranted the kind of bureaucratic review that had characterized the long saga of California conflagrations, and it was unlikely that each by itself could spark the kind of deep reform necessary. Yet it was possible that the trifecta might. Barely had the smoke settled from one than the next had flared. Between October 2017 and November 2018 California had created an internal refugee population out of those who had lost their primary shelter. The scene mutated from media spectacle to genuine catastrophe. One fire was an anecdote; three were a trend.

Yet over the course of a century California had shown an exceptional capacity to absorb big burns. Nationally, mass burnings seemed to join mass shootings as something a jaded, if not calloused, populace accepted with a shrug. The smart money would probably bet on the favorite, which is to say, more of the same. Seasons might grow longer; responses would grow to meet them. But serial conflagrations—three in 12 months—are different. If anything might change the calculus of response, this might, and if it fails, it's hard to see what kind of catalyst would.

• • •

It's too much to expect that California can reform overnight, even if it chose to. But it could begin nudging into better alignment with the National Cohesive Strategy. In recent years it has revised standards for new construction, worked to reduce hazards selectively, mapped and remapped areas at highest risk, and passed laws to allow CalFire greater responsibilities, but none of this is at scale, and there is a natural reluctance to enforce codes that do exist.

What California needs first is to deal with fire as fire. Because wildfire is so graphic and visceral it too often gets hijacked to advance some other agenda. The response to fire as *fire* gets lost or thinned into the minimum needed to distract public attention while partisans argue over logging or climate change. Second, while CalFire has most of the authority it needs in the form of basic legislation, it appears to lack the social and political legitimacy to enforce them. It needs a political charter that explicitly authorizes it to pursue the needed reforms. The third response is to accept that suppression alone won't work. Californians and California fire agencies know this. The authorities just don't know how to admit to the public that they have been wrong, or at least misleading, and that reforms are not just nice things to do, but necessary. California needs a statewide fire master plan to do all the things that modern fire management, not just suppression, require.

All this matters to the rest of us because what happens in California won't stay in California. It never has.

• • •

Postscript. Since writing this essay, California has experienced another windpowered fire bust, far less damaging than its predecessors, yet a media conflagration. Power lines remain a prime ignition source; in response to some $30 billion in claims, PG&E declared bankruptcy, began rolling blackouts to several million customers, and faces threats of being taken over by the state. Smoke has become a public health issue. The destroyed housing has created a population of environmental refugees. A score of bills blow through the state legislature. A billion dollars spread over five years has been granted to CalFire to begin mitigation measures.

California has been through such trials before and will rebound. Its response looks to be both big and serious. What actual form it will take is still unclear, but those responses will likely help direct the future of American fire.

NORTHERN ROCKIES

Fire's Call of the Wild

And from the fortieth winter by that thought
Test every work of intellect or faith,
And everything that your own hands have wrought.

—William Butler Yeats, "Vacillation"

O N JULY 2, 2012, a Cessna 182 climbed off the tarmac at Hamilton, Montana, for a fire recon flight over the Selway-Bitterroot Wilderness (SBW).
That morning homeowners in Colorado Springs were being allowed back
to neighborhoods incinerated by the Waldo Canyon fire, the most devastating
in Colorado's history; that afternoon, an Air Force C-130 Hercules dropping
retardant on a wildfire in the Black Hills crashed, killed four crewmen, and
seriously injured two others; and large fires were ripping up parts of Montana
to the south and east. The day previous lightning storms had passed over the
Selway, and fire officers worried about fireworks on the upcoming Fourth of
July, although errant bullets, careless campfires, and nature's own pyrotechnics
were enough to keep the land aflame.[1]

But the Cessna observers were not looking for new smokes. They were tracking the enduring aftermath of old ones. They were on a fire recon through
history. Forty years previously the first natural fire in U.S. Forest Service history
had been allowed to burn above Bad Luck Creek. One of the two leaders of that
brash experiment, Bob Mutch, was on board, as was Dave Campbell, ranger of
the West Fork District of the Bitterroot, whose domain embraced that patch
of the Selway-Bitterroot Wilderness. Their loop west and back was a journey
across time written into geography. The land below bore the proud insignias of
four decades of prudent aggression to reinstate natural fire in one of the storied
landscapes of the American fire scene.

If Americans had a National Register of Historic Places for fire, the Selway-Bitterroot region would rank among the early entries. The mountains had been part of the Big Blowup of 1910. They had hosted the 1932 pack trip in which the leading lights of the Forest Service debated over evening campfires how to cope with big burns in the backcountry and, without accepting let-burning or loose-herding, had proposed minimum suppression. Then, the Pete King-Selway fires of 1934 had prompted a formal review of Forest Service policy that led to the promulgation of the 10 a.m. policy. Having battled helplessly against both historic conflagrations, Elers Koch of the Lolo National Forest argued that fire suppression across the Bitterroot summit had done more harm than good, and the agency would be better off letting fires burn and not tearing up the land with roads to improve access for firefights it couldn't win. But the struggle went on and a paved road punched over Lolo Pass and by the 1960s others began to appreciate Koch's prophecy. The Bitterroots became a national war zone for the protest against destructive logging those roads made possible. Despite improvements in firefighting and a smokejumper base at Missoula, fires, occasionally big fires, continued.

• • •

What changed the calculus was the Wilderness Act of 1964. The Selway-Bitterroot Wilderness was among the first slated for inclusion. Still, someone had to connect wilderness with fire. The 1967 fire bust sent smokejumpers into the backcountry and bulldozers through Glacier National Park. Bud Moore, then on a fire inspection tour of Region One from the Washington office, reaffirmed the value of hard initial attack. "The strength, speed and effectiveness of Region 1 initial attack and first reinforcements actions needs study with the objective of identifying all possible means to improve them." Shortly afterward, he returned to Missoula as regional director of Fire Control. Yet a sense gnawed at him that the old ways couldn't continue in fire any more than in logging, that "we had taken the shovel and pulaski folks about as far as we could." By 1970 he had translated a conviction that fire belonged in natural systems into a plan to test that vision. The restoration of fire might even seem an act of penance for what he had unthinkingly done to mar the Lochsa. Besides, granted the language of the Wilderness Act, he concluded that suppressing fires in such places was "almost illegal."[2]

He sought to identify possible sites, selected two, and then lit on the White Cap Creek area of the SBW, near where he himself had first experienced the wild. He found allies in Bill Worf, regional director of recreation, and Orville Daniels, supervisor of the Bitterroot, who agreed to host the trials. Then he recruited an agency planner, Dave Aldrich, and a fire researcher, Bob Mutch, to translate notions about the proper role of fire into formal documents that a bureaucracy could act on. In 1971 the White Cap Five hiked into the SBW and met over campfires, an eerie echo of the 1932 pack trip; they completed their task in the early summer of 1972. By then Sequoia-Kings Canyon and Yosemite National Parks had experimented with "let-burns" and Saguaro National Park had invented the concept of a "natural prescribed fire." On August 17, 1972, Chief Forester John McGuire signed off on the White Cap Project as an exception to the 10 a.m. policy. On August 18 lightning kindled a snag on south-facing slopes above Bad Luck Creek within the approved zone. Mutch and Aldrich scrambled to the scene, but no one raised a shovel against the smoking snag, which smoldered in a brush field that claimed the site of an old burn. Fire on fire—the old fire changed the conditions for the new one. The Bad Luck fire burned an estimated 648 square feet.

The test came the next year when the Fitz Creek fire started on August 10. This one spread, creeping and sweeping in spasms, and it sparked a similar rhythm of responses. On August 13 and 14, as the fire moved uphill toward the Bad Luck Lookout, Moore ordered a B-17 to drop retardant to shield the facility. When lightning kindled 85 fires on the Nez Perce National Forest, the plume of the Fitz Creek burn attracted a crew, which assumed it was part of the bust, and Daniels had to intervene to pull them away. On August 18 the fire spotted across White Cap Creek, putting it outside the approved zone. In half an hour the escape, now named the Snake Creek fire, bolted up the slope, an independent crown fire, to the ridgetop. Daniels ordered an overhead team to suppress the escape (it was Mutch's team, for which he served as Fire Behavior Analyst). But Daniels mandated control only for the Snake Creek fire, not the Fitz Creek. One fire could call out two responses. The pluralism that would characterize the fire revolution had its microcosms as well as its macrocosms.

Before it ended the Fitz Creek fire ran to 1,200 acres, found a chronicler in Don Moser for *Smithsonian* magazine, and could claim bragging rights for the best quote of the new era. It came from a red-eyed crew boss called in for the containment. "I'd like to meet the son of a bitch who said we had to let this one burn." Since the White Cap Project had an approved plan and had followed it, the Washington Office backed it up. The next year the Forest Service retitled its Office of Fire Control to Fire Management. Bud Moore retired.

The program had what it needed: a celebrity site, anointing by a Grand Old Man who had himself risen from the ranks, a stiff wind from the intellectual climate of the times, a trying demonstration, timeless anecdotes, room to expand, and the raw stuff for a legend. The Sierra sites were self-contained, and often overshadowed by the story of fire in the Big Trees. The Gila could blow up and no one might notice. Whatever the SBW's storied past for the fire community, the ongoing controversy over logging assured that the Bitterroot would be in the public eye. Its smoke plumes would be seen. It might even be that the naturally burned land might compensate for the mechanically scalped.

• • •

That was how the program ignited. To propagate, it needed room to spread and authorization to do so. Both took time and a stubborn patience. Suppression remained the reflex response, the default option, the fallback or rally point when times turned tough. There was, in brief, nothing natural about natural fire. It happened because fire officers and district rangers chose it.

While the mountains might have a geologic unity, they were divided between two states, four national forests, and two Forest Service regions, all proud of their separateness and all granted considerable discretion to act differently. Expanding the land base, however, particularly through the 1981 Central Idaho Wilderness Act, put wilderness next to wilderness and nudged neighbors into common practices. After the 1988 Yellowstone brouhaha, every forest had to rewrite its fire plan, which encouraged a push toward common solutions along shared borders. The 1995 Federal Wildland Fire Management Policy helped legitimize natural fire as an option. The program spread from the legally wild to adjacent wilderness study areas, roadless areas, and even general forest wildland where fire behavior challenged survey lines.

The number of acres burned scaled up accordingly, never as much as had been taken out during the previous half century, and mostly keeping up with the expanding land base and branding the land in distinctive ways. There was enough fire to keep the program's momentum high and enough success to keep it pointed in the right direction. The numbers back up the anecdotes. In the decade preceding the Bad Luck fire, when wilderness ignitions were still being fought, the combined Selway-Bitterroot Wilderness and Frank Church-River of No Return Wilderness burned 4,989 acres. In the 1970s that figure scooted up to 6,838 acres, as a few fires began to take larger bites of the countryside. During the 1980s the figure increased fivefold to 38,026 acres. In the 1990s it doubled to 78,981 acres. In the first seven years of the new millennium it swelled by half again as much to 122,982 acres. Suppression shrank to point protection and loose-herding along the redline borders that wildfire could not be allowed to cross.

What mattered were the large fires, or the big seasons that held most of the big burns. They did the ecological work, and they began to rack up enough acres that successor fires burned into them in ways that tweaked the larger landscape into a kind of self-regulation. The administration adapted accordingly. Where, at Fitz Creek, a 1,200-acre burn had strained equipment, attention, and political will, by 1998 the SBW was accommodating multiple big burns; they reckoned they could handle 10 to 15 at one crack. The problem was negotiating outside the lines—with neighbors, even where the land was also wilderness, with non-wilderness national forest, with private lands overflowing with houses, with smoke in the deep valleys.

Then the contemporary cycle of big burns set in. While elsewhere in the Rockies suppression forces struggled to counter bust after bust, the West Fork District accepted more and more fires at once, subjecting them only to surveil-lance, and by so doing liberated scarce resources for more threatening wildfires elsewhere. The 2000 season shattered any lingering complacency about the power of suppression to contain wildfire in the mountains; after 90 years the big burns still rambled more or less at will. But it also granted the new era an image of perhaps equal power to R. H. McKay's iconic photograph of the Nicholson adit where, at the height of the 1910 firestorm, Ed Pulaski had held his crew at gunpoint. John McColgan's *Elk Bath* photo from the Sula Complex in the East Fork of the Bitterroot National Forest suggested a fire as much at ease in the woods as elk. Whether formal wilderness or not, the East Fork was coming to resemble the West Fork. The argument that the fire management practices of the wilderness might be injected into general wildland gained a powerful visual logic.

A climax came in 2005, the centennial of the Forest Service. The West Fork District accepted 50 fires that season in the SBW. At one point four large burns were visible through a single pair of binoculars from the Hells Half Lookout. Managers relied on lookouts and webcams to monitor the scene. The fires began to merge, new burn into old burns, to break up, smolder, reform, blow up, creep into extinction. The complex went on for 60 days. When the season finally ended, Dave Campbell, who had been on the district since 1996, received a Chief's Award for Wilderness Fire Management. The next year he received another for Line Officer Wilderness Leadership.

Over the course of a century the Selway-Bitterroot Wilderness had gone from a place where wildfires threatened the political existence of the Forest Service to where fires, left to roam in the wild, promised its redemption.

• • •

Each season seemed to confirm the principles that, like some bureaucratic pyromancy, had been divined in the flames of Bad Luck and Fitz Creek. Or as that dour philosopher Arthur Schopenhauer once put it, "The first forty years of life give us the text: the next thirty supply the commentary." By the 40th anniversary the commentary of the fire program was being sculpted into the SBW's conifers.

One was that a fire program had to function as a system. It could not bounce from one emergency to another. It required a lot of fires over a large landscape bounded by similar practices on similar lands. A second was that one fire could be treated many ways. Some parts could be accepted, even applauded, even as other parts might need herding or outright suppression. A third was that, to restate Yogi Berra, sometimes you can see a lot just by looking. During the Fitz Creek blowup, Dave Aldrich blurted out that he had "learned more in a few days watching that fire than I did in the preceding 15 years fighting fire." If you want to manage free-burning fire, you have to understand how such fires burn, and you don't do that with your shoulder pinned to a shovel. You have to watch, ponder, and watch some more.

There is little doubt that the development of the Rothermel fire model, published the same year that White Cap Project received official approval, suggested to wary administrators that they would now have the power to predict fire behavior with mathematical rigor and that those forecasts could undergird the decision to let natural fires roam. Edgy rangers could back up decisions with the seeming imprimatur of physics. Yet such forecasts were far beyond the model's capabilities (and still are). The only valid empirical basis had to come from the land itself, not a lab. It had to come by watching real fires, and that meant allowing real fires to run over woods and shrubfields, creek bottom and ridgeline. Even with ever-more-sophisticated decision-support packages, that still holds true.

The fourth finding could only emerge out of that actual experience. This was the way fires over time affected one another. Particularly in landscapes prone to stand-replacing fires, each big burn acted to buffer and check the next. The SBW began to fill out with burns, and even to reburn. Physics might appear to underwrite fire behavior, but it was history that underwrote physics. Fire regimes were historical constructions. They emerged from decades and centuries of fires burning, scorching, reburning, scouring, and just plain skipping around; of fires that burned in patches, and of big burns that burned out patches; of fires that blew themselves out when they struck recent burns; of fires riding over the troughs and crests of droughts and heavy snow years and winds capable of blowing down whole woods. Such patterns could not be reconstructed from first principles over a season or two, or even a decade or two. They required a long commitment. Probably they would demand centuries.

The Selway-Bitterroot Wilderness was an ideal site for such an experiment. Its remoteness and, for commerce, its emptiness had kept damages to a minimum; it had resisted all-out suppression as much as Florida had. Fire control did not begin to gain traction until after the Civilian Conservation Corps arrived, and then not until the postwar era let fire suppression take to the air. Probably serious suppression involved no more than two decades before the White Cap Project began to exchange tolerated natural fire for suppressed wildfire. The disruption of fire regimes was far less than for montane forests or parkland forests that had burned every 2–5 years and had excluded fire for perhaps 80–100 years, or missed 15–20 fire returns. Still, for a place that had been declared the test site for the 10 a.m. policy, a natural fire program was a bold decision. The progeny of the White Cap Project had chased suppression to its lair.

There was plenty yet to do in the SBW. Much of the landscape had still not burned even once, while a genuine regime required reburns. But the dynamic was unmistakable. Every fire suppressed made it more difficult to suppress the next fire. Every fire allowed to burn made it easier to allow future fires. There was no neutral position. You let fire do what it needed to do, or you broke the rhythms once more and added to the ecological debt burden. Moreover, the best way to simplify the process was to continue to expand outward over all the wilderness and as much quasi wilderness as possible. Restoration needed room and time. The SBW still lagged. There persisted a misbegotten legacy to overcome; the forest borders, bristling with new houses, were becoming more intolerant of fire; as personnel cycled through the many districts (four forests and two regions), newcomers had to be initiated into the regime when they passed through the portal. Not least, suppression lingered as a rootstock, ready to send out shoots from below the graft of natural fire management. Still, the burning was accelerating.

The SBW absorbed lightning-kindled fires with little expense and scant risk unless one discounts, which one can't, the sweaty palms of higher administrators. Yet the contrast was undeniable. To one side of the Bitterroots the future pointed to more ravenous fires, ballistic costs, and a vicious spiral of self-reinforcing suppression. To the other, the future pointed to ecologically enhancing burns, self-limiting costs, and a virtuous spiral of self-contained fires. The present, as always, was a muddle, and it was often hard to determine at any moment what cycle was spinning.

• • •

On May 31, 2012, Deputy Chief James E. Hubbard dispatched a memo to regional foresters, station directors, area directors, International Institute of Tropical Forestry director, deputy chiefs, and Washington Office directors laying out guidelines for the coming season. The text opened with a boilerplate reaffirmation that fire management was "central to meeting the Forest Service mission" and that, among the many tasks of that mission, the protection of life and property was primary.

Then came the bad news. The season promised to be a tough one, suppression costs were forecast to exceed the 10-year average (and allocated budget), and money to finance the firefight was tight and would have to come from other Forest Service programs, which would dampen the chances to do other tasks and alienate various and oft-vocal publics. The two poles that had torn wildland fire over the past 40 years—the wild and the urban—threatened to rift rather than meld. Congress wanted houses protected more than it wanted wilderness burned, and no one wanted anyone to die for either cause. Certainly some of the worst wildfires of the new era had originated as prescribed burns or wildland fires given a long leash that then went feral. Rumors circulated, too, that if the Forest Service busted its fire budget this year, responsibility for fire would be handed over to FEMA, a move that would destroy every vestige of good achieved during 50 years of the American fire revolution. Accordingly, a risk decision framework was specified to guide decisions. It directed that any choices that might involve allowing fires to free-burn would be taken from the hands of local officials and put into those of the regional foresters. The bottom line was—the bottom line. To protect costs aggressive initial attack, everywhere, would be the first response.

The memo came with a series of talking points that assured readers that the decision was only temporary, a reaction to the unholy collusion between a worsening climate and a weakening federal budget, and that its authors acknowledged, and emphasized, "that such an approach is not sustainable over the long-run." Critics had long recognized that firefights were like a declaration of martial law, able to put down an ecological insurrection or the occasional riot but not a means by which to govern the landscape. The historically minded might note that just such economic logic had guided the 10 a.m. policy, which Chief Forester Gus Silcox had announced as an "experiment on a continental scale." The big fires had run up the big costs, and the simplest way to prevent big fires was to kill (and overkill) all fires while they were small. Now, it seemed, the old experiment was being rerun. Fire and line officers wondered just what, in practice, the new orders meant. The simplest solution was surely to suppress everything. No one would be faulted for aggressive initial attack.

The sympathetically minded might note that, perhaps, this really was a temporary blip. But such mandates, like temporary taxes, tend to persist. The memo sent a clear message that the safest approach—safest for careers if not for life, property, and land—was to suppress every start. The sympathetic might also argue that the guidelines could be interpreted as the final nail in the coffin of the old prescribed natural/wildland fire use/resource benefits fire category, that all fires, as wildfires, could be managed in ways to promote ecological enhancements however they were labeled. But that is not what the interlinear text said. It said "aggressive initial attack" except where unsafe or imprudent to do so. An ecological insurgency, however, can be suppressed by sending in a fire constabulary for a while, but eventually it will boil over and consume the regime. Was the fire scene like the financial meltdown of 2008, requiring a temporary if unpalatable bailout? Or was it like an Arab Spring in which the longer repression remained the more likely an outright civil war would result?

There was, in July 2012, no way to say. What a century of fire suppression in the Northern Rockies had taught, however, was that the big fire years led to the big burns that did the biological work required and that helped to contain the blowups of the future. Allowing fires except in those critical years was a charade. Worse, what the Predictive Services emanating from fire sciences was forecasting was not just an upcoming season of exceptional demands, but an upheaval of climate that might make the temporary blip into a new normal. Suppression could, at most, create only an illusion of security, and in any event would soon blow through its budgets.

So the edict would go to the court of nature for review. There are ample examples in which, however desperate for fire restoration, officials decided that this particular fire at this time and place—amid an extreme season, next to major highways or housing developments—was not the one to loose. It might be that the Forest Service was making an equivalent choice for an entire season. There are equally many examples, from life as well as the woods, where such decisions are simply a false caution that is indistinguishable from a failure of nerve or outright bureaucratic cowardice. First establish order, then reform is the cry of counterrevolutionaries everywhere. The effort to impose that order becomes itself a process of further disordering that typically concludes with open revolt.

• • •

Dave Campbell said nothing about the national directives. The Selway-Bitterroot Wilderness has been so long in evolving that it was almost unthinkable to imagine a return to the ancien régime. Yet it was also clear that its power as an exemplar might be held to the scrawled borders of legal wilderness and the peculiar standing of the SBW in the national fire saga. It was possible that no tablets of instructions—what is ultimately an expressed ethic—might emerge from its burning bush. Yet for two hours it had been enough to savor the hardwired testimony of the Selway.

As the Cessna touched down on the tarmac at Hamilton, Bob Mutch observed, "We've come full circle." It remains to be seen which circle will best express his words.

The Embers Will Find a Way

The mind is not a vessel to be filled but a fire to be kindled.

—Plutarch

B EGIN WITH THE FUNDAMENTALS, which is where a fire science that wants to think of itself as applied physics likes to begin. The first question is how a fire gets started and the second how it spreads. Know that and the rest follows. The same might be said about fire scientists.

In 1950 when Jack Cohen was born, Tucson had a population of 45,000, and the family homestead was in any event closer to Mission San Xavier del Bac. The landscape was open high desert split by the riparian cleft of the still-flowing Santa Cruz River. When he left 18 years later, having graduated from high school at 16 and having spent two years at the University of Arizona, the city had swollen to 263,000 and in its search for water had pushed wells so deep that the Santa Cruz had become a ditch. It was an object lesson about sprawl and its environmental disruptions, but Jack was not one to look back. By then he had become the equivalent of a firebrand lofted into the prevailing winds, looking for a suitable place to alight.[1]

What gives that simile some verisimilitude is that he spent his youth starting fires. He was the one to kindle whatever fire someone needed set. He burned off the lawn in spring. He burned leaves in a burn barrel. He ignited the barbecue. He shot firearms, which in an age before ear protection ruined his hearing. If someone needed a cigarette lit—this was before the surgeon general plastered warnings on packages—Jack volunteered to make the flame. The family had a summer cabin in the Santa Rita Mountains amid the Coronado National

Forest, a place readily sparked by summer lightning storms, and prone to fires, although aggressive firefighting held nature's electrostatic match to sparks and snags.

Such were the basic facts, but like principles they needed some way to combine.

• • •

His youthful fireworks were all play; there was no malice, and no one got hurt. But it was not magic, because Jack Cohen was already an inveterate scientist who wanted tests, numbers, and proofs. For years he struggled, however, to find the right science. In high school he turned to chemistry and majored in it at the University of Arizona; and in important ways he never abandoned it. Meanwhile his interest in outdoor recreation grew. But nothing gelled. He did well in subjects he liked, and poorly in those he didn't. He took a year off, worked on drill rigs in New Mexico, and learned some practical geology; the rigs destroyed whatever remained of his hearing. He realized what he did not want to do with his life.

In 1971 he decided he wanted to be a forester and transferred to the University of Montana. He soon became disillusioned with forestry, took the minimum courses, and loaded up on botany, physics, and math. The critical moment of his Montana sojourn came in the summers of 1972 and 1973, when he worked as an assistant to professor Bob Steele, who had contracts for prescribed fire research. Jack had to learn everything: not just the science, but how to build line, work with hand tools, and set fires. That's what did it: direct contact with flame. For someone who thinks visually and tactilely, not verbally, it was a baptism by immersion that merged idea with act. Jack Cohen found himself touched by fire in ways that jostled together the various parts of his unsettled life into something resembling a coherent pattern.

He went to Colorado State for a master's degree, although, still bored with forestry, he enrolled in the atmospheric sciences program. But he craved a fire connection both for hand and mind. He found one by working his first summer on the Pike-San Isabel hotshot crew, which expanded his practical understanding of fire operations but left him again twitching with intellectual impatience. He resolved that issue by switching departments to biometeorology and bioclimatology, which led him to contacts with Forest Service researchers slowly assembling a National Fire Danger Rating System (NFDRS), a grandly synthetic program that brought everything known to bear on when fires might be expected to start and how vigorously they might spread. His 1976 master's thesis processed the fire climatology of the Colorado mountains.

What had been mostly playful now turned professional. He was immediately hired by the Missoula lab to help develop the NFDRS, which was released in 1978. And he switched from cutting fireline seasonally to analyzing fire behavior for going fires, graduating from the third fire behavior officer class. The NFDRS was about doing fundamental science but around a conceptual core that mattered to the field, and fire behavior officers took the best of that understanding and spoke to the overhead teams on big fires. The boy who grew up eager to light fire and easily bored had found a suitable fuelbed to alight on.

•   •   •

Fire was his field—that was confirmed. But fire burned in many settings, and Jack found himself blown from one to another. For the next 20 years he cycled through topics and Forest Service fire labs, at one point resigning altogether for 18 months. Throughout, there were two constants, his curiosity about ignition and the need for direct contact, to connect what was known with what was done in the field.

In 1979, a year after the NFDRS was released, he transferred to the Riverside Lab in Southern California. Here he met new topics like Santa Ana winds, Mediterranean climates, and chaparral, but they were also variants of old ones; the fundamentals were the same. What was most novel was the urgency of unraveling how live fuels burned and how fire actually behaved in the fractal frontier between houses and wildlands. The need for experimental fires led him to certify as a Lighting Boss on the San Bernardino National Forest. He reconnected with fire teams as the South Zone fire behavior officer. The need to put knowledge, the right knowledge, on the line persisted.

But fire had a way of crossing lines. It crossed from wildland to urban fringe, which also breached methods of fire suppression. The fabled foehn winds over the Transverse Range flung embers for miles, which mocked fuelbreaks and firelines. So, too, flames on the mountains broke through the parameters of models developed in the smooth beds of laboratory wind tunnels. Conducting field burns and doing tours with overhead teams mixed the administrative borders between research and operations. When in an internal report he criticized the prevailing Rothermel model of fire behavior, now the nuclear core of the

NFDRS prediction program, as unable to forecast the kind of fire behavior he was witnessing in the field and as propagating only "an illusion of understanding," he was reprimanded.

By the mid-1980s fire research was on the ropes. The Reagan administration sought to reallocate government science from civilian to military purposes and, buoyed by a bout of wet years in the West, chose to believe the fire problem was solved. Funding was gutted, and since the Forest Service had a virtual monopoly over fire science, that sent the entire field reeling. In many respects, the agency never recovered; it could sweep up the pieces but never quite reassemble them. The Forest Service tried to salvage what it could with a major redefinition of "the fire problem" and a massive reorganization announced in July 1986 that tended to anger and unhinge its staff. The reorganization proposed to send Jack to the Macon lab. "Uncompromising" family concerns, however, led Jack to resign and remain in Riverside. For 18 months he worked as a research engineer for Dunham-Bush Inc., a company that manufactured heat exchangers.

In August 1988 he returned to the Riverside lab as a full-time temporary employee and was fully reinstated a year later as he completed his transfer to the Macon lab. His assignment was to help adapt the NFDRS to the Southeast. But by the time he arrived the applications mission was being superseded by a new quest for fundamentals. The American fire scene was itself undergoing a continental-scale transfer of themes and attention as the ignition of houses became as critical as the kindling of wilderness. The wildland side had an NFDRS to forecast ignition and spread probabilities. The urban side needed an equivalent.

• • •

It was an inauspicious time for a research initiative, but desperate to find a topic compelling to the administration, Fire and Atmospheric Sciences Research identified what was awkwardly termed the wildland-urban interface. This was a place where an implacable urban sprawl met irrepressible wildland fires and hence collided with public safety and politics. Those in fire science hoped that it might give handholds for a research organization otherwise stranded on a cliff face.

For 25 years, since the Bel Air-Brentwood conflagration had swept over the Hollywood Hills, the problem had spread like blister rust, infecting the American landscape but without much visibility until it occasionally broke out in epidemics. The worst were in California. In 1985, however, the California plague went bicoastal. On the West Coast fires ripped into Baldwin Hills, Los Gatos, and Ojai; Jack self-deployed to the first and went to the others on assignment. On the East Coast a fire savaged the community of Palm Coast, Florida.

Suddenly, what had been a peculiarity of California exceptionalism looked like the next new thing in fire protection. The wildland-urban interface fire—"a dumb name," one critic noted, "for a dumb problem" because technical solutions existed—had become a national issue. The next year the National Fire Protection Association (NFPA), alert to the emerging crisis, partnered with the Forest Service to sponsor a conference, Wildfire Strikes Home, that became the basis for a program termed Firewise. NFPA took the lead. The hope that the topic might relaunch fire research faltered until the NWCG accepted a wildland-urban interface initiative in 1995. Jack Cohen served as Forest Service advisor.

At first blush it was a topic that seemed remote from wildland fire. But like the houses that crowded against wildland boundaries, its themes shared a common border in combustion science. The critical questions were identical: how did structures ignite and how did the subsequent fires spread? Better, it was a subject around which various Forest Service labs could rally. By 1989 the contours of a research program had become visible through the smoke. The pivot was the Forest Products Lab in Madison, Wisconsin, which took an interest because wood frame houses were the principal fuel. It wanted to understand ignition properties of wood as it did the strength and durability of construction timber. Initially it sought to create an expert system to identify the critical issues for research but soon discovered that, as Jack forcefully declaimed, there were no experts. Instead, understanding would have to come from fieldwork amid incinerated communities and from laboratory research into the fundamentals.

The Macon lab, excelling in combustion basics since the days of Wallace Fons and George Byram, took on topics in heat transfer and ignition. The Riverside lab had the problem in its backyard—an uncontrolled field experiment—ready for empirical analysis. The Missoula lab, having sponsored the prevailing model of fire spread, should have been as vital to the WUI issue as it was to the NFDRS, but fire behavior in suburbs was not a topic that could be tested in its wind tunnels, and doubts among the knowledgeable grew about the validity of its core model, which assumed that fire spread by continuously radiating fresh fuel from an advancing flaming front. Real fires seemed to burn in patches and belched in billowing plumes; there was little continuous about Rocky Mountain forests or subdivisions sprouting on a once-rural countryside; and some blasted communities had their houses burned to concrete stubble while surrounding trees remained green. Still, however dense the institutional combustibles, the WUI as a research topic did not self-combust. It needed a firebrand. It found one in Jack Cohen.

The disruptions from sprawl, of unthinking development destroying its own habitat, he had known from his childhood. But throughout his career the flaming fringe had followed him like a peripheral glow. He recalled as an undergraduate watching a documentary on the Bel Air-Brentwood fire, *Design for Disaster*. Then he saw equivalents burn without the filter of editors. As a new hire at the Missoula lab, he had observed, as a volunteer fireman on the ground, the Pattee Canyon fire leap through Missoula exurbs without, as he recollected, much of a flaming front. A crown fire had dissolved into a firebrand blizzard. As a fire behavior officer in Southern California, he had considered houses as another cache of combustibles, and again after the 1980 Panorama fire he had reconstructed the dynamics of ignition as wooden roofs broke into flame half a mile beyond the fire's perimeter. Such oddities of ignition had been, as it were, beyond the borders of his professional concern. Now they began to gather into a coherent agenda of questions.

It fell to him to develop a prototype model. What he discovered did not fit existing concepts but did follow his understanding of how science worked, by swarms of ideas searching for tests. Ninety percent of science, he decided, was asking the right question. He learned that windows fractured from radiated heat before exterior walls ignited. He found that fires did not rush across fringe suburbs like a tsunami of flame but blew over and through them in blizzards of sparks. Downstream of the plume or flaming front there were thousands of sparks, and even a mile away there might be scores. With so many swirling firebrands in an ember storm, it was unlikely any point of vulnerability would escape. Like the rain that falls on the just and unjust alike, the embers fell on everything. If any way were possible to kindle a propagating fire—on a wood stairway, in a collar of pine needles around a yellow pine, amid a town dump— the swarming embers would find it. By 1993 he had consolidated his insights into a prototype, the structure ignition assessment model, that shifted emphasis away from the flaming wildlands and onto brand-receptive houses; that also shifted the power of control from wildland agencies to what was in fact, if not acknowledged, an urban fire scene.

The insight was so counterintuitive that it seemed to defy common sense. As he came to dissect it, the whole WUI was a trompe l'oeil. The problem was not about the soaring flames but the swarming embers, not about the wildlands but the urbanizing fringe, not about the flammable woods but the combustible houses. To see it required a gestalt-like switch: what had once seemed like a vase became two faces. So, too, in the mechanics of this new fire scene the lowly (or obscure) trumped the obvious (or celebrated and iconic); and that could be said equally about fire models and researchers. But then, to Jack's mind, it had always been the role of fundamental science to overturn common lore with positive knowledge. The great ideas would always blow across the lines scratched in the duff by institutions.

His discoveries did not immediately fit into fire protection schemes that were, after all, and had been for most of a century, directed at halting the spreading flames. But neither did the agenda of the Macon lab fit perceived national needs. In 1995 the Forest Service, still consolidating its fire research programs, like a body in shock pooling blood away from peripheral limbs and into vital organs, ceded the building back to Georgia and shuttered the lab. Jack returned to Missoula. The structure ignition research program returned with him.

Throughout, he visited and investigated in detail the major WUI conflagrations; Baldwin Hills, Painted Cave, Los Alamos, Rodeo-Chediski, Hayman, Aspen, Angora, Grass Valley, Myrtle Beach, and Fourmile Canyon. Initially he had been invited to accompany teams fielded by the NFPA. Then he began self-deploying: he needed to see firsthand where the flame made contact. He became, in effect, an unofficial fire behavior officer for the WUI. More and more, the story of conflagrations proved to be a tale of tiny kindlings that grew monstrous. Then the International Crown Fire Modeling Experiment offered an opportunity to field-test theories of radiant heat transfer and house ignition. The trials confirmed that radiant heat—so obviously seen and felt—was a minor contributor. What mattered to exterior walls was flame contact. And what mattered to the textured structures that made up a house were niches to accept embers. As a colleague phrased Jack's argument, the embers will always find a way.

In 1999 Jack presented a paper on his findings to the International Conference on Fire Economics and sparked an uproar of controversy when environmentalists realized that the growing emphasis on wholesale fuels treatments on public wildlands, proposed as a means to protect exurbs, misdirected attention away from what Jack had termed the "home ignition zone," located on private property. His research challenged the scientific testimony behind what, following the epochal 2000 fire season, became the National Fire Plan. He felt he had become persona non grata for appearing to question the evidence behind evolving presuppression and fuels programs.

Still, in 2002—as record fires romped across Arizona, Colorado, and Oregon and crashed into classic sprawl—he released a popular video produced by the NFPA-Firewise program (*Wildfire: Preventing Home Ignitions*) that distilled what he had learned. The core problem, he continued to insist, was protecting the structure at risk, not suppressing the fire in the wildland. Wholesale treatments of forests, expensive and dangerous mustering of crews and engines— these could be useful but only if they did not misdirect attention from the home ignition zone. It was the point of contact that mattered. The rest was a distraction, and he worried that it might distract the wildland agencies from actively managing fire on the land. That September, thanks to the National Association of State Foresters and despite rumored grumbling by his own agency, he was awarded a Golden Smokey Award for fire prevention.

At some point what ignites burns out unless it can find fresh fuel to rekindle. Jack Cohen had always believed that research that failed to connect lab with field was unusable; now he found that research that contradicted the field might be unwelcome. The only solution, he felt, was to return to fundamentals.

• • •

The nonessentials fell to the wayside. He was asked to leave the National Wildfire Coordinating Group working party; he pulled back from counseling Firewise, for which the National Fire Protection Association had assumed responsibility, and after the Angora and Grass Valley fires, he quit touring WUI sites, from which he concluded he had little more to learn. Instead he joined colleagues like Mark Finney in a quest to unravel the mechanics of crown fire behavior, which most fire officers regarded as the big fire problem of wildlands, the one that threatened fire crews, inflicted the gravest damages and ran up costs, and, rightly or wrongly, were perceived to power megafires. Once again, he was starting over.

The research questions were the old ones: how did such fires start and how did they spread? But there were wrinkles involving live canopy fuels, which did not burn like saturated dry fuels but more like popcorn kernels (one reason crown fires roared). And there were surprises regarding the difficulty of getting pine needles to burn at all through simple radiant flux. Like fins on a radiator, needles accepted heat and also dispersed it. What mattered was direct flame contact. But then Jack Cohen knew that. It had been the story of his life.

Ignition had informed his scientific biography and whatever else of his life underwrote that narrative. He knew that big burns could start from a single ember, that great research programs could kindle from a solitary spark of insight, and that a reformation of practice could spread from the glow of an indomitable personality. Of course he understood well that the real power of fire was its power to propagate, and that a start could only spread if it had a suitable setting. But what he learned over the course of his career was that propagation was itself a process of rekindling, and that a lot of embers—or a really determined one—will always find a way.

How I Came to Mann Gulch

In my story of the Mann Gulch fire, how I first came to Mann Gulch is part of the story.

—Norman Maclean, Young Men and Fire

WHEN NORMAN MACLEAN was writing *Young Men and Fire* he says he was on the downhill side of his Biblically allotted three-score-and-ten years. When I came I was on the uphill side, with a few years yet to go, though I was aged enough to allow for a pilgrim's staff. I had mistakenly forgotten my maps but didn't really need them because I was not there to re-create the fire and because the landing at Meriwether Canyon was closed due to recent postfire debris. I was dropped directly at Mann Gulch. There was only one way to go.

The place has a preternatural quiet that shimmers between the sinister and the hallowed. The hum of flies and bees, the chirp of an occasional bird, the hush of the wind; there is nothing more. Completely absent is the secondhand babble that engulfs modern life. It's been said that exploration without an intellectual purpose is just adventuring, that fasting without prayer is just going hungry. So solitude without contemplation is just lonesomeness. Mann Gulch is a place that invites thought.

It is a pilgrimage site, one of the few the wildland fire community has. There are memorials and crosses where the 13 smokejumpers died and a path that remains visible though not formally maintained. Visitors leave small piles of stones, rather like Buddhist pilgrims tying a piece of cloth. Mann Gulch is remote and obscure enough to filter out the casual tourist and curiosity seeker. It is common for fire to be associated with some spiritual interest. The oddity at Mann Gulch is that the spiritual interest is the fire itself.

You hear often enough from the fire guild about the lessons to be learned at Mann Gulch, as with all tragedy fires. You will sometimes detect sneers from hardcore firefighters that the 1949 jumpers made mistakes that proper training would never allow today. You may be lectured by fire behavior analysts and fire scientists about what caused the blowup that consumed the side-canyon and how those on site might have evaded their tragic fate. But trekking into Mann Gulch to discuss fire behavior is like taking the pilgrim's trail to Santiago de Compostela to discuss catechisms. It's an error of literary, and even moral, judgment far more egregious than anything the jumper crew putatively did on August 5, 1949.

My reason for coming was not to honor those who died, although I would do that, nor to ferret out or puzzle through missing mechanisms of the fire, although the 360-degree panorama of the site made crystal clear what had happened and why. Rather, I came to pay homage to Norman Maclean. I wanted to know how you create a literary blowup.

• • •

Where now is our Tolstoy, I said, to bring the truth of all this home. . . .
Must we wait for some one born and bred and living as a laborer himself,
but who, by grace of Heaven, shall also find a literary voice?
—William James, "What Makes a Life Significant?"

Big fires require many coincidences, or what Maclean in his crusty woodsman's persona called screwups. The day was hot, the air unstable, the fuels ripened over a long summer. The winds of the Missouri gorge met those of the gulch and the plume of the building fire. So, too, tragedy required other screwups that put the wrong number (or right) in the wrong place (or right) at just the wrong (right) time. The crew had to arrive neither too late, nor too early, which meant they had to parachute in, and then they had to find themselves too high up to get below the fire and too far down to outrace it up the slopes. Fire and fire crew had to converge with only one possible outcome.

Yet the same holds for great works of the imagination, those that are not merely clever but that exhibit the fundamentals of the human spirit as big fires do the physics of heat transfer. They, too, involve a convergence of screwups. The place has to fit the story, the story has to fit the theme, the text has to reconcile style with subject, and then the book has to bond with the larger culture. It was

Maclean's genius to find the perfect place and story, and then to blend them in an evocation that only the most obtuse, literal, and positivistic could find objectionable. He understood that blowup fires could be tragedies, and that tragedies behaved, as a literary creation, much like big fires. They blow up. It was his triumph to have that work subsequently speak to the American fire community and then to American society in ways no other writing about fire has.

· · ·

How this happened is neither obvious nor inevitable. Mann Gulch had been in the chronicles for 28 years before Maclean took up the cause. Prior to 1992, when *Young Men and Fire* was published, Americans had had a century of forest reserves, on which they fought fires, without creating more than the odd doggerel and Sunday supplement journalism. Until Maclean saw the "It" through the smoke hardly anyone appreciated the literary potential of fires in a Montana mountain. But it is equally true that it took Maclean, who claims to have visited the area while it was still smoking, a long time to appreciate the possibilities. For decades he tried to write what would become the informing themes of *Young Men and Fire* by using the massacre at the Little Bighorn. He couldn't make it work. At Mann Gulch everything—all the little screwups of setting, symbol, and story—converged.

Consider each part in turn. First, the place. Mann Gulch presents a self-contained world, though one inverted from the usual conception of mountain ravines. It is wide at top and narrow at bottom. During the fire no one could get into it from below (the one man who tried passed out); and only two got through openings at the top (for reasons they could not explain). The sweep of rounding ridges along the crest top form a broad amphitheater, and Maclean sketches his jumpers as insouciantly imagining themselves performing for a crowd. Geographically, the Gulch is a perfect microcosm. It contains all the action.

Next, the story. It, too, is a miniature of a life cycle. The jumpers are born out of the sky, leaving the womb of the C-47 and breaking the umbilical cord of their static line as they descend. They live briefly on the slopes as they move to what they assume is a normal destiny. Then the fire blows up below them. They die, save for three. There is a remarkable unity of place and action, as Aristotle might have put it. The story is as self-contained as the place.

Still, it would have been unusable, as the Custer massacre was, without survivors. The smokejumper corps believe themselves an elite, which Maclean recodes as Calvinist elect. They are set apart from the run-of-the-mill populace who shun fires or find themselves recruited off barstools. Two smokejumpers survive for reasons that no one can discern other than chance or predestination. The Mann Gulch blowup, as scientists from the Missoula lab prove with mathematical rigor, was a "race that couldn't be won." The natural order determined to clean out the gulch. Firefighters, even those who thought themselves spared from the mundane and the mortal, had no more say than squirrels and deer.

The critical action turns on Wag Dodge and his "escape fire." And an escape it truly is. It allows him to evade the foreordained fates for all those seemingly destined for the flames of perdition. Through his own hands and wits he saves himself, and it is Maclean's challenge to show that in saving himself he did not doom others. Dodge is the hinge on which the drama turns. A story of action, however graphic, cannot carry much moral burden. A story without choice or agency is natural history, not literature. Without Dodge's escape fire the saga of Mann Gulch is an accident. With Dodge, it is tragedy.

Still, the text did not write itself. Tens of thousands of people have been on firelines, and thousands knew about Mann Gulch. The after-action studies led to reports and anecdotes and a Board of Review. But nothing like literature happened until Maclean did it. He moved from campfire anecdote to parable and then to universal tragedy. He saw in the episode what others did not. He wrote the equivalent of an escape fire.

The rules for nonfiction are simple enough. You can't make stuff up, you can't leave out something that truly matters, you have to match style and subject. Maclean's purpose is story, but narrative of a peculiar sort, because, in the end, the story is about storytelling. There is no formal outcome, tested as true and confirmed, a data set akin to a laboratory experiment. In fact, Maclean tries one form of explanation after another and they all fail except for simple storytelling, of the search as itself a story. So the design wanders, or appears to (Maclean is too savvy for a genuine random walk). What will make or break the tale is the voice of the storyteller.

Shrewdly, Maclean adopts the persona of an old woodsman, a man who knows fire and the countryside and the people who work it. In truth, he hadn't been on a fire since at age 17 he had labored on a Forest Service trail crew. So he's not exactly a Tolstoy, narrating omnisciently, but neither is he a laborer who got educated into literary theory. The fancy concepts get buried, as the old woodsman allows Maclean to indulge in plot diversions and slang and a colloquial perspective that keeps the storytelling front and center. The upshot is not the true story of Mann Gulch but a true quest for the storytelling, which is the best someone can hope for. So place and event merge, like a reverse prism in which multiple bands refract into a single white light.

That leaves a bond with the culture because a book does not contain its own meaning but like a fire is a reaction that derives its significance from its setting and whose power lies in the power to propagate. For a blowup book there must be some triggering fire whirl of meaning. Maclean had a leg up because he was already a celebrity writer with a national audience thanks to *A River Runs Through It* and because his pursuit of the Mann Gulch story was widely known among the cognoscenti. In a perverse way his death before completing the text added piquancy. Like his portrayal of the jumpers who lurched ahead after the flames had overtaken them, *Young Men and Fire* seemed to carry Maclean's last wishes a final lunge after death had overtaken him. The book became a bestseller and won a National Book Critics Circle award.

Public attention, however, was framed by two events. In 1988 Yellowstone had burned (and burned and burned) and alerted the American public that there might be more to fire than a day or two's flash of flame and prepared it to move from fires etched into a celebrity landscape to a fire from a celebrity author. The defining event came two years after publication. The South Canyon fire in Colorado seemed, with unsettling fidelity, to reenact the tragedy at Mann Gulch. For decades fires had burned over crews, and if not indifferent, the culture had little mechanism to interpret those losses beyond a regrettable industrial accident. Other than a spread in *Life* magazine, little came from Mann Gulch. It shot up a photogenic plume, which then blew away with the next day's news. Thanks to *Young Men and Fire*, however, the public had in its hands a Rosetta stone by which to read the hieroglyphics on the slopes of its successor at Storm King Mountain.

The fire, part of a ruinous season that killed 34 firefighters and burned through nearly a billion dollars in suppression funds, galvanized the fire community. Within a year a new common federal wildland fire policy was announced along with protocols to help ensure the safety of crews. The book and its times bonded, or as Maclean might have put it, the little screwups of the inside world that was the book began to fit with the screwups of the outside world that was American society until they became the same. Without *Young Men and Fire* the reforms would likely have come piecemeal and more haltingly. And without a recapitulating fire to confirm it, *Young Men and Fire* would have passed, not unlike the fire of August 5, 1949, without much purchase. The downside has been the propagation of cloying imitators, like a field of thistle, and the preciousness that has attached to everyone who dons Nomex. None of that can be charged to Maclean.

What he can claim is a singular success at changing how Americans thought about wildland fire and firefighters. The fact is, the Mann Gulch fire had barely registered beyond the Northern Rockies. It was even possible, until 1992, to write the fire history of the country with hardly a mention beyond including it in the roll call of tragedy fires from 1937 to 1956. The tally of what it did not do is longer than what most partisans believe. It did not redirect fire research. A reorganization was already underway dating from 1948, and what galvanized research into fire behavior was not a blowup fire in a small side-canyon of the Rockies but the explosion two weeks later of an atom bomb by the Soviet Union. (Almost contemptuously the fire claimed the life of Harry Gisborne, the leading regional fire researcher, who collapsed while tramping around the site to investigate what had caused the blowup.) It did not refashion training and fireline operations. That came after two more, larger tragedies in California. It did not implant fire as a plausible theme within the larger culture despite a photo-essay in *Life* magazine and a Hollywood movie, *Red Skies of Montana*, obviously based on it.

For American society wildland fires remained a curiosity and freak of western violence, like a grizzly bear attack. The impact of Mann Gulch remained stubbornly regional, agency-specific, even personal; the smokejumper corps, Forest Service Region One, and a handful of local oracles kept the memory alive; it inspired no memorial services, it sparked no literary legacy, it came and went like the fire that drove it. There are hints that the Forest Service, at a national level, was content to have it forgotten. Nor did Mann Gulch focus a national reconsideration of how the country ought to engage wildland fire. In 1949 the United States had endured a horrific world war, had entered a cold war, and, though unaware, was poised to enter another hot war in Korea; and since firefighting seemed to be war by other means, casualties would happen. No one wanted them, no one planned for them, they were just part of the business or what the pretentious called battling the red menace.

The fact is, the Mann Gulch fire was trapped in an obscure side-canyon of history until Norman Maclean jumped on it. It was his literary imagination that stirred and scattered the embers into a blowup. Almost single-handedly *Young Men and Fire* established a literary genre, shouldered its way into the consciousness of the fire community, and more astonishingly bestowed onto American fire what it had never had before, a kind of Tolstoyan saga that gripped the public at large. It gave the literate public a kind of reading glasses with which it could see the 1994 South Canyon fire as something other than yet another blurry entry in a long fuzzy chronicle to the status of cultural catalyst. No other piece of literature in American fire history has come close.

In a story compounded of paradoxes, this may be the most telling: over the course of the fire revolution the most significant publication—one of the revolution's decisive moments—did not come from forestry schools or fire labs but from a professor of Renaissance literature at the University of Chicago. Other fire intellectuals knew how to put numbers on the phenomenology of fire. Maclean knew how to create meaning out of it.

Ultimately the Mann Gulch fire did all that its recent chroniclers have claimed for it. Those effects just happened not in 1949 but in 1994, two years after Maclean explained why we might care.

• • •

But why did I come? The Mann Gulch fire occurred almost exactly five months after I was born, and *Young Men and Fire* had been published twenty years in the past by the time I trekked up the slopes on July 9, 2012. A decade earlier, in *Smokechasing*, I had published a critical essay on the book as literature and had little to add. What brought me was the sense that American fire needed a new narrative to explain what was happening on its lands and to its culture. I wanted to know how a narrative became great. I had earlier written about the Big Blowup with that quest partly in mind, and now I wanted to understand better what alchemy had made *Young Men and Fire* possible.

The issue was not academic. In 2007 the Meriwether fire burned over the entire Gates of the Mountains Wilderness, including Mann Gulch. The river passage to the mouth of Mann Gulch was lined, on its eastern flanks, with burned hillsides and scorched trunks and trees with blackened catfaces along the river's edge. The new fire had reburned not only the scene of the 1949 tragedy but the entire range. In a real way it had also burned over the story of *Young Men and Fire*; so while scars from the 1949 fire yet remain, the scene has been erased and written over like the ink in a palimpsest. The book endures as art, but the template it added to the narrative cache available to the American fire community may not apply 15 years later.

The Meriwether fire began when lightning kindled a fire on the ridge between Colter and Meriwether Canyons on July 17, 2007, as part of a regional bust. There were more compelling starts than one in a wilderness, and the specter of Mann Gulch still haunted the place and the tragedy's host forest was not about to send crews in to fight it. For several days they watched and occasionally drenched the creeping burn with helibuckets while they threw crews and engines at fires scattered like errant comets around Helena. Somehow—it's possible that one of those water dumps or the rotor wash from a Skycrane blew sparks down the mouth of Colter Canyon or roused the slumbering flames to fury—the fire blew up on July 23, a mechanical re-creation of the whirl that had perhaps eddied from the shearing winds between the Missouri's gorge and Mann Gulch. The flames surged over Cap Mountain, then galloped across the entire Gates of the Mountains Wilderness.[1]

This time there were no crews strung out along the slopes and daydreaming or taking photos. Before it ended, the fire blackened 47,000 acres, nearly twice the size of the legal wilderness, and threatened a subdivision along the forest's south boundary, which was spared in part because fuels work had been completed earlier that spring. But no one died. No one trekked over the ridges pulaski in hand. No one subsequently scrutinized the fire's behavior. No one wrote up so much as an official narrative. The fire had passed over the Gates with no more cultural consequence than had it been a flock of mountain sheep.

The Meriwether fire is now the dominant ecological story of fire at the Gates of the Mountains and Mann Gulch. But it doesn't fit any of the existing narrative templates. It was not a disaster story, it was not a classic firefight, and it was not a young-men-and-fire tragedy. It is to our times what the ten o'clock fire was in 1949. If it is to have cultural traction, it will need a narrative that walks with caulk boots up the slopes that matters to everyone outside the fire community.

It won't match *Young Men and Fire* until it again records what William James called "the sight of the struggle going on," of "human behavior *in extremis*," and that will require putting people back into the scene. But the story needn't be tragedy. It just needs to find some way to engage with the larger culture in a morally compelling way and be granted a narrative to carry that meaning. That task will need the right place, the right story, the right imagination, and then the luck of the draw, the right timing to say what the sustaining society wants to hear. The Meriwether fire won't do that, but another fire might, and we can only hope that before too much longer a new Tolstoy will peer through the smoke and find it.

Written on the ridge separating Rescue Gulch from Mann Gulch

What Makes a Fire Significant?

T ISN'T SIZE. The 1871 fire that incinerated Chicago and provoked panic on the New York Stock Exchange tallied 2,100 acres. The 1906 San Francisco fire burned 3,000. But using urban fires for an index is comparing pulaskis and shovels. Surely, wildland fires have their own dynamic of meaning for which size matters. Well, yes and no.

What makes a fire important is that it connects with the culture in some way that allows the burn to transcend its local setting. Those fires must matter. They must interact with a receptive society. They must get institutionalized, they must be recorded and pondered, they must hybridize with literature, art, politics, the sensemaking of those who live with and around them. Urban fires are positioned where they can produce casualties and be reported, which is to say, they become stories with currency in the larger culture. Wildland fires that become significant are more likely to be big because a big burn is better situated to find some points of contact than a small one, though not all big fires are significant, and not all significant fires are big. A region like the Northern Rockies is likely to claim a higher percentage of transformative fires because it has so many big ones. Even so, vast landscapes can be culturally empty except for those values, like wilderness, associated with emptiness. What matters is where fires and culture meet.

Today, we typically equate "significant" with "celebrity." For a fire to get public buzz it has to burn houses, kill people, or involve celebrities (a celebrity landscape will do). Otherwise, a large fire in a remote, unsung landscape is just a big smoke in the woods. They come and go with no more moment than elk in migration. Four hundred acres around Santa Barbara is worth more than 400,000 outside Winnemucca.

But what about in the past? Big fires didn't begin in 1910. The 1825 Miramichi fire across Maine and New Brunswick burned as large a landscape as the famed Big Blowup. There was enough of a jolt that it entered into historical archives and even inspired a folk song or two. The monster fires of 1871, 1881, 1894, 1902, 1903, 1908, and 1918 that gorged on the slash left by industrial-strength logging all found their way into the master chronicle because someone wrote about them or consciously created a memorial to them. It helped that whole communities were consumed, that hundreds died, and that these fires might be linked to other historical moments such as the Chicago fire.

What they share is that they all left a story in the ash. They tapped into older plots, contemporary memes, or prevailing metaphors—they illustrated themes and issues of interest to the public, which is to say, an audience. The Northern Rockies has a disproportionate number of such storied fires because it has so many blowups, because the U.S. Forest Service has such a commanding presence, and because the fires wormed into the cultural software of the agency. Once established, the lore of big fires became its own referent.

In the 19th century the significant fires raged through the Lake States. In the early 20th they appeared in the Northern Rockies, and in the postwar era, in California, which came with its own media megaphone. The big-fire era in the Lake States ended when fire and axe gutted its forests; fires would return only insofar as the woods did. In the Northern Rockies the big fires merged in the postwar era into a larger discourse about wilderness. But big fires had happened long before. As soon as the mountains had shed ice for trees, they burned, probably on a massive scale. Where are they?

• • •

They exist in the historic record, but with few exceptions not much in the cultural one. The obvious geodetic marker for comparison are the fires of 1889. The 1871 Peshtigo fire had joined America's best known fires because its curators have celebrated it as America's Forgotten Fire, which it hardly is. Researchers who have traced the geography of the 1889 fires have attempted to make it also famous for being overlooked. The burning that droughty season certainly ranged far. A historic survey identified some 35 region-spanning "fire episodes" in the Northern Rockies from 1540 to 1940, which left about 12 years between eruptions. There were plenty of other big burns, though they tended to be localized on a mountain range or two, not sprawling over the northern cordillera. Of the whole roster the "circa 1889 fires" yielded "the most compelling evidence of extensive burning."[1]

Mostly, the newspaper record of that year chronicles the vicious urban fires that ravaged large chunks of St. Louis, Buffalo, Philadelphia, Boston, Chicago, Savannah, Seattle, Spokane, and even New York City. The fires that made page eight were those that rambled in New Jersey, Virginia, Wisconsin, Michigan, Minnesota, Oregon, the Dakotas, western Kansas and Nebraska, and those embedded in the Rockies and the Pacific Slope. California suffered major fires from Marin County to San Diego. Bakersfield burned. San Diego County burned. The Santiago fire, under howling Santa Anas, was estimated at over 300,000 acres. Flames went "to the very gates of Portland." The fires in Yellowstone National Park offered the first serious challenge to the U.S. Cavalry as a fire control agency (it failed, if gallantly). Elsewhere—all the elsewheres of the region—the fires flared and bolted. Place after place found itself under a pall of smoke, with ash and cinders drizzling from the sky like snow. In the Lake States the fires burned in May and again in October. In the Northern Rockies they burned as widely as the 1910 fires; their far-ranging smoke smothered Denver. But those shouting flames had no echo from the culture. They were part of the ecological background noise of the Northern Rockies. That didn't change until 1910.

As a natural phenomenon only the regional scale of the 1889 fire season distinguished it. The worst outbreaks occurred where the land was most disturbed by mining, railroading, logging, and settlement generally. The causes were the usual suspects: lightning, railroads, prospectors, brush-clearing farmers, sport hunters, and the always-suspect transients. (In southern Oregon "tramps," if not treated as well as they wished, were said to set revenge fires. Outside McLean's three were caught and lynched but cut down and badly beaten before they died.)[2] On the Dakotas prairie fires were reportedly set by gangs collecting the bleached bones of buffalo to sell for fertilizer. There was no prevention program. There were no agencies committed to controlling fires before they brushed against towns and mills. Then companies and city folk would rally and send hundreds of men out to hold the fires back.

There were no effective institutions for fire at all. Both Idaho and Montana were still territories. Montana's statehood convention met in Helena while "great black clouds of smoke hang over the country, and for six days the sun has not been able to pierce the darkness."[3] Two months after snow and rain extinguished the fires, it was granted statehood. Idaho became a state the next summer. Until then it was overseen by appointed rather than elected officials. Idaho's Governor Shoup telegraphed Secretary of the Interior John Noble for aid to keep fires out of Boise, and received $500. After fires ringed Ketchum Secretary Noble authorized another $500.

It was laughable, of course, but there were no mechanisms by which to apply any funds. Revealingly, that same year the American Forestry Congress convened and campaigned for a system of forest reserves and professionals to administer them. They looked of course to Europe, but also to Europe's colonies. The *New York Times* reported a rhetorical question asked by many at the congress: "Why should we not take the same precautions for protection against fire that are taken in India?"[4] That would require government bureaus. The creation of forest reserves was two years in the future. An organic act for the reserves was 8 years out. The creation of the modern Forest Service was 16.

There were few mechanisms for reporting on the fires. No one added up the area burned, the men on firelines, the dollars spent or lost to damages. No one really knew where the fires were other than they were burning over the "whole countryside." Most reports came from random dispatches delivered to random newspapers. The most complete accounts on the California fires appeared in the *Lawrence Daily Journal* from Lawrence, Kansas. The reports on Secretary Noble's outlays for firefighting were reported in the *Bismarck Weekly Tribune*. An extensive survey of Montana fires, both prairie and mountain, came out of the *Fisherman and Farmer* in Edenton, North Carolina. The story of Portland's battle was published in the *Roscommon Constitutionalist* in Boyle, Roscommon County, Ireland.

The Missoula *Daily Gazette* might declare, as it did on August 16, that "from the northernmost boundary of British Columbia to the Mexican line, the broken mountain range is all ablaze. . . . The latter part of the dry season upon the Pacific Slope is always marked by forest fires more or less extensive. The entire absence of rain over long periods presents a predisposing condition. This year the destruction is unparalleled." They could at least see smoke through their windows and walk on the ash that sifted down on the city's streets. The *New York Times* had to rely on the report of a "gentleman just in from Northern Montana" as its source. A fire in the Adirondacks was in the *Times's* backyard. A fire in the Rockies might as well be across the Pacific. (In fact, a fire reportedly in Suchow, China, received as many column inches.) For nearly everyone the 1889 fires that mattered in the greater Northwest were those that burned Spokane and Seattle. However much big burns might rattle the region, they failed to register on the nation's cultural seismographs. They were a snag that burned and fell in the woods with no one to hear it.

In retrospect it surprises, though it shouldn't, how much the 1889 fire season fits the template that prevailed in the Northern Rockies between the advent of mining rushes in the 1860s and the onset of the national forests in the 1890s. The same season of fire arrivals and departures. The same fire causes. The same droughts and high winds. The same fire rhythm of creeping and sweeping. The hastily assembled crews of miners, loggers, cowboys, railroad gangs, and vagrants. The towns circled by flames. The isolated cabins burned over. The stories of desperate flights, refuge sought in mine shafts, trains laden with evacuees crawling through dense smoke and flame. The whispered rumors of isolated deaths. The same reporting, at once breathless, mundane, and macabre that could speak with one gush about whole mountains ablaze and whole communities arraigned to struggle against them, suns blotted out by smoke, scores of miners missing, a Swedish family maimed as they fled, and "telegraph lines somewhat demoralized by burning poles." The pieces were interchangeable with big fire years across decades.[5]

Previously, before the arrival of westering Americans, another set-piece encounter had prevailed. And later, after institutions were created to preserve and manage a public domain, yet another template for big-fire narratives came into play. Throughout, big fires persisted. Undoubtedly they changed as they feasted on the slash of colonizing Americans, threatening a timber famine; then on wildlands made wilderness, overgrown with combustibles by being denied fire; and then as the exurbanites recolonized the rural scene and as climate inflected into a new normal. With each phase change the meaning of those big fires for their sustaining society morphed, along with how they would interpret the big burns of the past.

The fact is, outside of devoted fire researchers and the occasional antiquarian, no one has much cared about the 1889 season. At best the 1889 fires were a dress rehearsal for the 1908 fires, which were a prelude to the Great Fires of 1910— and these were the fires that mattered. They were not the largest or powered by the most horrendous drought or the highest winds, but they are the ones that burned with the greatest severity for American society. They established, for the first time, the terms of the American way of fire protection in wildlands. They traumatized the agency—the U.S. Forest Service—that more than any other laid down the infrastructure for nationalizing their lessons. They inspired more archives and gathered more records than any wildland fire previously. They wrote the creation story. We are still living with their aftershocks. We aren't with those of 1889.

•   •   •

Wallace Stegner once observed that "no place is truly a place until it has had that human attention that at its highest reach we call poetry." That's as true for fires as for waterfalls, and mountain peaks, and sandstone gorges. A compilation of big fires is little more than a scatter diagram of history; without context it is a data set, not a narrative. Its Cartesian points are so many snags that burned and fell in a forest and no one heard, at least in a way that anyone else cared about. The Northern Rockies fire scene has repeatedly conjured up enough big fires with a rough approximation of backwoods poetry to influence national policy and define a Platonic ideal for what fire protection in the form of a set-piece firefight might look like. Yet there is a problem with the kind of meaning those big fires created.[6]

The American fire community is not prone to citing William James. But as they seek to turn fires into data, and data into narratives of meaning, they might consider one of James's essays, "What Makes a Life Significant?," in which he argued that what animated events, and lives, was conflict and struggle—a desperate contest, uncertain in its outcome, pivoting on the will and strength of character of the actors. This certainly holds for wildland fire as well, and it is a primary reason why the firefight has proved more compelling than the prescribed burn. The firefight against monster flames can serve as the Hollywood script of a desperate struggle, its outcome uncertain. The prescribed fire, by definition, is a controlled process whose conclusion has, in principle, been foreordained. A firefight has drama. A prescribed fire should have none.

No fire regime has only one kind of fire, and the Northern Rockies is no exception. But it has a defining fire. The big blowup in the backcountry shapes the region as much as Santa Ana–driven conflagrations do Southern California or near-annual surface burns do the Flint Hills. So, too, no regime if it is to be robust can depend on one story or triangulation marker. The challenge for the Northern Rockies will be to bond its new avatar of the big burn, the megafire, with cultural clout. In 1910 it did the trick by aligning with another classic essay from William James, "On the Moral Equivalent of War," which was published in *McClure's Magazine* the same month the fabled Big Burn roared over the Northern Rockies and blew through a Forest Service that still had the dew on it. The timing was mostly coincidental, though not entirely. The firefight as battlefield here found its philosophical apologist.

What the Northern Rockies will do in the coming era will determine how relevant the region remains to the national and even global saga of fire. A big fire can become a great fire only when the rhythms of blowups match the cadences of a colluding culture. If the Northern Rockies hold only with big smokes, bravura firefights, and young men lost to the flames, if it continues to recycle its memes as it does its big burns, pumping them into larger sizes without reconfiguring their meaning, it will fade into a historical palimpsest, as the once-burning Adirondacks and North Woods have. If, however, it can twist the prism of meaning to catch more fashionable themes, as for example with wilderness or sustainability, surely a desperate struggle of our times, it might refract those brilliant flames into new significance.

The Other Big Burn

G LACIER APPEARS like a patch of First-Day nature still fresh from the Pleistocene. Everything about the place screams monumental, a vast sculpture garden from Nature's Romantic age. Its fundamentals seem far removed from human meddling. Snow, ice, streams, lakes; water hard, water soft, water flowing, water still; chiseling, caressing, abrading, covering—here water in all its forms acts on an upswelling of rock to make what most viewers would regard as an ur-scene of natural grandeur. Even amid the Rockies, the massif is striking, like a glacial erratic left from the ice age, moved not over space but through time. So immense are the domes and peaks that the cliff-covering forests appear like a veneer of lichens and those in valleys like moss. They frame the core scene like vignettes in an illuminated manuscript. Once the place was discovered, the founding question was never whether it deserved to be a national park, only when and how.

Yet Glacier's story is equally one of fire. The park was birthed amid the Big Blowup, its fires have repeatedly influenced the National Park Service, and its future may well be decided by the Big Burn of industrial combustion.[1]

· · ·

When Major William R. Logan arrived to assume his post as superintendent of the newly authorized national park, Glacier was burning. It was an outlier of

1910's Great Fires, but the fires threatened the park's entry and core. The Forest Service assumed control, assisted by the Great Northern Railway. Perhaps 100,000 acres burned. Another 50,000 acres within the park burned in the regional fire bust of 1919.

Its fire regime has a northern rhythm. A big burn comes every decade or two, and the administering agency finds itself with too large a fire organization in the cool years and too small a staff in the hot ones. What has aggravated the scene is the feudal nature of the national parks, which is less a system than a coalition of baronies, each park separately created and governed. Not until 1916 did the country create a civilian agency, the National Park Service, to assume responsibility from the U.S. Army. With its scattered holdings, the NPS found it hard to match fires and staffing. There were few buffers and backups; unlike the Forest Service, it did not have access to emergency funding for firefighting; and the administrative apparatus could crack when its dispersed political structure, which could not muster much collective response, met a northern fire economy in which the extreme years drove the system. In this way Glacier both miniaturized and exaggerated the national conundrum. As a big fire could upend Glacier, so a big-fire year at Glacier could unbalance the National Park Service. Just that happened in 1926. In 1910 the park had essentially no staff when the fires began; in 1919, it was still a fledgling, finding its wings as a civilian agency; but in 1926 Glacier was a relatively mature park and one of the system's crown jewels. Several fires began outside the boundary, and one burned across the western border on July 7. Public complaints led the Department of Interior to dispatch Horace Albright, assistant director of the service, to handle them. Then on July 31 a fire started from an exploding gas tank on a salvage logging operation within the park that swelled around Howe Ridge into the Lake McDonald fire, and high winds on August 5 sent another fire, the West Huckleberry, over the north slopes of Apgar Mountain and into the Lake McDonald burn. The two fires merged, blew up, burned over 50,000 acres, and cost $230,000 to fight (nearly $3 million in 2012 dollars). Without emergency funds to draft from, the firefight caused the Park Service to shutter some parks to pay the bills. The 1926 fire season was, by almost any measure, a disaster—financially, administratively, politically.

The big burn worked on the NPS much as the Big Blowup had the Forest Service. It made a local issue into a national one, a move boosted the next year by the establishment of the Forest Protection Board to oversee the federal government's commitments on forested land and fire protection. The Board required fire plans, of which the NPS had none. Planning demanded someone knowl-edgeable to write plans and administer them. The service recruited John D. Coffman from the Mendocino National Forest, a man toughened by fights with local ranchers over light-burning and familiar with Coert duBois's concepts of systematic fire protection, which had been devised for California. He focused on a national scheme to satisfy the Forest Protection Board with a report released in 1928, and the creation of a fire organization at Glacier to prevent a repeat of the 1926 debacle and furnish an exemplar for the system overall. The outcome could do for the national parks what duBois's treatise had done for the national forests. In early 1929 he and chief ranger F. L. Carter wrote that fire plan.

Still, in 1929, as Albright became director, the NPS's national fire organiza-tion amounted to Coffman, a solitary fire guard at Sequoia National Park, and the planned operation at Glacier. By August fires had again invaded the park. The Half Moon fire, started outside the west boundary in logging slash, burned over 100,000 acres (50,000 in the park) and cost the service $120,000 to con-tain. Coffman again appealed to Forest Service techniques, this time a style of postfire analysis, to critique the Glacier fire program. The review became, like the park's fire plan, a model for the service.

Then the fires began to fade away. The fulcrum was the Civilian Conserva-tion Corps, begun in 1933, which at Glacier fielded 1,278 enrollees organized into nine camps over the next eight years. The CCC erected much of the infrastructure of the park, including its fire program. Enrollees even removed thousands of acres of unsightly snags killed by the 1929 burn. The 1935 and 1936 drought years racked up most of the burned acres, 5,456 and 7,722 (the Heavens Peak fire claimed 7,600), respectively; but together they amounted to a third of the area burned by the Half Moon fire. The 1936 outbreak destroyed park facilities at Many Glacier, led to mutual-aid agreements with the Black-feet Indian Agency, and inspired a service-wide review of every park's fire pro-gram, the promulgation of national guidelines for protection, and accelerated training for CCC crews. In the 1940s, despite wartime drafts on manpower, the largest fire, the Curly Bear of 1945, burned only 289 acres. By the end of the war aerial fire control replaced the lost camps of the CCC. Smokejumpers were attacking fires by 1946; in 1953 aerial reconnaissance was supplanting all but a small handful of lookouts. Meanwhile, visitors started far fewer fires, and the national forests were better at holding fires before they could break over the boundary. The 1950s had only one bad year, 1958, when 33 fires accounted for 3,000 acres.

In 1963, as the Leopold Report introduced the fire revolution to the National Park Service, only 12.3 acres burned. Over the next years even that annual acreage fell by two orders of magnitude: 0.0 acres, 0.1 acres, 0.2 acres. Glacier National Park had become what Albright and Coffman intended—a model of fire's exclusion. When Glacier published a new fire plan in 1965, it retained the 10 a.m. policy as its lodestone. The park's residual fires seemed as much a relic of a former age as its mountain glaciers.

•   •   •

The illusion vanished in 1967. Big burns returned to the Rockies, and two of them slammed into Glacier. On August 11 drought and dry lightning kindled 20 fires; other storms ignited another 15. Two grew large, and both swept through lands that had burned in earlier conflagrations. The Flathead fire chewed through the Apgar and Huckleberry Mountains, the scene of the 1926 burn and, whipsawed by the wind shear of cold fronts, spilled outside the park boundaries. The Glacier Wall fire burned across the region of the 1936 Heaven's Peak fire. Between them the fires consumed 12,330 acres. The park threw everything it had at them.

Glacier was as well endowed for fire control as any unit of the National Park Service, but it could hardly cope with burns on a scale it hadn't seen for 30 years, and since the region was up in smoke with more than 30 fires over 1,000 acres, there was little reserve it could call upon from allies. The park responded by hiring what labor and equipment it could (3,500 men), it brought in overhead teams including the best fire officers in the NPS, ordered helicopters and National Guard units, authorized whatever means were necessary, expended $2.5 million, and after the smoke cleared hosted an elaborate postfire review. The siege went on for 62 grueling days.

What prompted the review was political attention. Local inholders primarily gave Senator Burton Wheeler a club with which to beat the NPS, while the more sympathetic Senator Mike Mansfield had to respond to constituents as well. The Washington Office had to reply. The intent of the two-day review was to demonstrate that the Park Service was serious about fire suppression ("fire is today, without a doubt, the greatest threat against the scenic grandeur of our National Parks"); that it had done everything possible to beat back the two big burns; and that Glacier remained a crown jewel not only of scenery but of effective administration. A primary conclusion from Superintendent Keith

Neilson was that the major failure lay in public relations, in not getting the proper information to the press and public.[2]

Yet through the course of the discussion there were queries, mostly a subtext, about the wisdom of sending in a phalanx of bulldozers, particularly against the Glacier Wall fire. Les Gunzel voiced the most emphatic concern. The Flathead fire threatened developed areas and private cabins and had been burned over at least twice in park history; the dozers were warranted. At Glacier Wall they only wreaked havoc, and left scars far more vicious than burned snags. He wanted explicit guidelines. Others, not in the firefight, looked from the Leopold Report to the costly wreckage and wondered why in the remoter areas the fires were being fought at all.[3]

The next year the service published its famous Green Book for the administration of natural areas, which replaced the 10 a.m. policy with options drawn from Leopold Report recommendations, including the need to restore fire, through natural means where possible, through prescribed burning where otherwise necessary. The effects were felt most immediately in the Sierra parks, particularly Sequoia (a proponent of light-burning in its early years) and at Saguaro (where Gunzel was chief ranger). Yellowstone rewrote its fire plan in 1972. Glacier was slower; after all it had been a centerpiece of Park Service suppression for almost half a century or nearly all the existence of the NPS. The new policy nudged forward in 1974, and a revised fire plan in 1978 put forth some tentative feelers while still granting suppression primacy. In 1980 the fire organization drew a zone of 100,000 acres in upper elevations for natural fire; it was a safe strategy in part because there were no fires. From 1983 to 1984, in association with researchers at the Missoula lab, Glacier experimented with a few prescribed fires. Other parks, many much smaller (and perhaps more nimble) or with fewer fire problems, were experimenting boldly with the new options. The era pointed to Wind Cave, Saguaro, even Rocky Mountain. But Glacier—a big status park with a history of big burns—was a laggard. When the 1988 season rolled over the Rockies, it attacked all starts. Even so, some 27,520 acres burned, the largest since 1936, although the Yellowstone fires mesmerized the national audience and deflected political attention away from Glacier's blowup as it did the Canyon Creek fire in the Bob Marshall Wilderness, which shot out of the mountains and overran tens of thousands of acres.

This time serious reform followed from the postfire reviews not of Glacier's fire bust but of Yellowstone's. Every park with any aspiration of incorporating natural fire had to rewrite its fire plans. Glacier completed the task in 1991. It zoned patches for prescribed natural fires and conducted some low-complexity burns in Big Prairie, an isolated ponderosa pine savanna. But the push was on for more black: the program needed to show some burned acres. Then, as so often before, a fire in Glacier rippled throughout the system. The year was 1994, a season notorious mostly for running up the first billion-dollar suppression budget and for the catastrophic South Canyon fire that engulfed a crew outside Glenwood Springs, Colorado. Those events, as emergencies are wont to do, overshadowed at the time what happened at Glacier, which pivoted on a managed fire, not a crisis. By sucking attention like stars into a black hole, they granted Glacier and the NPS some political space.

The Howling fire began on June 23, 1994, from a lightning strike on the North Fork of the Flathead River near the park's western border. At the time Glacier was one of a handful of parks with the size, funding, and clout to allow prescribed natural fires. A large burn, the Starvation Creek fire, which burned across the Canadian border, was under the direction of an incident management team. Suppression resources were strained. To oversee the Howling prescribed natural fire the Park Service assembled an ad hoc team to predict the fire's behavior and plan for contingencies. The Howling fire did not blow up, and by late August it was blocked to the east by two other fires, both wildfires, but both controlled through a confinement strategy. The three fires burned together. After the Starvation Creek fire was controlled, the Howling fire overhead team assumed control for it as well. For 75 days the team stayed with the fire until responsibility was ceded to the park. The fire burned for 138 days.[4]

By accident, necessity, and daring the Howling fire experience demonstrated how to cope with long-duration fires by substituting fire behavior knowledge for heavy machinery, trading land for options to maneuver, and adapting opportunities presented by nature to the strategic purposes of the park. They began by making the best of an awkward situation. They ended by inventing a new mode of operation. The 1995 Federal Wildland Fire Management Policy received its momentum from the shock of the South Canyon tragedy and the season's cost in lives, dollars, and burned area. But the outcome led to a steady liberalization of how to handle wildfires that dissolved the prescribed natural fire in favor of wildland fire use or resource benefit fires. Out of the Howling fire team emerged the idea behind fire use managers and modules. To the architects of those ambitions the Howling fire was the proof-of-concept test. Once again, without drawing national attention, Glacier had helped reform how the country would tend its fires.

Still, not until Glacier amended its fire management plan in 1998 did the park fully leap into the new era. Within a year it was managing the large Anaconda fire as a wildland fire-use burn. During the 2000 season, which ravaged so much of the Northern Rockies, Glacier suppressed the fires it received, but selectively, letting many blow out into rocks. The next year it accepted—it had little choice, really—the Moose fire from the Flathead National Forest, then 43,000 acres and let it burn itself out for another 28,000 acres until it hit the site of the old Anaconda burn and expired. The breakout year, however, came in 2003.

The fires grouped into seven complexes for a total of 145,000 acres, or 13 percent, of the park; two of the complexes, the Robert and the Wedge Canyon, burned 45,659 acres in adjacent national forests. The Robert rambled over the Apgar Mountains and Howe Ridge, the scene of the 1925, 1926, and 1929 Half Moon burns. The Trapper fire reburned much of the 1936 Heavens Peak and the 1967 Glacier Wall fires. To protect against the Wedge Canyon fire, Parks Canada bulldozed a line 100 feet wide and 10 miles long just north of the international border. Fighting the fires would have been hopeless—there were too many fires in the region and too few suppression resources to throw at those in Glacier. The park felt instead that it could manage those it had, use the techniques it had pioneered on the Howling fire, keep the burns out of headquarters, let them do their ecological work in the backcountry. Its neighbors were still evolving toward such notions. They all had different standards and National Environmental Policy Act approvals. In 2003 Glacier burned more acres than in any year since its creation.

Glacier had gotten its black, and then more. In 2006, despite attempts to suppress it from the onset, the Red Eagle fire blew up to 19,153 acres in the park before blasting out the park's eastern borders for another 15,050 acres on the Blackfeet Nation. By the time the 2010 fire management plan had coded the new strategies into bureaucratic language, some concerns were being voiced by resource management officers that there might be enough fire for a while, that the park could use a few years to assimilate what it had, that it might think about protecting some rare or old-growth patches in the name of biodiversity regardless of whether free-ranging fire, however natural, might be prepared to take them.

To some observers the value of a pause pointed to the future, that fire management needed to serve goals beyond getting the black, that this was not Big Burn National Park. To others, however, the real concern was not the fires recycling lodgepole pine or cleaning out western red cedar groves. It was that other big burn, the one that was driving out the glaciers.

• • •

Much of the Glacier National Park story is about boundaries. The park itself was created by slashing out a big chunk of the Blackfeet Reservation. Its northern perimeter traces the international border with Canada. It's flanked elsewhere by private holdings and national forests. These are not borders that matter to fire.

For its early years the park's fire problem was to keep fires that started outside from burning in. More recently it's to keep fires that start within the park from burning out. As national forests like the Flathead move away from extractive industry to recreation and wilderness, their aligned goals have allowed for more mutual fire plans. Park and forest can accept fires from one another and redraw their management strategies to work with fire behavior considerations rather than against them. The northern border delimits different thinking about how to restore fire. Under guidelines to enhance ecological integrity, Parks Canada favors prescribed fire, even high-intensity burns. Under notions of wilderness and "primitive America," the National Park Service leans toward letting natural fires, or loose-herded wildfires, do the job. The eastern border with the Blackfeet Nation is tougher since prevailing winds will drive fires—crown fires—from the park outward. Proposals surface from time to time to construct a fuelbreak along the perimeter, but it would take a swath a mile wide to break the chain of wind-borne spots. Today, every entry into the park is burned. What a visitor first sees at Glacier National Park is a scorched land.

The most interesting border is along the south, particularly the main ingress at West Glacier. The entry itself, through a small town on private land, is unburned. Behind it, however, at the visitor center and concession area, lies Apgar Mountain and the backdrop to Lake McDonald and Howe Ridge, and these were burned over in 2003. Yet the most significant fire scar lies outside the West Glacier complex: it's an overpass for the BNSF Railway, which visitors must drive under to enter the park. Its predecessor, the Great Northern, was instrumental in promoting the park, or in other words, the park as park is partly an artifact of steam. It's not simply academic muttering to suggest that this industrial gateway too is a portal of fire.

The eponymous glaciers for which the park is named also represent a blurry border, in this case of climate or of climate inscribing geologic epochs. The end of the Pleistocene, and the onset of the Holocene, is usually reckoned at between 10,000 and 11,500 years ago when the latest interglacial became undeniable. But the fundamentals behind the ice ages did not change then. The Sun radiated as it had before; the Milankovitch cycles, with their stretching, tilting, and wobbling, continued; no replumbing of Earth's heat transfer machinery occurred equivalent to Pleistocene-onset shutting of the Isthmus of Panama. By geologic standards the Holocene is simply another warm period within a 2.6-million-year-old climatic wave train of chillings and warmings. The ice should return. In fact, by some accountings, it should have already begun to mound up. Instead of still-shrinking glaciers we should be witnessing spreading ones.

We don't, because the ice didn't. The Little Ice Age did not ripen into a full-blown glaciation. Instead, the planet has warmed. It continues to warm in defiance of geologic history and astronomical rhythms. The evidence grows that the cause for the warming is humanity, or more precisely, a change in how Earth's keystone fire species conducts its business. We shifted from a primary emphasis on burning surface biomass to burning fossil biomass. That pyric transition has taken vast amounts of stored lithic carbon and released it as gases. The Earth could adapt to changes in the rhythms and scale of burning grasses, shrubs, and forests; it has done so for over 400 million years. It could not adjust to the sudden upheaval in planetary combustion, or rather, it is undergoing an accommodation that will take centuries. So remarkable and recent has this revolution been that many observers believe the contemporary scene deserves its own geologic epoch, the Anthropocene, which begins in the late 18th century when fossil fuel combustion becomes more than locally significant. (For that reason it might as well be termed the Pyrocene.) For fire history this redrawing of the geologic timescale is akin to the arbitrary geographic bounding of places like Glacier National Park.

The park's mountain glaciers are shriveling. The USGS estimates that the original site held 150 glaciers in 1850, and that in 1910, when the park was formally founded, all were still present. In 2010 only 25 glaciers larger than 25 acres still existed. Some prognoses suggest that all the permanent ice will be gone in another 50 years. As a scenic spectacle and geologic presence, the collapsing glaciers are being replaced by conflagrations. The ice age is yielding to a fire age.[5]

The two epochs differ, and not just in their temperatures. The propagating fire is not simply an inverse of the melting glacier. Glaciers rise and fall according to purely physical conditions. For fires to shrink or spread the ambient temperature, or climate generally, must refract through ecological systems and human societies. Still, ice and fire can't coexist. As the ice recedes, plants will reclaim that land, and as conditions favor, they will burn. There are few ways to stanch the retreat of the glaciers. There are many ways to intervene in the prospect of burning woods. The future of fire management at Glacier will likely hinge on how to understand that pyric border and manage across it. Wildland fire management can do little with the deep drivers; it can no more halt fossil fuel combustion than it can the sprawl of development behind the wildland-urban interface. It can, however, cope with the consequences.

• • •

The fire economy of the Northern Rockies is one of episodic big burns in which capabilities are almost inevitably out of sync with needs. There are too many pulaskis on hand during the slow years and not enough during the peak ones. Recurring crises push institutions to move responsibility up the chain of authority; as with wars or emergencies, big fires tend to centralize power. Yet the issue is no longer how to stop such fires but how to live them and in places restore them in ways that help the habitat without burning down human settlements. The thrust is toward landscape-scale management, a geography larger than the biggest burns.

In the future, the park must think similarly across temporal scales. The real economy of fire, like the dual portal at West Glacier, must include internal combustion as well as free-burning flame. The combustion of fossil fuels is the big burn that will shape the future. How to draw boundaries around it is far from clear.

GREAT PLAINS

Seasons of Burning

Prairie Fire in American Culture

THE GREAT PLAINS lie within the center of the country, and close to the historic core of its pyrogeography, yet today they reside on the periphery of a national fire culture and its accompanying narrative. The paradox is even more peculiar because the plains gave the nation its only school of fire art and one of its two major fire literatures, educated foundational figures in what evolved into fire ecology, and became an exemplar in the debate over private and public conservation. The easiest explanation is that prairie fire, like the western plains, was met and then leaped over, and by the time the country reengaged those fires were receding and the country's major institutions set. Even as the Great Plains were being pioneered the country had moved on.

How the region went, culturally speaking, from fiery core to burned out cinder has its curiosity as a study in historical contingency. It was largely a matter of timing. It seems that in history and culture as well as in nature, season of burning matters.

IN ART

Some countries create a fire art, some do not. It helps if fire is a part of quotidian life, or fuses with a sense of the national sublime, but fire by itself does not make art; art does. Russia produced a Urals school of forest fire painting in the late 19th century. Canadians, with a similar boreal landscape, painted nothing. Australians have an unbroken legacy of fire painting and poetry. South Africa has almost none. For art to emerge it has to speak to cultural interests and have on hand suitable genres to express them. America's flaming prairies were ideally situated for both.

Painting was morphing from the Grand Manner—large canvases that illustrated enduring moral lessons—to landscape in which natural history replaced human history as a source of inspiration and emulation. America was especially receptive to this shift since it had no monuments from antiquity, no Colosseums or Parthenons, but it had an abundance of wild nature. It had Yosemites, Yellowstones, Niagara Falls, and Grand Canyons. None, however, captured the congested drama of battles that served so often as the backdrop for the Grand Manner. Prairie fire did. The flames divided the field into competing realms: they forced action: they drove herds and people. For artists increasingly fascinated with light, as the 19th century was, fire offered a glorious palette.

An art drawn to nature in the wild met an expanding frontier aflame. Artists joined other explorers and journalists in reporting on the national estate and the saga of pioneering that was claiming it. But they could also make a reputation as an artist from such scenes; and many did, or having painted scenes from the American West generally, they added the burning plains. Once established the genre perpetuated itself until the prairie fire was no longer novel and, with the advent of modernism, art found art itself more interesting than the things it had formerly represented.

The chronicle began with George Catlin, who painted several famous scenes from his 1831 voyage along the Missouri River, each of which established a trope. One showed roiling smoke and flames along the horizon, with a Native American family watching from the foreground in nervous alarm. Another is a close-up of rolling prairie with the fiery front wending like a rivulet of flame. "Where the grass is short, the fire creeps with a flame so feeble that one can easily step over it. The wild animals often rest in their lairs until the flames touch their noses. Then they reluctantly rise, leap over the fire, and trot off among the cinders, where the fire has left the ground as black as jet." Such fires, Catlin notes, are "frequently done for the purpose of getting a fresh crop of grass for grazing, also to secure easier traveling," but they have an aesthetic no less than utilitarian outcome. "These scenes at night are indescribably beautiful, when the flames, seen from miles distant, appear to be sparkling and brilliant chains of liquid fire hanging in graceful festoons from the skies, for the hills are entirely obscured."

A third trope moves the action to the center as a mounted band flees before an approaching flood tide of monstrous flame. Catlin jotted breathlessly into his journal that "there is yet another character of burning prairies . . . the war, or hell of fires! where the grass is seven or eight feet high . . . and the flames are driven forward by the hurricanes, which often sweep over the vast prairies of this denuded country. There are many of these meadows on the Missouri, and the Platte, and the Arkansas, of many miles in breadth, which are perfectly level, with a waving grass, so high, that we are obliged to stand erect in our stirrups, in order to look over its waving tops, as we are riding through it. The fire in these, before such a wind, travels at an immense and frightful rate, and often destroys, on their fleetest horses, parties of Indians, who are so unlucky as to be overtaken by it."

The two polarities Catlin painted established the genre. That hellish onrushing fire, with the rolling smoke like a thunder cloud and the flashes of flame like lightning, became the dominant design as artist after artist painted surfs of flame that drove bison, elk, antelope, wolves, sheep, cattle, and people before it. William Hays split the scene in half with a roiling fire driving vast herds before it from left to right, as though the canvas were a map. Gustave Doré had panicked sheep rushing toward the viewer. A variant imagined the scene of a small band under threat. Charles Deas painted such flames bearing down on a mounted trapper, and again threatening a wagon train. Alfred Jacob Miller showed trappers and indigenes responding by setting backfires and swatting out the fires on the downwind flank. Nathaniel Currier and James Ives dramatized for their popular lithograph series the more common response in which the party kindles an escape fire into whose rapidly burned patch they would move to ride out the fast-encroaching front (see A. F. Tait's *Life on the Prairie: The Trappers Defense, Fire Fights Fire*).

Well into the 20th century similar artwork appeared even from the hands of the country's most celebrated painters and illustrators. Frederic Remington painted Indians setting grass fires and ranch hands quenching them with "beef drags," and desperate cowboys driving herds ahead of the consuming flames. Charlie Russell showed the Crow tribe burning the Blackfeet range; bison and antelope crossing a river to escape stampeding flames; and assorted campfires, signal fires, and cooking fires. The larger list goes on to include locomotives belching sparks while wild flames and bison race alongside, whole galleries of fire-driven stampedes, and many, many copies and colorizations of the classic images.

More recently, the revival of enthusiasms for environmental matters—and hence for burning prairie—has rekindled the interest of contemporary artists, although, like most art, their deepest instincts are to imitate the masters or to work within the genres of their time. That, after all, is what makes it art.[1]

IN LITERATURE

In the same way that art comes from art, literature comes from literature. The pivotal figure seems to be James Fenimore Cooper and his Leatherstocking Tales. The first, *The Pioneers* (1824), opens in upstate New York and ends with Natty Bumppo and Chingachgook trapped in a forest fire. The sequence concludes geographically with *The Prairie* (1827), in which the Leatherstocking, now identified as the Old Trapper, guides a group onto the prairies. Along the way the sagacious frontiersman saves his party by kindling an escape fire. The push of American settlement, originating in the northern woods, concludes when it strikes the grassy plains. The torch passed from Adirondacks to Great Plains.

The Leatherstocking Tales are widely recognized as a kind of creation story for the American experience, the beginning of its fascination with the frontier. Cooper was himself the Old Trapper of the western novel. Thereafter literary men traveled to the plains for the same reasons artists did. Washington Irving followed with his *A Tour of the Prairies* (1835); explorers, travelers, and later newspapermen added to the roster. An increasingly literate populace, or at least one that carried a tradition of belles lettres among their baggage, created a popular body of written work on fire. With the prairie fire the expression "illuminated manuscript" took on new connotations.

Cooper's oeuvre foreshadows the two regional literatures of American fire and the way they are linked through the flow of settlement. The Lake States have a literature of forest fires overrunning newcomers, a grand narrative that ends with the land cleared and the fires removed. The Great Plains have a comparable tale of homesteaders and small towns resisting or being overtaken by grass fires that likewise concludes with the flames banished by plowing, roading, gardening, grazing, and replacing. (Even classics such as *Little House on the Prairie* have an obligatory set-piece fire.) For both, the blowups of the past help measure the heroism of the pioneer. But prairie fire has become a story from the vanished past, along with cholera epidemics and locusts. The rootedness of settlement had no place for something as free flowing as wind-driven flame. Wildfire went from being a recurring theme to a literary trope.

What is striking is that there is no similar literature from the southern plains, or for that matter from the southern pineries. Settlement proceeded at an earlier time and from a culture less bound to formal art and literature. While settlers in the north were recording prairie fires within a narrative of endurance against the elemental plains, the classic accounts from the south were about droving, and although wildfire was a threat, the fires that populate accounts like Andy Adams's were campfires. The southern narrative is about moving: prairie fire is, as it were, just another variant of the plains' tendency toward the fluid. The northern narrative is about staying put, which makes fires a threat. Like its authors the literature was rooted in a place.

Outside the Great Plains, however, or for that matter outside the Great Lakes, it did not spill over into a distinctively American genre, any more than the art of prairie fires did. Instead the grand narrative of American fire turned on the creation of a permanent public estate: these were the landscapes that mattered to the nation overall. The Great Lakes conflagration is replaced by Great Fires and Big Blowups. Eventually, a culture of fire emerged out of that experience, passed from generation to generation; and much later a literature, though one made de novo and curiously divorced from its antecedents.

It would not be far off the mark to date the reemergence of such a literature to 1992 when Norman Maclean posthumously published *Young Men and Fire*. That book did for contemporary fire, grounded in the public lands of the Far West, what Cooper's Leatherstocking Tales did for the trans-Appalachian frontier. The central action of Maclean's meditation turns on an escape fire lit by foreman Wagner Dodge, which Maclean regarded as an unprecedented invention. It took a colleague to remind him that Cooper depicts just such a fire in *The Prairie*; but there was no continuity between the techniques frontiersmen had long used in tallgrass prairie and Dodge's daring act amid the piney savanna of Mann Gulch, and no literary continuity apparent even to a professor of literature such as Maclean. The written legacy had broken, had been left with the sod house and the mule-drawn plow. The prairie fire had passed from the national narrative and a regional variant would have to be reinvented.

IN SCIENCE

With its fractal patches and edges America's unsettled rural countryside was ideal for wildlife and inspired many a youth to become a naturalist. Not a few of those who figure in postwar environmentalism (think Aldo Leopold) and in the fire revolution (see Herbert Stoddard and E. V. Komarek) came from the Midwest. The tradition of the autodidact who later acquired formal learning is an old theme, especially calculated to warm the cockles of America's populist heart. But the plains did more.

Its northern naturalists recast their experiences into formal learning, just as artists and litterateurs did. The upshot was an American school of botany, which segued into a school of ecology. With uncanny fidelity, its life cycle meshes with that of the pioneering era, which is to say, with the epoch of quasi-wild prairie fire. As their chronicler observes, "at the core of their community was the shared experience of the prairies and plains." The first generation grew up on the frontier, studied botany as farmers commenced serious sod busting, and "created the science of ecology" at the same time as American philosophers, also responding to the vigor of nation building, fashioned Pragmatism. The second generation—their students—could, as Ronald Tobey describes, still "experience isolated fragments of pristine prairie" and "feel the entire presence of the original plains as a deepening echo in their lives."[2]

Once again, a contribution cannot be separated from time and place. The "first coherent group of ecologists" in the United States emerged from the study of grasslands. The model of the natural world that they constructed exactly mirrored the experience of change that they had grown up with. Nature, it seemed, was also a pioneer, whose progress led to a civilized and stable climax, and while periods of rapid transition were disruptive, those disruptions only reset the clock. The Great Plains were too vast and implacable for humanity to deform in any serious way; at best, they could substitute domesticated grasses for wild, and livestock for bison and pronghorn. Most felt, as John Weaver, leader of the second generation, put it, that the prairie "approaches the eternal."[3]

The project began with Charles Bessey, who moved from Iowa State to the University of Nebraska in 1884, introduced laboratory methods into the classroom, wrote the dominant textbook on botany, and inspired legions of students. Bessey was a pragmatist (what we know is what we can learn from experiment) and a pioneer (we have to do in order to know). His two students, with whom he constituted a first generation, Roscoe Pound and Frederic Clements, split that inheritance. Pound helped bring Pragmatism to jurisprudence. Clements introduced the quadrat to ecology, which was a practical means of breaking down the undifferentiated vastness of the plains into study units—the quadrat as the sodbuster plow of prairie science. Having devised methods, Clements ventured into other landscapes, notably, the Rockies. In 1910 he produced America's first formal paper on forest fire ecology, "Fire History of Lodgepole Burn Forests." In a tidy reversal, the grasslands did not derive from the forest—were not a biota with the trees removed; they were the model for how forests worked, at least when subjected to fire. Fire behaved in lodgepole pine as it did in tallgrass prairie.

The difficulty came when Clements and his successors translated their data into a paradigm of succession in which ecology could explain phenomena by placing them into various cycles and epicycles of progressive change. It was, in truth, the ecological equivalent of William Morris Davis's model for geomorphology, which also underwrote an American school. The concept spread like red brome. The foundational observation was, as Tobey puts it, "the central fact" about the prairie was "its natural stability and tough perseverance."[4]

Then came the 1930s. Drought crushed even native prairie, which presumably had long ago adapted to plains climate. Drought and depression, moreover, not only stopped pioneering, they undid it. But not before the record of untrammeled sod busting had ruined the capacity of the prairie to prevail. The Clementsian model could account for neither nature's nor humanity's effects. Intellectually, the theory went bankrupt; and just as damning, it could not say how to intervene and correct the errors that had contributed. Students interested in grasslands migrated into range science. The theory persisted well into the 1960s, but primarily because its old members, ensconced in universities and research institutes, continued to publish within its framework and because it was easy to teach. It resembled a massive tree that could be felled but not uprooted, and whose stump was left to rot away. As the adage puts it, science advances one funeral at a time.

When the American fire revolution burst onto the scene in the 1960s its partisans were dismissive of Clementsian ecology, which could only envision fire as a disruption of natural progression, not as a necessary process of continual renewal; and they were skeptical of academic science, which had so long supported it in defiance of common experience. When fire science rebuilt, it arose out of government labs and looked to the conifer forests of the Northern Rockies, the chaparral and sequoias of California, and the longleaf and rough of Florida. Prairie fire ecology became a subset of range management. And like prescribed fire, which struggled to find enough extra grass to support itself, the science of grassland fire scrambled to find sufficient intellectual fodder and space to free-burn. Instead it found refugia like the Curtis Prairie and Konza Prairie and the burning brush around Texas Tech.

For a while grassland fire provided the model for an American science. For much of the past century, however, it has struggled to find an adequate place for itself within science at all.

IN POLITICS

When drought fused with Depression, the consequences were felt beyond farms and schools of ecology. Critics of an American way that had ripped up the national economy as it did prairie sod had many examples of environmental havoc whose costs were too great for laissez-faire responses; but none rivaled the Dust Bowl. With subcontinental dust storms blowing to the national capital, the Great Plains returned to center stage of a national debate about conservation and the relative merits of public and private economies. The outcome stood the saga of pioneering on its head.

The Great Drought was not an era of massive fires on the plains because generous fuels require abundant rain. The big burns migrated west—Matilija (1932), Tillamook (1933), Selway (1934). Their smoke palls were the forestry equivalent to the enveloping soil squalls. Only indirectly, by mobilizing national attention to the wastage of natural resources, and by refuting Clementsian ecology, did the plains contribute to America's fire narrative. *The Plow That Broke the Plains* had its counterpart in the unfilmed, but widely understood, axe that broke the woods. Smoke and dust were their dark-double offspring.

If anything, the Dust Bowl experience (as synecdoche for the 1930s Great Plains) moved the region out of the nation's pyrogeographic focus. In 1947 a quarter million acres burned in Maine and another quarter million in South Dakota. The Maine fires folded into other alarms, helped nudge the country toward a formal civil defense program, and led directly to the Northeastern Forest Fire Protection Compact. The South Dakota fires blew away with the wind. The Maine outbreak looked to the future. The South Dakota conflagrations echoed the past, as though a macabre reenactment of pioneering days.[5]

Yet the Great Plains were not through with fire, and the national fire narrative had not fully abandoned the plains. The mechanisms for return were two, and they differed in their focus, their methods, and their means of propagation. One was interested in restoring prairie. It had powerful cultural capital to draw upon and a deep well of empirical evidence that fire was elemental to any attempt at restoration. Whatever prevailing theories intoned, it was instantly clear to those up to their elbows in big bluestem and black-eyed Susan that prairies could only thrive if burned. As more prairie was protected, the more fire had to return. The model of controlled burning to advance grasslands, however, ran into stiff headwinds. However much humanity's firepower—for all our existence as a species, really—had depended on grasslands, however long hunters and herders knew that game and livestock gathered around burned patches, which exerted a gravitational pull akin to Jupiter, moderns had lost that connection or had immigrated from places for which substitutions had been devised or had so committed to maximal production that they left no grass sufficient to carry fire or simply feared the wild extravagance. The preserved prairies needed fire, and they needed legitimacy for the idea of fire. They propagated that message through a civil society of environmental groups, especially the Nature Conservancy.

The second mechanism was the academic refugium created by Henry Wright at Texas Tech. The Wright group also promoted fire to restore grass, but the target at the end of their driptorches was brush. Their leverage came not from appeals to prairie but to paychecks. Fire was a relatively cheap, effective, and environmentally benign tool to drive back the mesquite and general scrub that overran pasture. The Wright staff resembled a prairie dog village that, through offspring, colonized new lands, particularly in the southern plains. Henry Wright codified his knowledge in his co-authored text, *Fire Ecology*, but its message was more often spread through extension agents. It appealed to ranchers: it made fire a legitimate tool in the barn. Across fire's fulcrum it balanced the ecological enthusiasms of prairie restorationists with the economic pressures of ranching. Later, the concept of patch burning helped bridge economic interests with ecological benefits.

Prairie fire thus entered into America's fire revolution. That reformation needed more exemplars than working landscapes in the southeast; and prairie restorationists needed intellectual heft from an emerging science of fire ecology and practical instruction from experienced burners. The Curtis Prairie at the University of Wisconsin and the Konza Prairie under the direction of Kansas State University evolved from research plots to provide some of the fundamental science and announce demonstration sites. Fire and prairie alloyed with special force, however, with the emergence of the Nature Conservancy as a presence. Thanks to Katharine Ordway, TNC acquired significant holdings, from which they were reborn, after a period of hesitation, into apostles for burning.

TNC only became a national player in fire after the organization moved into Florida, where prescribed burning was not simply an implement for restoration but a means of survival. From its Florida chrysalis a national fire program emerged, bulked up, hardened, and committed to National Wildfire Coordinating Group standards. That was an investment most private landowners, and even most TNC prairie managers, thought unnecessary. But with marginal federal holdings on the plains, some other institution (or institutions) would have to overcome the tendency of life on the plains to disperse and attenuate. Something had to thicken what the unbroken vistas thinned. Flint Hill ranchers and wildlife refuges held fire in niches, but TNC became the institutional fulcrum by which to leverage prairie fire into national awareness. TNC accepted that role, set up fire learning networks and training exchanges, and stiffened prescribed fire associations.

Even as his idealized model was faltering, Frederic Clements had argued vehemently on behalf of New Deal reforms and government intervention. That didn't happen—couldn't happen. The Shelterbelt soon deflated, even as its counterpart, the Ponderosa Way in California (a 650-mile-long fuelbreak), cut across the Sierra Nevada. The federal government bought abandoned or degraded lands and created national grasslands, which it placed under the administration of the Forest Service. But such lands were not large enough to reshape the economy of the plains nor to tilt the national narrative. Prairie fire was a local, not a national, event. Fire management would require alliances among small and dispersed sites under all levels of government and mixed ownerships.

By the time prairie restoration had become popular, the country had moved ahead. It had rechartered its fire programs and rewritten the country's master fire narrative. Burning prairie was not a primary means to change national thinking on fire; rather, national thinking had reformed from other causes and was now applied to prairies. So, too, its long and expansive association with burning prairie had granted the Great Plains a distinctive fire culture that in places had persisted and, at certain seasons in the almanac of American history, had contributed to a national culture of fire. No other region could make so broad a claim.

Yet, in the end, with so much land lost to fire from plow and hoof, and with so little symbolic or charismatic value beyond the region, the plains found themselves outside the triangle that defined America's pyrogeography. Florida, California, the Northern Rockies—what happened here had national ramifications. What happened in the plains remained in the plains. Like a whiff of smoke that triggers deep memories, the Little Burn on the Prairie could invoke a wave of national nostalgia, but little more.

Pleistocene Meets Pyrocene

U NNUMBERED, PRAIRIE POTHOLES dot the northern Great Plains. Most are pinpricks or blotches. Many are ponds, and some, lakes. They dapple the moraines and loess lands of North Dakota like a geologic tickertape recording, for nature's economy, the rise and fall of Pleistocene ice. Eastern South Dakota has perhaps a million, sprinkling the landscape like oregano. Minnesota's prairies have their 10,000 lakes as much as the north woods. Since they first appeared in the wake of the Wisconsin ice sheet, those potholes in the prairies have been an indispensable waystation in the great flyway that seasonally sends flocks of mallards, blue-wing teals, blue and Canadian geese, ibis, egrets, cranes, pelicans, gadwells, and others, north and south in a biotic echo of the advance and retreat of the ice. For the traveling flocks the potholes are not bumps in the road: they are the caravanserai that make travel possible.[1]

It might, at first honk, seem odd that amid speckled wetlands there should be a role for fire. But enmeshed in tall- and mixed-grass prairie there is no way they could not burn along their edges since the surrounding prairie will burn routinely, and when dry spells drop the shoreline, that they would not flare and smolder into the sedges and organic soils piled there, and when deep drought empties the waters and drains the peaty bottoms, that the fires would not from time to time gnaw into and even scour out those cauldrons of peaty combustibles. Yet just as the birds' habitat requires both upland and wetland, so fire and water become the yin and yang of the pothole province. Those innumerable water pockets make a pyric scatter diagram: through it one can trace the curve of an unusual multivariate relationship inscribed across the American fire scene. They are fire's flyway over the plains.

The upshot is a tableau of fire and plains. The way that fire interacts with climate and grazing to make prairie. The way conservation must mingle public purpose with private lands. The fusion of the ancient with the evanescent, as a biome primed for almost explosive change rests atop a relic geology fashioned by the movements of ice sheets across tens of millennia. The dispersed character of the potholes, so scattered that mere fragmentation pales beside it and cohesion only appears in spasms (even on the wing). Coherence demands relentless exertion. Amid this starry field of potholes spangling the prairie firmament, patterns emerge over wide expanses, among them the biotic constellations traced by the migration of herds and flocks. In the prairie potholes fire management builds on the outwash of the Pleistocene.[2]

•   •   •

Arrowwood National Wildlife Refuge (NWR), 30 miles north of Jamestown, North Dakota, is large, old, and characteristic.

Settlers moved in during the 1870s as the plains tribes moved out. It was a harsh, remote land, given to a mixed farm-ranch economy. More settlers filled in the blank spaces during the early 20th century, but much native grass remained. When the 1934 Taylor Grazing Act ended the transfer of public domain to private hands, drought and Depression were ravaging what biomes had survived. The federal government began acquiring through purchase, tax delinquency, or resettlement lands for the protection of the Great Plains flyway—a part of reclamation programs that throughout the plains included the better-known Shelterbelt. It was not enough to cease the ruinous practices of hilly sod busting, overgrazing, and overhunting that had damaged the prairie: it was necessary to rehabilitate them.

In 1935 Arrowwood Wildlife Refuge was established out of the former Riebe ranch, homesteaded in 1882. Together, the Works Progress Administration and a Civilian Conservation Corps Camp 2774 put in dikes to help stabilize the wetlands, erected housing for staff and equipment, planted shrubs and trees (including chokeberry and Russian olive), and controlled predators. They built trails and roads that doubled for fuelbreaks; and when fires broke out, they fought them. World War II returned management back into custodial status.

The postwar era saw a revival of applied science, an agronomic model, to maximize waterfowl production.[3]

But the system was not quite right, and those close to the land realized that the number of ducks shot was not a measure of habitat health any more than board feet measured forests or "animal use months" grasslands. They needed more land for coherent habitats, and they had to think in ecological terms. The land issue was addressed through purchase and a program of easements on private land for wetlands and grasslands. Owners could still farm and graze, but they couldn't break new sod or drain off the waters. The upshot was a density of mixed-ownership protected sites unlike anything elsewhere: North Dakota has 63 of the 561 units in the National Refuge System (the nearest rivals are California with 34, Florida with 30, and Louisiana with 23). But when combined with auxiliary easement wetland management districts, and waterfowl production areas, they render the Missouri and prairie coteau into a veritable Milky Way of protected sites. Arrowwood was an administrative complex that included three fee-title refuges, six easement refuges, and three wetland management districts.

Still, more land would be ineffective unless it was managed properly, and not draining or overshooting was not by itself sufficient management. The wetlands were only as good as their adjacent uplands, which were habitat for many species other than waterfowl (such as prairie chickens), and that biome was in bad shape. Something was still out of whack. In 1963 Arne Kruse and Leo Kirsch, later joined by Ken Higgins, began experiments at Woodworth Study Area (subsequently the Northern Prairie Wildlife Research Station) to see if fire could defibrillate a habitat not only decadent but becoming infested with exotics like smooth brome and Kentucky bluegrass and woody plants that intercepted ground water and broke the hydrology of wetlands. They used Arrowwood refuge for their beta field trials. They found that fire was inextricable—and indispensable—to good prairie and hence to the long-term health of waterfowl and upland birds. Their trials thus paralleled the fire revolution, and in 1972 Kirsch and Kruse addressed the 12th Tall Timbers Fire Ecology Conference with their findings. The gist was, fire's removal "has done untold damage to prairie wildlife." What was good for the prairie was good for birds.[4]

In this way plains waterfowl were inscribed into the honor roll of firebirds that had, over the course of the 20th century, forced a reconsideration of landscape fire. The collapse of red grouse in Scotland, as grouse replaced sheep and protected covert replaced widespread burning, had inspired several parliamentary inquiries. The decline of bobwhite quail in the coastal plains of the southeast had sparked the Cooperative Quail Study in the 1920s. In more recent decades the endangered red-cockaded woodpecker has brought another fire-drenched habitat to public attention. Between the prewar quail and the postwar woodpecker came the birds of the plains—the migratory waterfowl and the prairie chickens primarily. They all—every study—paired sagging populations with deteriorating habitats, and they each, every one, recognized that the loss of birds followed a loss of fire. In its final 1911 report the Grouse Commission noted that the diseased grouse were also a harbinger of ill sheep, which is to say of sick heath, and it concluded that landowners needed to burn big, perhaps a third of an estate annually. Its successors agreed.[5]

This was, however, a hard sell to an agency historically committed to maximizing game birds because it meant a short-term loss of ideal conditions, which for ducks meant cover and nesting conditions, for the longer-term health of the habitat. The wildlife community thus aligned with ranchers who refused to lose any forage to fire, even seasonally, if it meant less stocking. But the value of the wetlands depended on the vigor of the upland prairies, and that meant serious burning.

The U.S. Fish and Wildlife Service had plenty of fire experience, but like its lands, that knowledge was local, diffused, and specific to units. The agency had no national fire program, not even a coordinator. That began to end after fires in the Kenai NWR became explosively expensive; after fires in 1976 at Seney NWR in the Upper Peninsula—one natural, one set—went wild; and then yielded to a series of fires that were either botched or resulted in deaths. The climax came in 1981 when two technicians died fighting a fire at Merritt Island NWR. Congress raked the agency over its mismanaged coals. But out of that ordeal by fire the FWS got marching orders and money, and despite the agency's inertia it began to integrate with the national fire community, which meant adopting national standards.

One unexpected outcome was that, in the prairie pothole region, the FWS became the largest federal fire presence. The U.S. Forest Service had a few national grasslands and remained the conduit to the states, but in places like North Dakota the FWS could do with fire what it had done through easements and collaboration with wildlife protection. What migrating flocks did to bind the pieces together ecologically, the FWS had to attempt institutionally.

Bonding to national trends brought a more professional fire staff, but it also welded refuges to national cycles of fire management. These boomed after 1994 and boosted again after the 2000 National Fire Plan. Then they collapsed as Congress and the Great Recession applied fiscal tourniquets to the public sector. In 2011 the region had a permanent fire staff of 58; by 2014, 26. Funding collapsed, which meant projects needed partners, and with most grasslands in private hands, that meant public-private collaboration at a time when rural populations were collapsing and there was little economic incentive to burn. Few ranchers would sacrifice immediate forage for sustained grassland health, even as cool-season invasives like Kentucky bluegrass and smooth brome shrank pasture. There was no Conservation Reserve Program for fuels. There was no endangered process act for fire.

Even so, and after rising and falling from the 1970s, burning became a routine practice: it was reestablished along with big bluestem and sharp-tailed grouse. The Dakotas prescribe burn 40,000 acres a year; Arrowwood alone burns an average of 8,000 acres annually, or roughly 10 percent of the total land under its administration. Yet this is no more than a tithe on what the prairie needs, and the acres burned are a weak index of the mix of burns, dormant- and growing-season, back-to-back and spread out, surface flushing and deep scouring that the land needs, and all this must be syncopated with grazing to achieve the best results. Prescribed burning and prescribed grazing are the prairie's winning tag team.

After the flush from early burns in the 1970s, the ecological effects have worn out: nothing persists long unchanged in the prairie. Fires that had come no more than 4 years apart must now wait 44. Amid declining budgets and staff, and worse, political instability; with a warming climate shuffling northward and exotic weeds seizing the better, formerly disturbed soils and massed at the borders ready for invasion; with surrounding buffer grasslands put to the plow to grow corn and soy, and a new era of sod busting accelerates as crop insurance encourages putting Conservation Reserve Programs and marginal lands into production; and with critics finding more imaginative causes with which to restrict the scope of burning, whether it be smoke or an eccentric butterfly, refuge managers are struggling to keep enough good fire on site, even as it becomes ever more apparent that fire is the strong nuclear force that bonds the biotic pieces into something like ecological integrity on a land that needs as much resilience as it can get if it is to absorb the still greater blows to come.

• • •

The problems can seem as dense and diverse as the botanical mix of grasses and forbs, microbes and mammals, insects and birds, that constitutes prairie. But boil that pitchy brew to its essence and it leaves a hard distillate called scale. The glory of the prairie pothole landscape—its expansive range, its role as dispersed way stations of a flyway that spans North America—is also its curse.

It includes the problem of small units that need burning—a question of edge to area. It can take as much effort to burn 200 acres as 2,000. The diffuse character of private and public land ownership means it isn't possible to subsume hundreds of upland-wetland units into a single burn, or to muster the varied entities and folk into common set-piece undertakings. Without whole landscape burning fire officers can't muster favorable economies of scale. They can't mass critical capacity and infrastructure.

Scale means that they can't trade space for time, or time for space. Few sites have sufficient bulk to both patch-burn and let bison or cattle free-range. They can afford to burn land that temporarily reduces forage for livestock and nesting sites for birds only if they have enough land elsewhere to compensate. They see fire as a tool, not as a necessary process that must be fed and tended. They can't accommodate the quirky needs of the Dakota skipper butterfly, which thrives in older prairie, with the needs of plovers, which flourish in recent burns. Instead of expanding, and absorbing diversity, each new demand shrinks the room for maneuvering, so that the cost of operations becomes exponential, and the price of failure irreversible. The grasslands move quickly: without room to roam, it's hard to buffer and balance conflicting needs. Arrowwood NWR has perhaps 16 days a year available for burning. Today, the rate of change, always sudden, is becoming explosive.

Scale means it isn't possible to maximize management for one value or another. The parts will only flourish within the context of the whole, but the whole demands more space to accommodate its complexity. To acquire the resilience needed to survive, you have to manage for the habitat overall, and that requires enough land to accommodate the peculiar needs of all the parts and processes. The Great Plains as a biota flourished because there was space enough to move and seize the particular opportunities of varied sites and seasons. Always, some parts were out of sync, yet the whole was hale. When those once-expansive habitat smorgasbords shrank to the size of teacups and teaspoons, the capacity to absorb and respond became brittle and vanishingly weak. There is less margin for error, less capacity to counter threats, fewer opportunities to reclaim lost ground.

Nor is it enough to put fire as fire on the ground. Vigorous prairie needs lots of fires in many combinations, each with its own compounding interactions with grasshoppers and bison. Arrowwood needs a medley of fires to match its menagerie of waterfowl. It needs double burns in a year to stall Russian olive, repeated spring burns to hinder Kentucky bluegrass, fast burns in the fall to hammer snowberry, and burns year after year to promote bluestem and grama, and places and times spared flame for a spell. Yet it is far worse to have too little fire than too much. Almost any regimen for fire on prairie is better than no fire. And a fire regime once extinguished is, like an extinct predator, tricky to return. The same is true for a fire culture.

There is plenty of room for despair or just plain cynicism. But there is always hope. In *Prometheus Bound*, Aeschylus has his tortured but still defiant Titan declare that, along with fire, he gave humanity hope. Those who keep the flame are keeping the future.

• • •

Today the northern plains are increasingly split between two realms of combustion that symbolize two geologic epochs. One is a relic of the past, and its flame is kept to preserve the living memory of that past. The other is a harbinger of the future, and its flame threatens to remake the planet. The first powers migratory North American wildlife; the second, an ever-restless American society.

The prairie pothole region is a relic world, a geomorphic imprint from the Ice Age. For millennia flames swept the region, sparing only the wet blotches and potholes themselves, the speckled traces of the old ice. Today's residual burning is diminished, like dispersed candles compared to the sweeping conflagrations of the past. Yet the biota, and ultimately the potholes themselves, cannot survive without those surface burns. The flames migrate with the seasons like geese and bison.

The coming world is the glowing constellation of gas flares that map the Bakken shale being worked for oil, an emissary from a gathering Fire Age generally called the Anthropocene but which might be termed the Pyrocene. Drill holes belching natural gas speckle western North Dakota in an eerie counterpoint to the pothole country to the east—a dark pyric double. These fires burn day and night, summer and winter, without regard to surroundings; they propagate along a flaming front of fracking. They announce a future world shaped by fire as the past was by ice.

Those flares are not static. In reality, they are combusting downward into the rising gases, and in so doing they are burning down into the geologic past, while their emissions loft upward into the future. That industrial firepower is rewiring the American landscape, redefining what of it is a resource and recoding how those resources might be routed through a human economy. They are shifting climate, allowing for the northward spread of crop cultivation. They power the tractors that plow under the prairie. They run the distilleries that process corn and soy into commodities and then haul those products to market. They run the motor homes that send Dakota snowbirds southward for winter. They make possible a new biochemical era of sod busting that relies on Roundup and Roundup-ready seeds to replace native flora with genetically engineered culti-vars. Just as the upland burns at Arrowwood are a catalyst for a whole biota, so the fracking flares are for America's industrial society.

The prairie potholes are caught between those two fires. They desperately need more of the first and far less of the second. But how the prairie potholes and the prairie frackholes play out will determine what kind of future we might hope for. In this tension, too, they are a portal for realms of fire on for the Great Plains overall.

Konza

TALLGRASS PRAIRIE IS to the Great Plains what the longleaf forest is to the coastal plains, and big bluestem is as much a signature species as longleaf pine. Both biomes need fire. Both were so decimated by settlement that only 2–4 percent of the original biomes remain extant. Both have been the subject of intensive preservation efforts. Both have served as a poster child for fire science. Any survey of the Southeast—even the ecological equivalent of a planetary flyby—will include longleaf, particularly such residual patches as the Wade Forest in the red hills of northern Florida; and any reconnaissance of the Great Plains will find its orbit deflected by the gravitational mass of Konza Prairie in the Flint Hills.[1]

The northern flanks of the Flint Hills sit close—100 miles—to the geographic center of the continental United States. So, too, they crowd near the core of tallgrass. The Flint Hills were too rough to plow and too remote to pave. They survived the surge of agricultural settlement that elsewhere ground over the plains and that here turned instead to ranching. In the post–Civil War years the hills became an open feed lot from May to June for Texas cattle being shipped to Kansas City. The ranchers burned to freshen the glorious native grasses; they fired every spring, typically in April. That regimen put the hills in production and kept them burned. The prairie thrived. The Flint Hills had more tallgrass than anywhere else. They were to tallgrass what the southern Sierra Nevada was to sequoia groves. When the time came to protect that fast-dwindling habitat, the hills gathered into one place the whole tribe of reasons for doing so.[2]

• • •

There were many purposes behind prairie preservation, and repurposed ranches furnished the means and scale necessary. Because the land was grazed (and burned) and not plowed under (or its fires extinguished), the basics of the original biotic matrix survived. Ranching persisted, and burning with it, most spectacularly along the hills' blurred terraces. Both Tallgrass Prairie National Preserves are here—the National Park Service's outside Strong City, Kansas, and the Nature Conservancy's in Osage County, Oklahoma. Konza Prairie Biological Station triangulates the northern lobe of the hills, south of Manhattan, Kansas.

Few preserved prairies lack a research component, targeting whatever special feature of the landscape led to its being set aside. Some survive because, as with ranching, they had an economic rationale. Some were created for historical and cultural reasons—the restoration and preserved memory of presettlement conditions. Others promote biodiversity or tourism; there are even pay-to-burn festivals in which visitors can experience the thrill of setting the prairie aflame. Ranchers want fire research to help boost productivity. Preservationists support science to better understand how to protect extant patches and rebuild prairie out of cornfields. The tourist industry wants visitors to experience *Little House on the Prairie*, or the wonder of cross-continental trekkers on their way to Oregon, California, or Utah, and they need fire to sustain something of that original scene. Biodiversity interests, too, yearn to know what regimen of fire or of fire in association with other processes best promotes particular species or biomes. Each looks to research to create data and turn that data into practice. But only Konza is dedicated to full-spectrum, basic science.

Its 3,487 acres are managed as intensively as any commercial farm. Konza is not a natural wilderness, left to its own devices. It is not a tourist destination, open to hikes, picnics, and recreation. It is not committed to maximizing commodity production. It is not outfitted with gift shops, simulacra of sod homes, or faux pioneering museums. It's a field station set up in 1971 and run by the Biology Department at Kansas State University; a long-term ecological research site under the auspices of the National Science Foundation since 1980; and the primary facility for understanding the dynamics of an intriguing, deceptively complex, and vulnerable ecosystem nearly as old as the Holocene and one threatened by the cumulative abuses of the Anthropocene. The research agenda at Konza studies every aspect of the biome. Not least, it is one of the few research enterprises anywhere created from its origins to study fire. It is to tallgrass what Tall Timbers Research Station is to longleaf.

If the Flint Hills have become the homeland of tallgrass prairie, Konza has emerged as the founding lab for prairie science. It provides the strong nuclear force of data and concepts that hold collective understanding together.

• • •

Why Konza? Why fire? The story opens, like a door, on two hinges. One is the recognition in the postwar era that tallgrass might disappear if it was not actively protected, and that its disappearance would hurt science along with other prairie values; the need existed for a field station dedicated to research on land that more or less retained its critical properties and processes. The other is the personality of the project's prime mover, Lloyd Hulbert.

Observers early recognized the association between prairie and fire, but they disagreed over whether the fires were an epiphenomenon that just happened or whether they were integral or even informing, without which the biome would vanish. They knew that if fire were removed, brush and woods would overrun the grasses, but they did not know if fire as biocombustion was necessary or if some other, mechanical means to clean out the thatch and repel hardwoods would work as well. They could agree with Aldo Leopold as he eyed the Wisconsin prairie and noted the "ancient" ally the grasses had in fire as a flaming brush cutter to beat back burr oak. The Curtis prairie at the University of Wisconsin Arboretum decided in the late 1940s that it had to include fire in its efforts to return farmland to prairie. The first prescribed burn by the Nature Conservancy occurred in the Helen Allison Savannah outside Minneapolis in 1962. And while everyone could see that tallgrass thrived amid the annual burning that lit up the Flint Hills every spring, they did not know how those grasses might flourish if fire were removed. They lacked the disciplined knowledge that could only come from controlled experiments. They burned because burning worked—had always worked. They relied on tradition, folklore, and operational pragmatism.

Something else, however, had to tweak that perception into a research agenda and to distinguish Konza from other places that studied grasslands. What happened was Lloyd Hulbert, the genius loci of Konza, who for several years during

World War II led an improbable life as a smokejumper in the Northern Rockies. Fire did for him what it did for the big bluestem that he preferentially studied.

Hulbert was born at Lapeer, Michigan, in 1919 and raised a Quaker. He took a degree in wildlife conservation from Michigan State in 1940. When the United States entered World War II, he signed up for Civilian Public Service as an alternative to the military draft. He spent a year and a half in a camp in Michigan, then, in one of fate's happy quirks, the program offered a chance to serve in the fledgling smokejumper corps, which was just going operational when Japan bombed Pearl Harbor. He spent the next two years in Missoula, jumping on Northern Rocky Mountain fires, and during the off-season doing rangeland reseeding through the Intermountain Experiment Station. He was discharged in April 1946.[3]

For Hulbert the experience meant a sudden immersion into a very different life—mountains instead of plains, sweeping forest fires instead of field burns in stubble, practice rather than research, all amid a daring occupation barely beyond its beta phase. If only by osmosis, he encountered a distinctive fire culture. It was one committed to fire's suppression, but it put fire at the core of life and understanding. His rangeland labors for the Forest Service, though not all of it specifically in fire, still put him in an organizational culture for which fire was an enduring obsession and a bureaucracy whose research mission was tent pegged by a network of experimental forests and ranges.

For the next few years he taught as an instructor at Montana State and Minnesota. In 1953, he went to the State College of Washington (now Washington State) to study under Raymond Daubenmire, the doyen of grass biology, where he worked on *Bromus*, another pyrophyte. (He's credited with working out the extraordinary root system of cheatgrass that makes it so effective an invasive.) In 1955 he joined the faculty at Kansas State University. Here he found himself proximate to another fire culture, this one tethered to the Flint Hills and working under a dean, Frank Gates, especially keen on academic field research stations and a department that had considered one since 1953. He plunged into research on prairie grasses and grass-tree interactions, initially emphasizing soils, then fire.[4]

Over the next decade Hulbert campaigned for a network of protected areas in the state that could also serve as control sites for research; the notion was broadcast as "A Plan for Natural Areas in Kansas" by the Conservation Committee (then under Hulbert's leadership) of the Kansas Academy of Science, and enacted as legislation in 1974. The crown jewel was the Konza Prairie Research

Natural Area six miles south of Manhattan, which served as the Kansas State Biological Field Station. With funds from the Nature Conservancy, the ranches that became Konza were acquired from 1971 to 1977. Former ranch buildings morphed into labs and housing for visiting researchers. Burning for pasture evolved into systematic experiments on fire regimes. In 1987 Konza acquired a herd of bison to help round out its complement of grazers. It lacks a similar suite of browsers, and its scale is too small to accept the wolves and bears that were natural predators, to say nothing of human tribes, so scientists must serve instead to cull the herd and burn according to the prescriptions of experiments rather than the imperatives of hunting. As fires kindled by aboriginal firesticks had ceded primacy to wooden matches and ropes soaked in kerosene, so now they yield to the strike of a steely will on the flint of the hills.[5]

Since its founding, Konza has enjoyed a steady drumbeat of expanded ambitions and recognition. It was one of the six founding sites for the National Science Foundation's long-term ecological research program. It became a research site for NASA, for the USGS (National Benchmark Hydrologic Network), for UNESCO's Biosphere Reserve project, and for the National Ecological Observation Network. In a couple of decades it had gone from working ranch to world-class research center, part of a global consortium committed to understanding grasslands. Not everything known about tallgrass prairie comes from Konza, but no synopsis can ignore it, and it effectively synthesizes all the others.

When it began, Konza was not the only academic field station in the United States or even the only one dedicated to prairie. Agronomists fretted over fire and range, and agricultural scientists at Kansas State (and elsewhere) were burning plots by 1926. In 1933 the University of Wisconsin had acquired 60 acres of abandoned land for its Arboretum and began to reclaim it, with some assistance from the Civilian Conservation Corps. A decade later, under the leadership of John Curtis, it commenced formal experiments in restoring pasture to original prairie; some experiments included fire. The project was arduous and complex since the land had been plowed and farmed, then left for fallow, and finally used to pasture horses. Restoration from the gashes, feral cultigens, and the camp-follower weeds of settlement remained its raison d'être.

Konza sprang from another vision. Hulbert wanted a large site capable of multiple, long-term experiments on land that, while grazed by cattle, which grazed differently than pronghorn and bison, had remained prairie throughout. Unlike other experimental grasslands, this one would investigate the full gamut of prairie processes; would have a scale that made it more ranch than kitchen garden; and would include fire as a founding feature. Lloyd Hulbert never penned his intellectual autobiography or traced a genealogy from Rocky Mountain snag fires to the rituals of Flint Hills burning, but something had to pique his curiosity that did not happen to others, and it is likely that his tour as a smokejumper made fire, as fire, a topic for him worth pursuing. Certainly those who knew him thought so. Uniquely, Konza embraced both prairie research that involved fire, and fire research that occurred on prairie.[6]

• • •

Today, Konza organizes its research around the three themes that most inform tallgrass: climate, grazing, and fire—or what really matters, how they interact. In the spring of 2014 there were 165 active research projects on what is probably the most heavily instrumented patch of prairie in the world.[7]

The lesson that repeated, like a fractal, at all scales, is that one effect, by itself, matters less than the ways it interacts with others. To measure those effects requires a large landscape to accommodate at least some of those exchanges and their experimental manipulation. The separate parts get simplified into dichotomies—soil shallow or deep, bison grazing or not grazing, herbivorous insects or their absence, burned or unburned, drought or deluge. That is how reductionist science (and the human mind) tend to work. But those factors do not behave in the field like toggles that simply switch on or off: they are rheostats that modulate through a range of conditions. Some sites might be burned annually, and others once a decade; some patches are grazed routinely, others infrequently. The permutations among all the possible interactions probably equal the number of atoms that make up Earth. No science can test for them all, but if Konza cannot replicate the world's "blooming, buzzing confusion" (as William James called it) with controls, it can at least capture some of its richness with studies of second- or third-order complexity. Ideas, too, achieve their power from their interactions.

The results can be counterintuitive. What matters are the linked regimes of rain, grazing, and fire, the three-legged stool of prairie ecology. The right mix can enhance biodiversity and ecological stability. Fire doesn't control sumac, for instance, unless the burn is really hot, and probably comes midsummer, and it still needs browsers to work a one-two punch to drive the brush back. Herbivorous insects stimulate more budding, and the bud bank, not the seed bank, responds best after fire. Bison and burning, together in the right mix, increase forbs and dampen the trend toward monocultural grasses. And so it goes, fractal after fractal, from bison to grasshoppers, from flowering bluestem to buds, from landscape to patch. Grasslands offer opportunities for managed manipulation on an annual basis not available to forests or brush. And Konza offers the scale in space and time necessary to grasp a bit of the complexity, even if its insights seem little more than fireflies captured in the gloaming.

Its location allows it to investigate fire-grazing dynamics at all levels and scales. Its size—almost 60 times that of Curtis—means it can accommodate comparable, controlled experiments at a watershed level. Its commitment to the long view means it can track secular changes in climate and brush encroachment, and that it can test the possibility of ecological reversibility. Can a system that had remained unburned for 20 years be "restored," and if so by what combinations of fire, grazing, and climate? It can imagine fire as an ecological catalyst, as a synergistic process, as something embedded within a biological matrix. Its grasses are not simply fuel; they are also feed and fodder, and the rush and rhythms of biotic actors determines the character of the fires as much as wind and drought. These are unique features of the Konza infrastructure and why the National Science Foundation has renewed its status as a Long-Term Ecological Research site for the seventh time.

•   •   •

For Lloyd Hulbert it was the task of a lifetime. Honors eventually came, among them the Nature Conservancy's Oak Leaf and President's Stewardship awards. But Konza as a working lab was his real testimonial. There he personified the old adage that a single spark can start a prairie fire. In truth, he started hundreds, all set within a meticulous matrix of experimental designs and imagined over the span of a lifetime, and beyond. For Hulbert the long view was always the better choice. In September 1985 he wrote his sons that "most people try to ignore the subject [of death], which is the wrong thing to do. Death is necessary and sure for everyone. If there were no death, there could be no birth." Eight months later his own death came. His ashes were scattered over Konza's bluestems.[8]

People of the Prairie,
People of the Fire

T WICE OVER THE PAST 20,000 years the Illinois landscape has been destroyed and rebuilt. In the first age the agent of change was ice, mounded into sheets and leveraged outward through a suite of periglacial processes from katabatic winds to ice dam–breaching torrents. The ice obliterated everything, leaving as its legacy a geomorphic matrix of dunes, swales, moraines, loess, great lakes, and landscape-dissecting streams. For the second, the agent was iron, forged into plows and then into rails. Coal replaced climate as a motive force, and people pushed aside the planetary rhythms of Milankovitch cycles and cosmogenic carbon cycles as a prime mover. They left behind a surveyed landscape of squared townships.[1]

The first event worked through a geologic matrix; the second, a biological one; and they were equally thorough. All the state went under ice at least once; the last outpouring, the Wisconsin glaciation, pushed south from Lake Michigan and covered perhaps a third of Illinois. The frontier of agricultural conversion put nearly all of the state to the plow, or where rocky moraines prevented it, to the hoofs of livestock. When it ended, only one-tenth of 1 percent of the precontact landscape remained more or less intact. Less than one acre out of a thousand held its founding character, and that acre was itself minced into a thousand, scattered pieces.

In both ice age and iron age, however, life revived after extinction with fire as an informing presence—fire in the hands of people. The biological recolonization of the landscape after the ice had fire in its mix and expressed itself as oak savannas, tallgrass prairies, and grassy wetlands, stirred by routine burning. Fire was a universal catalyst; in particular, prairie and fire became ecological symbionts. The reconstruction of the second landscape has relied on industrial combustion, fueled by fossil biomass as fallow.

But those intent on sparing, or actively restoring, the former landscape must appeal to open burning. A fire sublimated through a tractor does not yield the same effects as one let loose to free-burn through big bluestem. The regeneration of such settings, so unstable and scattered, can be troubled. The reliance on fire is both essential and challenged.

• • •

The indigenes at the time of European contact, the Potawatomi, were known variously as the people of the place of the fire, or the keepers of the fire, because they maintained the great council fire around which the regional confederation of tribes gathered. But that fire did not stay within the council circle. It spread throughout the landscape, a constant among the diversity of grasses, trees, shrubs, ungulates, small mammals, birds, and insects that congregated around the informing prairie. In time the Potawatomi became known equally as the people of the prairie since the one meant the other. Remove fire and the prairie disappeared. Remove prairie and free-ranging fire lost its habitat. Remove the keepers of the fire and both prairie and fire vanish into overgrown scrub, weedy lots, or feral flame.

Restoration is a slippery concept. In some places it means mostly finding ways to preserve and enhance relics that have survived the battering. In other places it means an outright regeneration, or a reconversion of farmland to prairie. But at its core it involves sparing the pieces and saving the processes that connect them. In Illinois, once the prairie state, now a factory farm, prescribed burning is what connects those pieces, and prescribed burners are the agents that join them.

KANKAKEE

The unity of the Kankakee sands region lies in one of those convulsive geologic aftershocks of glaciation. As the Wisconsin ice receded, it melted, and the meltwaters ponded behind berms of moraine or lobes of adjacent ice sheets.

Eventually those dams themselves melted or were breached, and the impounded waters drained out. This often happened catastrophically in the form of floods or, in local parlance, torrents. At Kankakee the outrush left a scoured landscape of sand dunes and wet swales and incised streams. It became an archipelago of soils and landforms whose connection looks back to events 17,000 years in the past.

Each site took on additional characteristics as the result of its recolonization and, during the second—the settlement—torrent, the ways in which it was farmed, drained, grazed, or subdivided. Historically the lowlands were marshy and grassy, and the uplands more forested. But extensive draining converted the swales into corn fields, while routine burning kept the uplands into a woody savanna—the largest remnant of extant oak savanna anywhere. Critically, while grazed, the uplands were not plowed: their soil structure remained intact. And, exceptionally, they continued to burn.

The great northward migration of African Americans had an echo in a secondary outflow from Chicago, a city some found too alien and job poor, into subdivided lots around Pembroke. There they settled down amid old habits, including casual fire, and an absence of government services, not least fire protection. The lack of trash collection, in particular, meant they burned refuse, and these fires frequently escaped to kindle the countryside. The surrounding sand ridges burned roughly every 1.5–2 years. An area of extreme economic poverty became, paradoxically, a place of exceptional biotic wealth.

Today that miscellany of missed places constitutes an atoll of natural areas, some 32,000 acres in all, allocated among 33 designated sites, hopefully labeled the Greater Kankakee Sands Ecosystem. The archipelago includes Goose Lake Prairie State Natural Area, Des Plaines Conservation Area, Midewin National Tallgrass Prairie, Wilmington Shrub Prairie Nature Preserve, Laughton Preserve, Mazonia-Braidwood State Fish and Wildlife Area, Iroquois Woods Nature Preserve, Mskoda Land and Water Reserve, Sweet Fern Savanna Land and Water Reserve, Kankakee Sands Restoration Project, and Willow Slough Fish and Wildlife Area, and with those sites a roll call of Illinois conservation organizations that ranges from national agencies to state and county bureaus to nongovernmental organizations (NGOs): the U.S. Forest Service, the Fish and Wildlife Service, the Illinois Department of Natural Resources, the Nature Conservancy.

In all this—remnants scattered like lithic flakes, restoration projects sprouting from corn stubble, a variety of institutions as diverse and dispersed as their biotic relics—Kankakee is a cameo of the Illinois conundrum. No single site, institution, or vision contains it all or organizes the pieces. There is no commanding height—not topographically, not institutionally, not intellectually. A federal presence is muted, quarantined on checkerboard hills in the far south. There is no domineering private landowner—no Weyerhaeuser, no Ted Turner—to deform the space-time of land use. There is no counterforce to challenge the industrial plow. What the pieces and players share is a variously defined commitment to nature protection. They are, like the Potawatomi, peoples of the prairie, which means they are also peoples of fire.

They differ in goals. Some believe that the task demands a way to connect the fragments into a whole, at least conceptually; they seek out corridors to join the parts, or ideas to help identify which pieces should be protected in what order. Others believe that salvation depends on size. Unless the protected areas are large, unless they contain within themselves all the required parts, the whole cannot hope to survive against fragmenting forces of regional or continental scope, not to mention globalization. Yet the practical scale of either strategy is so small that the atolls they oversee may both be drowned in the rising sea of a modern economy. Chicago adds more rambling exurbs yearly than the state does protected preserves. Farmland converts to city, not nature.

Each site resembles a miniature, the ecological equivalent of a ship in a bottle. Its minuscule scale allows for some processes to persist, and for the abolition of known destructive practices. But they struggle to become a whole; the separate parts cannot absorb the roaming elk and bison (and successor cattle), or their predators, that helped define the historic scene. Their collective fauna is one that travels by air, and that is also tiny; the faunal diversity consists of birds and especially invertebrates.

This can cause troubles, however, because insects can be highly specific in their preferred habitats, and they can attract partisans that consider butterflies and leafcutters in old-growth prairie as the counterpart to spotted owls in the old-growth forests of the Pacific Northwest. The species triumphs over the habitat. In order to accommodate, even small plots might have to partition into micromanaged patches. A landscape that boasted white deer and wolves must shrivel to one for beetles and the regal fritillary. This matters because some management practices cannot be indefinitely shrunk, any more than Newtonian physics can scale evenly from quasar to atom. A butterfly and a bison demand different minimums of place.

So, too, does fire. In a miniature landscape it acts more like a blowtorch than a free-ranging wind. It behaves like an implement of horticulture, a clipper or hoe, no longer feeding itself as it propagates but consuming what it is served. The patches resemble cages in an open-air zoo, or to mix in a more benign metaphor, like rooms in a hospice. The ecology of a candle bears little kinship with that of a prairie aflame. No one knows the scaling laws for fire ecology that might join the nanoniches of a prairie refugia to a boreal crown fire. They only know they must have fire.

•   •   •

This, however, is the second element the system shares: a commitment to burning. Fire does nothing here it does not do elsewhere. It just seems more prominent because it is indispensable and the small scale of the operation makes it undeniable.

The remnants survived because they were burned. If flame leaves, woody plants will quickly swarm over and smother prairie and savanna. A handful of years is sufficient to let invasive shrubs and trees establish themselves to the point that fire alone can no longer knock them back. Like a boa constrictor steadily tightening its coil whenever its victim breathes out, the woods crush the grasses and forbs when the pause between fires lasts too long.

Questions of scale do not, however, abolish all principles of fire ecology, and one is that organisms do not adapt to fire in the abstract but to a fire's regime. You can lose a site as surely by burning badly as by not burning at all. The bouquet of sites around Kankakee argues for a bouquet of burns; and in the absence of particulars, a useful rule of thumb is a three-year rotation, which approximates the core cycle of postfire recovery and, at Kankakee, will accommodate almost all species if a site is large enough relative to the organism's demands. Still the threat of too little fire probably trumps the threat of too much. The premier relicts like those around Pembroke burned almost annually, or no longer than biennially; they burned as frequently, that is, as fuel existed to carry the flames.

Another principle is that fire is *bio*technology. Its flames do more than act as a fiery brush cutter; beyond merely mowing and mechanically rearranging, they transmute; they chemically change the biomass they consume, as grazers do. Nothing else provides their range of ecological services. Moreover, add to the roster of precepts that fire is an interactive technology as well. Fire's effects rarely result from fire alone, but from the cascade of interplays it sets in motion, among them grazing and browsing, both gone from the Kankakee complex except at the level of insects. That is why mowing or planting or shunting fuels around cannot substitute. You can't swap clipping for burning as you can an electric bulb for a candle.

All this argues for prudence, which here means perpetuating the regime that kept the sites intact. To the uninitiated the precautionary principle might seem to argue for the opposite: they might invoke it to halt burning until a full-spectrum ecology can be worked out. But of course a complete ecology will never be known, and while species-partisan Neros and academic researchers fiddle, Rome won't burn, with disastrous results.

• • •

From the grand perspective of fire's recession and recovery across the American continent, the kindlings around Kankakee seem quaint, a daub of mom-and-pop stores amid an economy of big-box retailers and multinationals. This is boutique burning, almost a farmer's market of handcraft fires. California's Cedar fire burned an order of magnitude more acres in one savage surge than the Greater Kankakee has under its collective protection.

But such metrics miss the point. It is not the number of scorched acres but the richness of their impact that matters. Acre for acre, probably more comes out of the Illinois burning than from all the firing of southern loblolly pine plantations and of generic western wildlands. Here, fire is the critical catalyst, without which the land cannot be defibrillated back to integral prairie and healthy savanna. Fire alone can't make that restoration work, but nothing done without fire can succeed.

NACHUSA GRASSLANDS

The Kankakee complex is a strategy of small parts in search of a larger context. Its conceptual counterpart is a large preserve that might contain the varied habitats of interest within its own borders. In Illinois its prime expression is the Nachusa Grasslands managed by the Nature Conservancy. But Nachusa is becoming large (in relative terms) not because it has preserved an ancient landscape but because it is rebuilding one.

Its core is a small ripple of rocky hills that escaped the plow and hence retained some native species. In 1986 TNC purchased 400 acres. But around that atoll lay a platted sea of row-cropped maize. Over the course of 20 years the preserve has expanded to 2,000 acres, all purchased from willing sellers at market prices. Those acres must be brought into the system through a laborious process of restoration. By 2009 some 93 projects, ranging from 2 to 60 acres each, had expanded the dominion of prairie. Half the labor has come from volunteers.

It isn't enough to leave the acquired land to nature's touch: it will grow weeds. Instead prairie must be cultivated with even more tending than commercial crops; and there is little natural about the boundaries, which must follow the square-surveyed townships of settlement and the economic rhythms of commercial agriculture. Until they have the wherewithal to begin the conversion they lease out the land for corn. Meanwhile stewards gather seeds from existing prairie.

They begin actual restoration—or more accurately, a reconstruction—by harvesting the corn, burning the stubble, and sowing a heavy mix of native seed, as much as 55 pounds per acre of 150–200 species. The next year assorted plants will have rooted, along with a street gang of weeds. Some weeds are an immediate concern and are clipped and individually herbicided. Others will succumb as the indigenes thrive. What matters is removing the nasty species and stimulating the desired ones, especially the grasses like little bluestem, since they will carry fire over the plot, and fire is what sparks the system to life.

The overseers burn as often and as intensely as possible. Where desired plants flourish they may overseed with more, and where some seem lacking, they may try again, and let nature determine the suitability of niches. Meanwhile they burn. By the third year of tedious culling a raw matrix of prairie grows on the site. When they determine the mix is more or less right managers can begin backing off annual burns and feel their way toward a suitable cycle. Fires spread across the surveyed borders and suture the larger quilt of patches together. In this way the restoration reverses the frontier inscribed under the parameters of the Northwest Ordinance of 1787; and with so much work done by volunteers the process resembles a kind of reverse homesteading.

• • •

The fire story at Nachusa is simple enough to state. Fire initiates the conversion, and once it has worked that alchemy repeated burning perpetuates the revived biota. Restoring prairie has meant restoring fire: this much is unexceptional, however quirky the process might appear to deep ecologists intrinsically wary of Roundup and flame. Rather, Nachusa's natural character resides in its present expression, not its history—or as William James famously described pragmatism, "By their fruits ye shall know them, not by their roots."

Yet there is a second narrative of fire restoration at work as well, in which fire is returned not only to the land but to the hand. The reconstruction of Nachusa reinstates fire to ordinary people. The volunteers who do much of the hard work of gathering and disseminating seeds, clearing invasive shrubs and weeding new acres, also do the burning. As much as reinstating big bluestem and lady fern, Nachusa has returned the torch to folk practitioners, the kind of fire-wielders who sustained the prairie peninsula through millennia. The people of the new prairie have become people of the new fire.

This is a story easily lost among the attention paid to the traditional big hitters of fire management, and it counters two trends. One is the grand narrative of Earthly fire by which industrial combustion has replaced open burning through technological substitution and outright suppression. This is why there is no fire on the still-farmed lands around Nachusa, why cooling towers from a nuclear power plant loom over the northern horizon from the rebuilt barn that constitutes preserve headquarters, and why quads and tractors rather than draft animals fill the sheds. Nachusa is putting fire back on the land.

The other trend is the systematic stripping of fire from the hands of the folk. The simplistic yet orthodox narrative for justifying the restoration of fire on public wildlands is that nature had set fires and misguided public agencies extinguished them, and the outcome is the shamble of present-day fire regimes. Such a narrative implies that restoration means no longer suppressing nature's fires. It means that people have to quit interfering with nature's logic. Nature will then begin deleveraging the landscape into it proper state.

Yet the record for virtually every landscape is that people had set most of history's fires, and this leads to the conclusion that the missing fires—those that have disappeared over the past century—are the result of people no longer acting as we have acted throughout our existence as a species. Less and less burning got done because there were fewer and fewer burners to do it.

To be sure not all of that erstwhile burning was prudent or systematic; some was abusive and promiscuous, and not a little simple fire littering. But in shutting down the excess fire became, in effect, a government monopoly, something so seemingly arcane and technical and intrinsically dangerous that ordinary citizens could not be trusted with its stewardship. In this narrative, restoration means getting people to burn again. What Nachusa adds is the return of the torch to private citizens, not solely to agents of government.

RECESSION AND RESTORATION

The ice age receded, across a span of 10,000 years, with a succession of geologic spasms like the Kankakee torrent. The recolonization of that evacuated landscape by life took several millennia, and after the climatic maximum of the Hypsithermal, humanity helped stabilize its dimensions and the resulting pastiche of prairie and savanna by regular firing.

The iron age ended with the bleeding soils from a thousand thousand cuts and with a slow smothering beneath a blanket of domesticated flora. Its regeneration will take centuries, if not longer, quarter section by quarter section, township by township, and it will act out against a fast-morphing climate, likely the byproduct of an industrial burning run amok. But it will happen at the hands of a humanity wielding fire.

This is not the kind of creation story or heroic narrative that American environmentalism has traditionally thrived on. But it is what must happen if nature's economy is to continue to produce the goods and services we want. It's a story in which the Hippocratic injunction to first do no harm means you will harm if you don't first do. And it's a story in which the people who want prairie must also become a people who want fire.

The Blackened Hills

T HE BLACK HILLS ARE an island, a cameo of the Rocky Mountains plucked onto the plains. In their geology, their biota, their history, their contemporary economy, the hills align with the West. But their fire regimes seem to have absorbed some features from the surrounding prairie. It's as though, here, the plains congealed into a patch of the Rockies.[1]

The mountains bring rain. Like an ocean isle the moisture is greatest on the windward (northwest) side and lightens to the lee (southeast). The rain grows forest amid what would otherwise be mixed- or shortgrass prairie. But while the trees come from the Rockies they behave like brush on the plains. The western yellow pine more resembles eastern red cedar than it does the ponderosa pine of montane Montana or the Mogollon Rim. It propagates like a woody dandelion. It regenerates spontaneously, gregariously, promiscuously: only the saw, beetles, and flame hold it in check. Save that a timber market exists for the larger trees, and that its massed forest serves as an amenity for the tourist industry, the pines might well be considered a noxious woody weed. And like the cedar its history is intimately intertwined with fire, which is to say, with how natural conditions and human history have met, mingled, clashed, and colluded.[2]

· · ·

The anomalies begin with its geographic location, as an outlier of the North American cordillera. What Midway is to the Hawaiian Islands, the Black Hills are to the Rockies.

Around the northeast border of Colorado, the central Rockies thin and pivot to the northwest. The Black Hills continue the old trajectory, like a Pacific island stranded by shifting plates, or in this case by the shifting rumbles of the Laramide revolution. Like such isles the hills are igneous: their nucleus is a bubble of granite that rose and bulged the surrounding strata, and then abraded down. The overlying and the peripheral weaker layers washed away, leaving a hard core and a crusty ring like a barrier reef. Both peak and reef were forested, while a lagoon of grass lapped between them. The resulting dome resembled an etched turtle shell. Even with thousands of feet of rock removed Harney Peak remains the highest point in the United States east of the Rockies. The Black Hills rise out of the Great Plains like Tahiti out of the Pacific. From every vantage point around it attracts the eye. It draws all to it.

It's a biotic island as well. Remoteness, however, can simplify as well as retain. Being more mobile the hills' fauna could reach the isle from elsewhere as well as remain while the post-Pleistocene climate staggered toward some degree of stability; its complement of mammals is astonishingly rich. The flora is simpler. The grasses of the plains wash against the slopes and wend through the interior valleys. The trees are residual, however, a forest ark from the Pleistocene. For the Black Hills National Forest, which largely excludes the grassy valleys, some 95 percent of its land is forested, but 93 percent of that is a single species, *Pinus ponderosa*. The other dominants typical of the Rockies are absent, save aspen and spruce; these claim niches, ecological relics like the mammoth bones in sink-holes outside Hot Springs. The informing principle of the biota is its exuberant pine, which binds everything else into a whole.

The hills both drew and held that most peregrinating of creatures, human-ity. The colonizers collected and concentrated almost every theme of western settlement. Their story, if staggered slightly behind most regions, claims it all in miniature: a sprawl of mining, railroads, mass grazing, widespread logging, and after the wreckage became unbearable, late reservation to public lands and administration under doctrines of state conservation. The Black Hills Forest Reserve was proclaimed by President Grover Cleveland in 1897, the year the forests received their organic act and 23 years after Custer's Black Hills Expe-dition set off a gold rush. The reserve's boundaries encompassed most of the hills that lay within South Dakota, but those borders incorporated swathes and packets of private holdings from mining claims and ranchers patenting water holes, scattered across the mountain like beetle borings. The U.S. Forest Service assumed control in 1905. Seven years earlier, Henry Graves, a future chief, had visited the hills and despaired for its future. Its woods, its waters, its very soil was disappearing as though down a sluice box.

Gradually, the USFS wrested the region away from its ruinous laissez-faire rushes and imposed a minimal order. It brought some regulation to logging, forest grazing, watersheds, and reservoirs, and of course it imposed fire protection. Still the Black Hills were, and have remained, overwhelmingly the dominant economic engine of the West River country, the impoverished stepchild and federal-laden half of a lightly settled state. (South Dakota is the 17th-largest state but has one-fourth of 1 percent of the nation's population.) The Old West economy of commodities spiky with booms and busts hung on surprisingly long. The Black Hills National Forest continued as catalyst and counterweight.

Eventually the economy of the hills, as with the West overall, turned to services and amenities—to summer homes, recreation, and tourism, much of it playing on an idealization of that abusive past, as though the Hickoks and Oakleys of logging and grazing had become tradable securities in bidding for public attention. Eventually the era of hard-rock mining symbolized by the Homestake yielded to the chiseled granite of Mount Rushmore. Still, whether as emblem or economic player, the old order persisted. The ponderosa pine continued to hold the parts together. The Black Hills remained a living keepsake of the Old West, preserved, as it were, in pine pitch.

In eerie ways, the Black Hills came to resemble Lake Tahoe. One was a hump on the plains, and the other, a hole in the mountains, but they shared a comparable power to distill and concentrate the themes of western history. Each created a bounded miniature. Each is a lodestone for its regional economy, and its evolution. Both began as commodity producers before segueing into a modern economy of recreation and service. Without a governing body to orchestrate use on all lands, the hills, however, have boosted schlock to an order that could make Tahoe blush. Its private lands are littered with the bric-a-brac of industrial tourism like beer cans tossed along a highway, a cacophony of niche theme parks, kitsch, specialty museums, and souvenir shops like baited hooks for passing cars. But where Tahoe centers on a lake, and overseers obsessed with its purity, the hills have an inland sea of woods, fecund to excess. Cutting in the Tahoe Basin practically requires an act of Congress. Cutting in the hills is essentially an act of survival.

The Black Hills has retained an old-model multiple-use forest of the sort that has become extinct elsewhere. It is both a working and a heritage forest. It stands to the national forest system as the rebuilt Deadwood does to mining; its evolution is like watching a frontier dance hall morph into a modern casino. Whether its continued high-volume logging is a salvation or an administrative equivalent to a bank of slot machines, generating extra revenue but at the cost of what Hal Rothman, historian of western tourism, termed "devil's bargains" remains to be seen; the most likely outcome may be both. Revealingly, the region has a state-legislated Black Hills Fire Protection District that aims to regulate all open fires and to attack any that occur.

What seems clear is that the Black Hills concentrates what comes to it. The good news is that it thus overcomes the plains' tendency to disperse. The bad news is, it concentrates the bad as well as the good.

•   •   •

Its fires are a hybrid; more powerful and persistent than prairie fires, less consuming than the classic Big Burns of the Northern Rockies. Fires can spread at rates more typical of grasslands: the Jasper fire clocked off 30,000 acres in six hours. But thanks to pockets of heavy duff and abundant dead-and-down trees they can hold for five to seven days, well beyond the one or two burning periods common for even monster grassfires. The pine, that is, behaves more like prairie brush than either woods or grassland alone.

Intriguing facts—but the big story is the magnitude of burning that must have shaped the presettlement forest. Thanks to W. H. Illingworth, who accompanied Custer's 1874 expedition as photographer, a documentary record exists of what the Black Hills looked like a century after the Sioux laid claim and on the cusp of an American invasion. Grassy valleys are free of woods; south-facing slopes are culled of dense forest, letting trees flourish in stringers and pockets; and north-facing slopes are dappled with aspen and pocked with canopy holes. A century later those same scenes were rephotographed, with astonishing results. Despite being logged over at least once, despite waves of bark beetles descending on the pine like locust swarms on corn, despite serious fire years, the forest had thickened at least threefold. South-facing slopes filled in. North-facing slopes shed their mottled texture in favor of a uniform canopy of even-aged pine. The only explanation for such a dramatic shift is an overall reduction in fire. The amount of presettlement burning—surface, crown, mixed; spring, summer, fall—must have been staggering to fight back the woody brush.[3]

The hills, while making a compact unity when compared to the surrounding plains, are a complex of fire regimes when examined in detail. Chronologies of fire-scarred trees suggest long waves in which burning overall waxed and waned with climate and human migration. Average return intervals are 16 years, with some sites as frequent as 2 years and others reaching 34 or so. The serrated texture of the geologic dome that defines the hills offers plenty of room for variability; averages across the hills probably mean little. Moreover the fuel arrays defy simple cycling because they are mediated by two fauna. Grazers, most spectacularly the bison, compete with fire for the grasses, while bark beetles work over the pines. What the record clearly reveals, in common with western landscapes generally, is that the old regimes broke down in the late 19th century. Somewhere between 1881 and 1893, the last landscape fires occurred. Over the next century big fires meant a single burn in the occasional year on the order of 10,000 to 20,000 acres.[4]

The bill of indictment cites the usual culprits. Logging shattered the structure of the old forest, grazing checked the grasses that might have helped keep fire on the ground, and the Forest Service committed to fire's exclusion. A classic photo of light-burning was taken in 1898, showing Henry Graves, the man who would lead the fight as chief forester, standing among Black Hills pine and smoldering surface burns. The exuberance of pine regeneration made a mixed fuel array not only the norm but a constantly primed tinder box. Having proclaimed the forest a poster child for all the ills of laissez-faire land use, the Forest Service had to demonstrate an alternative. It could not abolish mining, logging, or grazing, or throw out the thousands of homestead patents, or wave off mountain pine beetles. But it could attack fire.

It had modest success, perhaps from fire's exhaustion as much as its suppression. Then the Great Drought framed the hills with big fire years in 1931, when fire burned to the Rochford city limits, and 1939, when the McVey fire roared outside Hill City; three years of flooding followed. Between those dates the Forest Service adopted the 10 a.m. policy. The state legislature responded in 1941 by creating the Black Hills Fire Protection District to regulate all open fires. In 1959 a fire burned to the edge of Deadwood. In 1960 two large fires broke out near Wind Cave, and in 1964 the park experienced its biggest burn, the Headquarters fire, at 14,096 acres.[5]

The modern era commenced in 1985 as a new spiral of drought began circling the West. Big fires broke out in 1988, and again in 1990. Then the Jasper fire—part of the 2000 Northern Rockies constellation—ripped over 83,511 acres and burned a cavern in the center of the Black Hills. Thereafter large fires have struck almost annually. The southern ridge—the reef around the island peak—have caught the most; Elk Mountain suffered so many repeat burns it was no longer even black. Fire prevention succeeded in holding the number of human ignitions; but lightning, like coyotes filling the void left by vanished wolves, more than made up the difference. Then drought and beetles began mutating green forest into red slash and standing fuel, and brushed against a built landscape that, morphing into amenities communities, magnified the risks to life and houses. The forest of the Black Elk Wilderness upwind of Mount Rushmore is 85 percent bug-killed. Instead of sprawl, the Black Hills suffered from infilling—helpful in cities, but dangerous in wildlands. The collective landscape threatened to sink into a forested variant of gambler's ruin.

As fire risk went ballistic, fire protection scrambled to keep pace. Spooked by the Jasper fire, the various fire authorities in and around the Black Hills established the Great Plains Interagency Dispatch Center at Rapid City. In 2002 an executive order by Governor William Janklow consolidated the authority of the state forester over the Black Hills Fire Protection District. In 2006 North and South Dakota, Wyoming, and Colorado signed the Great Plains Interstate Fire Compact. The thrust to all these efforts was protection.[6]

The Black Hills National Forest ran a full-gamut suppression program. From 1982 to 2011 the forest annually averaged 120 fires that burned 9,709 acres; over the past decade, aggressive fire control brought those numbers down to 97 fires and 6,356 acres a year.

• • •

Behind both landscapes, the built and the wild, lay the ponderosa pine. It underwrote the paradox that while people sought to save the forest, they also had to save themselves from it, for even as fires and beetles took out mature stands, seedlings sprouted like ragweed. The forest seemed to grow faster than it could burn.

It was too dangerous to let wildfire and free-ranging beetle infestations remove the trees, not least because both left abundant fuels in their wake. The dead trees, once fallen, could stoke pilot flames into the maturing thickets of pine regrowth. Instead, in an almost textbook example, the national forest turned to silviculture. It would let prescribed felling control fire, and then let prescribed fire assist continued felling. Whether cut, chewed, or burned the pine seemed irrepressibly capable of regenerating, and only some kind of thinning program could prevent the revanchist woods from consuming the landscape. To the mind of the public forester the agency needed to regulate nature's laissez-faire economy as it had that of the settlement era.

The forest has turned to cutting to fight off beetles—sending mature timber to mill, cutting and chunking where patches are too tiny for commercial harvest, thinning ahead of infestation frontiers. It cuts to similar purpose around towns, monuments, and I-zones in the name of fire protection. It logs to create a mosaicked landscape, one better buffered against wildfire. It relies on variations of shelterwood thinning, not clear-cutting, so seed trees remain. In the 1990s the forest turned to whole-tree logging, which further diminished the visual impact of large harvests. The land appears to heal rapidly. Regrowth soon covers what traces remain. There is virtually no public protest.

Instead of holes, however, there are piles. A market exists for mature trees; none for the smaller diameter stems and tops, which are stacked into mounds the size of strip malls. Here is where prescribed fire enters the cycle: when snows prevent spread, the piles are ignited. Some broadcast burning occurs, but mostly burning serves cutting, just as cutting is intended to assist the control of wildfire. But even when stacked like cordwood (literally) the burning falls behind. After a dry winter or two in which firing is omitted the Black Hills might better be called the stacked hills. In the meantime an immense network of roads laces the mountain, which allows for rapid initial attack.

In principle it is a forester's dream. But logging hasn't stopped the beetle outbreaks every 30 years or so; the hills have been logged over several times, and the beetles keep coming. Nor has it stopped the fires; the roads, the thinning, the whole-tree harvests—none have stopped ignition or scotched the prospects for further Jasper-sized burns. The Jasper fire, after all, burned through landscapes logged several times. The past decade has even experienced crown fires at night, unheard of before the modern system came into sync.

At the most basic level the program doesn't go far enough. Timber managers estimate they need 50,000 acres treated a year, and fire managers want at least 35,000 acres thinned and burned. Timber operations reach about 5,000 acres, while fire has achieved about 4,400 acres annually over the past 10 years. A decade ago the Black Hills surpassed the collective national forests of Region 6 in timber production; now it exceeds Montana; and it achieves that output not with big trees but through the sheer volume of its smaller ones. The market for piled fiber, however, is very small. Without extraordinary effort, the prospects are that regeneration will romp through the hills, and that fire will rip through the regrowth.

With 3,500 miles of internal borders due to private inholdings, and with prospects for explosive runs, there is little opportunity to allow landscape fires much room to roam; the developed sites all seem to lie downwind from the relatively unbroken forests. For the same reason it is tricky—more art than science—to finesse and coerce prescribed fire to behave as desired and stay within its designated bounds, like training a grizzly bear to dance or a mountain lion to cuddle. For a forest that struggles to find the prescription windows to keep up with its pile burning and the operational latitude to conduct broadcast burning on the scale required, fire management must veer toward suppression and the hope that between beetles, saws, and burns, fire officers can gain enough space and sufficient time to catch the bad fires before they blast into communities.

As it approached the 2012 season the forest could certainly boast an enviable fire establishment. It had 14 engines, 2 hotshot crews, a 7-person helitack crew, a Type 1 helicopter, a tanker base at Rapid City, access to 3 dozers, and a peak-season staff of 150. The Black Hills Fire Protection District and Great Plains Interagency Fire Coordination Center gave it an institutional context for cooperation. The forest was the largest player by far in South Dakota. It had been fighting fires for over a century.

• • •

In triangulating the bulk pyrogeography of the Great Plains, the Black Hills suggest contrasts both to the eastern prairie and to the western forest. The hills are to the western plains what the prairie peninsula is to the eastern, putting trees amid grass instead of grass amid trees. The one puts the West on the plains, and the other, the plains into the Ohio Valley. What this means beyond an interesting intellectual exercise is unclear. What makes the Black Hills significant to wildland fire is that it displays an alternative history, what multiple-use fire suppression might look like had the nation chosen to stay on the old slash-and-suppress path during the 1960s.

Like a vintage car, the old model kept being rebuilt and upgraded as it added mileage. The forest had advantages not generally available. There are no threatened and endangered species to allow environmental groups leverage; only one token wilderness existed, the Black Elk, with an area probably equal to that of the Rushmore and Crazy Horse monuments; and the local communities supported the timber industry. Two-thirds of the forest is available for harvest, and accessible by road; and salvage logging is common. In a state of 824,000 people, almost all of them east of the Missouri River, the Black Hills remained a gold mine, even if it meant panning for tourists rather than nuggets. Most of the classic tools of forestry still lined the workshop walls.

But if the Black Hills National Forest concentrated the opportunities for continued multiple-use management, it also concentrated its ills. Summer homes sprouted like mushrooms, but did not make a sustainable economy. Logging cut more or less freely but could not, alone, halt beetle epidemics or megafires, and left mounds of debris like woody junkyards, and an economy that continues to decline. Deadwood was reborn as a casino city and TV celebrity but continues to wither nevertheless. Grazing persists unimpeded, jostling against fire and wildlife, yet it remains economically marginal. Drought drains streams along with the Pactola Reservoir. In 1984 the BNSF railroad shut down its last line through the hills. The mines that established the cycle of boom and bust, after a brief bubble, have collapsed once and for all. Galena lives on as a superfund site. The Homestake Mine finally shut down and bet its future on hopes that the National Science Foundation might turn its monumental shaft into a field lab for neutrino research. One can imagine no better illustration of the shift in the hills' fortunes.[7]

And fire? The Black Hills may evolve into another kind of experiment: a test on the ability of the multiple-use forest to achieve a level of fire management that has eluded (or been denied to) other models of national forest administration. The complexity of fire management here simplifies. New starts are suppressed, prescribed fire is mostly pile burning, and the landscape links flame to fuel, not to matters of ecological integrity. Legal, social, and political shackles that fire managers elsewhere blame for making their job impossible are here unfettered. Instead the complications reside in economics; in dispersed summer homes and tourism, weak markets, and the costs of remoteness. The isolation that makes the hills special also renders them vulnerable. The hills concentrate fuel and fire problems without allowing an equally focused response. The fires come, and they threaten to rush over the land with unremitting frequency and with perhaps greater severity.

Old-style forestry claimed it could manage fire if silviculture were granted carte blanche and fire suppression were staffed adequately. In the Black Hills it has been granted that wish probably as fully as any national forest in the West. What happens will thus affect not only the hills themselves but the ceaseless national conversation about why wildfires continue to metastasize and what might be done, at what costs, to stop them, and what of those fires might need to endure, even if a final reckoning may conclude that fire here is no worse than in other places, and no better, that it is simply another anomaly rising out of the plains.

SOUTHWEST

The Jemez

STAND ON THE RIM of Cochití Canyon and look over a blasted landscape. What once was a densely forested gorge, a sky island on its head, is now stripped clean, save for ghostly legions of trunks. Moonscape is a common journalistic trope to describe postburn scenes; but those landscapes return, often with more ecological vigor than before. Here even the ash is gone; and only in niches in Cochití can you find a patch of green. There are no squirrels or rabbits scurrying among the ruins. There are no bark beetles amid the boles. There are no birds; no juncos, no red-shafted flickers, and because there is no carrion, no vultures soaring on thermals. Even the last-survivor ravens have fled. There is silence ruffled here and there by wind. There is only rock, and the trunks like lithified pillars, and occasional pulses of floodwaters, the geologic matrix out of which an ecosystem might sometime return.[1]

It's a setting in search of meaning, or at least a metaphor. To natural scientists, it's a land being reorganized. To novelists, it's an apocalyptic, nuked landscape. To those who like their science and fiction fused, it's as though the Genesis Device, featured in Star Trek's *The Wrath of Khan*, had unleashed its fiery front and destroyed the preexisting world in "favor of its new matrix."

It's a place to contemplate the future. Are the serial fires that are blasting the Jemez Mountains, and where the fires return, compounding one burn with another, a new normal? Or are they a biotic outlier as peculiar as the Jemez is a geologic one? Is the Jemez a sacrifice zone of the Anthropocene? Or the first tremors of a global Götterdämmerung that signifies the end of the world as we know it?

• • •

The Jemez Mountains are one of two vestigial megavolcanoes in America. In broad terms they mark a spot where the Rockies and the Colorado Plateau meet atop buried rifts before spilling into the Basin Range to the south. Over the course of 14 million years the Jemez has spewed lava and belched tuff and gases and built up an immense pile in northern New Mexico. It remained active during the Quaternary. There were major eruptions 1.6 and 1.25 million years ago before, roughly a million years ago, the mountain blasted lava, ash, and pyroclastics on such a scale that it blew away its own top and the great mass collapsed into a caldera. A resurgent dome rose at the center, the crater filled with flowing rhyolite, and later bursts surrounded most of the central massif with a daisy chain of smaller domes, leaving a landscape of grassy plains and forested hills known as the Valles Caldera.[2]

Since then the Jemez Mountains have spilled biological rather than geologic fire. From the onset of the modern monsoon climate, the landscape has been saturated with ignition. When wet-dry cycles stoke suitable fuels and winds whip and slosh over the great conifer-coated massif, fires break out and ramble around the *valles*, over domes, across flanking mesas, and down gorges. The older record is chronicled in soil and lake charcoal; the newer, over the past centuries, in fire-scarred ponderosa pine; the most recent, in texts and photographs.

For millennia burn had passed over burn, leaving strata of ecological history. In some places the biota thickened, the equivalent of rhyolite domes. A few years had almost no spreading fires. A few, as in 1748, had many, or fires that crept and swept across the whole of the range, dodging rains and grasping for winds, over the long summer months. Some 23 times, back-to-back years managed to scorch virtually everything. But throughout that meticulous annal, the forests and grasses and patches of scrub persisted. Through most of the mountains the fires scrubbed the biotic grime of 5–10 years from the surface. Mixed-conifer woods trimmed the crest of the highest rims, and when they burned, the flames wiped out stands, but such bursts were little more than vignettes decorating the margins of the text. What the fires removed either survived or was effectively replaced. The material record testifies that fires, and fire years, swelled and contracted, but always within bounds. The biota persisted through layer after layer of surface fire flows.

The Jemez Mountains are among the most intensively inventoried landscapes in America. Few places have a fire history as thoroughly documented: there is little dispute about the character of the fires that populated the Holocene. Century by century, millennium after millennium, the cadence of the fires continued, like hot springs fed by a deep chamber of climate.

• • •

Then that rhythm broke. In the late 19th century an irruption of livestock stripped out the grasses and stopped fire stone cold. After 1883 the annals of fire effectively end. Afterward, although ignitions might sprout like lupines, they failed to spread. Their fuel went into sheep and cattle, while flames starved and expired. The record of fires simply stops, as though the recording needle ran out of ink. In 1905 the U.S. Forest Service assumed control over the forest reserves and instigated a program of systematic fire exclusion. People could no longer set fires freely, and fires of any source that did appear would be suppressed. Though the core of the Jemez, most of the Valles Caldera, was held privately under the Baca Grant and its successors, its fires beat to the new regime.

The rhythm broke a second time a century later when, beginning in 1977, savage, woods-stripping—with a severity unprecedented over the duration of the Holocene—fires began to blotch the mesas and domes. When such fires reburned the same site, they vaporized wood and shrub and grass, baking the caldera like a ceramic bowl. Such serial fires no longer inscribed a living record on the land but obliterated the record altogether. The Jemez began shuffling from one of the best-documented landscapes in America into one that threatened to become a tabula rasa. History wasn't stopped or even reversed. At places like Cochití Canyon, it was extinguished.

Behind this transformation lay new sources of power. The plateau had long been subject to seasonal transhumance of local flocks, though their size, and hence their impact, had been limited by the capacity of the local communities to absorb them. The market was local and, compared to the landscape, small. Shepherds burned, but along the transhumant corridors of seasonal travel and within established fields of fire. The flocks competed with fires, but there was ample room for both.

Then in 1878 the Atchison and Topeka railroad crested over Raton Pass. A railway reached Albuquerque in 1880, while a spur line brought locomotives puffing into Santa Fe. The next year a complete east-west connection across the continent was possible using those routes. In 1883 the Jemez experienced its last widespread fire season. (In 1885 the Atlantic and Pacific Railway completed a more direct route.) Those rail lines proved a chasm in combustion history as profound as the rifts underlying the massif.

The ancient isolation of northern New Mexico ended. Flocks and herds no longer had to accommodate into the tidy string of pueblos along the headwaters of the Rio Grande: they had the nation for a market. They expanded to meet that demand. The problem was that the fuels feeding the flocks was far more finite than the fuels stoking the locomotives. While the coming of the U.S. Forest Service in 1905 changed the patterns of indigenous burning, it did nothing to alter the onset of the new combustion regime, which had its sources and sinks elsewhere. Industrial combustion had a magma chamber thousands of times larger than that beneath the Valles Caldera. Over rails and roads it spilled out surface flows of internal combustion, while wiping the land clean of open flame.

The larger landscape began to adjust to the absence of routine fire. An ancient predator had been removed, and the population of conifers that it had held in check now exploded exponentially. The most dramatic changes occurred in the montane belt of ponderosa pine. Land that had held trees per acre that could be counted on a ranger's fingers and toes now numbered in the hundreds, and in places, the thousands. Locally, as around the forested caldera, abusive logging shattered the structure of the forest and left behind a scrub of conifer thickets. Combustibles stockpiled, now so congested that a flame in any particle could ignite a score of others, and they other scores in what might aptly be described as a chain reaction of combustion. Suppression could challenge the dragon only through good luck and a favorable climate.

• • •

In the 1960s a revolution in thinking, and then in policy, sought to reverse fire's exclusion with its controlled reintroduction. That protest flared most spectacularly in Florida and California; the Southwest contributed little directly; and here suppression remained the rule. By the 1970s the deteriorating scene inspired regional discussions, and by the 1980s some experimentation. Most of it occurred in the Mogollons, not the Jemez. If viewed from space, the Jemez would seem ideal for fire restoration: a massive island that if not wholly self-contained was an environmentally bounded unit. The reality was different.

Whatever the sentiment of reformist fire officers and creative ecologists, the freedom to maneuver at the Jemez was less than it appeared from afar. There were complexities hidden among the pixels and polygons. There were ruins on the surface, modern pueblos around the perimeter (the wildland-urban interface of the Jemez), and a deep distortion in regional pyrogeography known as the Los Alamos National Laboratory (LANL). The lab had pioneered, with the atomic bomb, the explosive invention of an alternative energy to combustion. In fact, much of the damage from weaponized nukes came from the fires that followed, and nuclear power never displaced combustion when tamed into reactors (or even came close). But the lab has acted in the space-time of the Jemez fire field like a white dwarf star, deforming everything around it. Among federal facilities probably only Merritt Island, which houses the Kennedy Space Center within a fire-prone wildlife refuge, approximates the peculiar dynamics of the Jemez.

With almost malevolent cunning the lab and the town that serviced it were sited for maximum risk. They rose amid a forest unburned since the 1880s, and then kept unburned, save for a patch here and there every 20 years or so. It sat on the eastern flank of the massif, where it could receive fires driven by the prevailing spring winds from the southwest. And its secretive nature and lethal pollutants made treatments unlikely and prescribed burning unwanted. It resided amid a natural dump of combustibles that could burn with the energy output of Little Boy and Fat Man combined, though released over hours rather than nanoseconds. It made political sense to house the project in a remote, secure location. It made no ecological sense. The lab sat amid nature's more benign version of U-235. The mechanical muscle of wildland fire protection, still bulked up with postwar military hardware, shielded the lab and its anomalous town from the consequences of the hasty decision to staff the Manhattan Project on the slopes of a combustible biovolcano. The lab was created by a wartime emergency. It was protected by a cold-war fervor for fire control.

• • •

Then on June 16, 1977, lightning kindled a fire on the eponymously named Burnt Mesa in the uplands of Bandelier National Monument. It escaped control and, with prevailing southwesterly winds behind it, fresh with spring vigor, it barreled over mesas laden with Anasazi ruins and a century of weapons-grade biofuels and bolted toward the Los Alamos National Laboratory.

The La Mesa fire announced a new era in the pyric history of the Jemez. The Park Service repurposed archaeologists as line scouts to keep bulldozers from flattening the ruins that dappled the mesas, insisting that there was no point in destroying the artifacts that were the reason for the monument's existence—more subtly, perhaps, challenging the unquestioned supremacy of national security needs over all other goals. A symposium followed, which put down markers for the monitoring of future change. And the fire, though halted before it reached LANL, effectively posted a notice next to the others along the lab's border. The Department of Energy signs warned against intruders. The La Mesa signs, in the form of blackened snags, dared the fence to keep fire out.[3]

The fires paused as the region went into a wet cycle. It was a grace period, but while Bandelier prepared plans and began to light prescribed fires to counter the vast imbalances accrued over the past century, the Santa Fe National Forest did not, and LANL appeared to dismiss the La Mesa outbreak as an outlier, as though it were the outcome of an experimental error. The wet cycle dried up. In 1996 a dry cycle returned with pent-up ferocity. The Dome fire, boiling out of an abandoned campfire on April 26, again roared toward Los Alamos. A spring fire, it burned hotter and faster. This time a thin red line of backfires held it on Obsidian Ridge within shouting range of Los Alamos, only because the winds had providentially spun to blow from the north. Had the prevailing southwesterlies held, the fire would have ripped through lab and town.

The numbers were starting to add up, as though the Jemez were venting a lost century of fires in explosive spasms. The La Mesa fire had burned 15,444 acres; the Dome, 16,516; but far larger landscapes, laden with aching fuels, lay unburned. More than that it was obvious that the Pajarito Plateau was in the firing line: it lay in a seasonal wind tunnel primed with explosive matter that only needed a spark, an ignition device, to set it off. Officials were looking at regime change. Two years later fire officers and lab officials met to discuss some perimeter barriers in the form of fuelbreaks. The treatments were still underway when a prescribed fire near Cerro Grande in the uplands of Bandelier National Monument escaped control on the evening of May 4, 2000.

The suppression strategy elected to back off to Highway 4 and burn out. That burnout was lost, and the fire bolted out of the monument altogether and soared over the Pajarito Plateau toward Los Alamos. A fumbled suppression strategy again failed to halt the flames or to actively protect the town. One head fire crossed untreated borders to blow into the lab; another crept through the western half of the town incinerating houses by the hundreds. The fire reached 47,000 acres—the largest forest fire on record for New Mexico. Final costs tallied nearly a billion dollars thanks to the Cerro Grande Fire Assistance Act; most of it went to rebuild the town. Blame scorched the air, but there was no escaping the fact that the prime mover was the botched prescribed burn.

The worst, however, was yet to come. On June 26, 2011, a dead aspen fell across a power line and kindled a fire amid drought, wind, and more than a century of accumulated combustibles. The plume twisted, held by horizontal roll vortices, as it rumbled across the caldera like a slow-churning tornado on its side. When the vortices faltered, a mushroom-cloud plume rose, and when the plume collapsed, as though it were a pyric thunderhead, it sent the ash, gases, and flames downward in a rush, a biotic flashover. The effect resembled a pyroclastic outflow that stripped everything. Where that rush channeled into gorges like Cochití, it poured through like a flood of pyric debris. Where it splashed over mesas, even those stocked with dispersed junipers little prone to burn, it immolated the landscape. Where it passed over lands fried by the Dome fire, it took everything down to rock. The Dome fire had stripped out the forests. The Las Conchas fire wiped away the locust and oak and exotic grasses that had grown in its wake. By the time the rush ended, Las Conchas had tripled the size of the Cerro Grande fire.

The latent horror was that such a fire would blast through the lab and liberate stored or buried materials including plutonium that would contaminate the land for decades and, worse, poison the municipal watersheds of downstream communities. But even conventional incendiaries were sufficient to render the Rio Grande unfit as a source of potable water for 20 days at Santa Fe and 40 at Albuquerque. Nature's nemesis had answered human hubris with a force majeure against which resistance was futile. It seemed as though the Jemez had again found new avatars of eruptive energy.

That ignored the deeper driver. Las Conchas had started when the old source of natural power, free-burning fire, had crossed the new one, industrial combustion. An older order had fallen on a newer one. The 115-kilovolt transmission line that sent the spark over the fallen aspen ran power for the Jemez Mountains Electric Cooperative, which received its energy from coal-fired dynamos throughout the greater region.

• • •

With its dramatic prominence amid the northern New Mexico landscape, with its remarkably long and complete record of fire history, with its array of ur-landscapes abutting ultramodern ones, with its metastable bonding of human institutions like Bandelier and LANL, with its sudden eruption of landscape-scouring fires, as though the slumbering volcano had mated with plutonium and spawned Loki-like freaks to sport with the forest, the Jemez Mountains easily qualify as one of the geodetic markers of southwestern fire.

It's less obviously a cipher for the future. Each observing group sees its own agenda in the flames and its prophesied future in the postfire ruins. Are the eruptive fires agents of irreversible change? Or are they simply synthesizing the changes around them? Are the Jemez forests a model for the Southwest's sky islands, the beta version for the regional future of fire? Or are the mountains once again an exception, a place of spasmodic violence, its megafires the modern outflows of a megavolcano?

The temptation is strong, in particular, to make fire a subplot in the saga of global climate change. But it makes at least as much sense to see climatic warming as a subplot in the epic of fire history. When the Earth's keystone species for fire changed its combustion habits and reached into the Earth's past for fuel to stoke its craving for more firepower, it committed the planet to a new future of burning. Ancient and modern fires are still seeking a working equilibrium.

In a pragmatic sense it matters little what the ultimate cause might be. For the next few decades, perhaps for the next few centuries, the reality promises more of the same as the Jemez moves toward an end point over which we will have marginal control. We can't muster a counterforce to halt fires of this savagery—we can't bomb them away. But lots of smaller actions, added together, might dampen and deflect the damage. The energy released by the Las Conchas blowup might rival several nuclear bombs, but only over the course of days. So, likewise, a counterforce might soften the fires' eruptive power by altering the landscape over many years until it reaches a calmer angle of repose. That strategy might work, or it might not. But the no-action alternative is clear. It's to stand on the caldera, as at Cochití Canyon, and peer into an equilibrium of emptiness.

Thinking Like a Burned Mountain

O N SEPTEMBER 18, 1909, a young Aldo Leopold, then a ranger with the
U.S. Forest Service, shot two timber wolves in Arizona's White Mountains. He noted the episode casually in a letter home. But the incident, like
embers in an old campfire, glowed in his mind, and in April 1944 he wrote one
of his most celebrated meditations, "Thinking Like a Mountain," in which he
described standing over the dying she-wolf and watching the "fierce green fire"
in her eyes die and wondered if shooting the wolf had helped unhinge the larger
landscape. Too much emphasis on safety, he thought, was dangerous. He quoted
Thoreau's dictum, "In wildness is the salvation of the world."

The essay, or more accurately moral epistle, became one of the founding
documents of 20th-century American environmentalism. It helped make the
wolf the living emblem of the wild, and wolf restoration a measure of ecological
enlightenment. About 10 miles southeast of Leopold's kill site, Mexican gray
wolves were reintroduced in 1998. But his insights also helped underwrite a
campaign of nature protection that focused on the preservation of pristine lands.
Leopold was the architect of America's first "primitive area," the Gila, located
in an adjacent national forest, which subsequently became the inspiration for
a National Wilderness Preservation System 40 years later. In 1984 the system
acquired the 11,000-acre Bear Wallow Wilderness, about 10 miles as the crow
flies southwest from where Leopold shot his wolf. Between them the three
sites form a triangle of environmental thinking transformed into action—the

deed into an idea, the emblem into a restored species, the wild into a legally gazetted preserve.

A century later a mammoth wildfire boiled out of the Bear Wallow Wilderness, blew over the wolf reintroduction site, and overran Leopold's vantage point above the Black River. The Wallow fire, kindled by an untended campfire, burned 50 times as much land as the wilderness held. An idealistic green fire met an all-too-real red one.

• • •

The contrast almost overflows with symbolism, but two themes seem most useful. One speaks to nature protection, and that preserving the wild is perhaps not just a paradox but an example of a misguided urge toward safety, in this case the security of nature, not unlike Leopold's shooting a wolf. "In those days we never heard of passing up a chance to kill a wolf." Fewer wolves meant more deer, and no wolves meant "a hunter's paradise." So, too, it has seemed self-evident that removing the human presence would mean a healthier land, and no people would mean paradise.

The other theme is fire. At the time Leopold killed the green fire, he was also swatting out red ones. Fire control was the among the most fundamental of ranger tasks; to ignore fire could be cause for dismissal. Interestingly, posters from the era even equated fire with wolves: the fire wolf running wild through reserves was a ravenous killer that needed to be hunted down and shot. Over time this belief, too, yielded to the realization that fire's removal, like the wolf's, could unravel ecosystems. The difference was that fire was renewed annually, if not through human artifice then through lightning (the American Southwest is North America's epicenter for lightning fire). The spark is always there: if wind and fuel are aligned, fire can spread.

But the deeper story was that the sparks decreased and the fuel was stripped away. Lightning fires were attacked and extinguished at their origin. People quit setting tame fires to substitute for nature's wild ones. And overgrazing slow-metabolized on a vast scale what fire had formerly fast-burned. Cattle and sheep cleaned out the country's combustibles. Flame might kindle in the isolated snag; it could not easily spread. Over decades, however, the removal of predatory fire allowed a woody understory to flourish, akin to the metastasizing deer population that blew up after the wolves were extinguished. Both yielded a sick, impoverished landscape.

So a campaign to restore fire ran parallel to that for reinstating wolves. Their histories are oddly symmetrical. The population of neither wolf nor fire has reached its former levels, and the landscape teeters on a metastable ridgeline. The issue is that success requires not merely the presence of wolf and flame but a suitable habitat in which they can thrive. The power of fire resides in the power to propagate, and that sustaining setting was gone. Fire, however, had other properties wolves lacked, notably a capacity not simply to recycle but to transform. A single spark could transmute thousands of acres almost instantaneously.

On Memorial Day weekend, May 2011, flames returned. This time they came as feral fire. It was certainly not a tame fire—not a controlled burn or a prescribed one suitable for wildlands. Neither was it a truly natural fire; it started from a slovenly kept campfire and burned through decades of forests whose structure had been destabilized by logging, of grazing that had destroyed their capacity to carry surface fire, and of doctrines of fire exclusion that had prevented nature's economy from brokering fuel and flame. The Wallow fire could no more behave as it would have in presettlement times than could a wolf pack dropped into a former hunting site now remade into a Phoenix shopping mall.

Probably fires had burned as widely in the past, but through long seasons in which they crept and swept as the mutable comings and goings of local weather allowed. Undoubtedly, in the past spring winds, underwritten by single-digit humidity, had blown flame through the canopies of mixed-conifer spruce and fir and left landscapes of white ash and sticks. But it is unlikely that earlier times had witnessed a similar combination of size and intensity. The Wallow burn was not what forest officers had in mind when they sought to reintroduce the ecological alchemy of free-burning flame.

• • •

The Wallow fire has not destroyed Leopold's parable, any more than its flames have destroyed the forest in which it took place. But in burning over story and landscape it has transformed them. A burned mountain might think differently than an unburned one. What emerges out of a meditation at the kill site today is a modern parable of the Anthropocene. It describes what has happened when the Earth's keystone species for fire changed how it did business.

The Wallow fire hinges on two paradoxes. One is that industrial societies— those most ravenous of natural resources—are also the ones prone to create nature preserves on a significant scale. The curve of formally protected nature reserves traces closely the curve of fossil fuel combustion. The other is that a relatively unbridled capitalist society, in the full flush of the Gilded Age, set aside roughly a third of its national estate to shield against the ravages of its own economy. A significant fraction of that (80 million hectares) America has committed to national parks and wilderness.

When humanity opted to acquire our firepower by burning fossil biomass, we quit burning surface forms, and we used our enhanced fire engines to suppress what fires nature or accident kindled. Apart from technological substitution, the shift was a deliberate policy intended to protect natural places. The founding legislation usually identified sites as sanctuaries "from fire and ax." The immediate outcome was success: forests remained standing and conflagrations halted.

The longer consequences took several decades to become apparent. The shift from external to internal combustion shocked the system as surely as overgrazing or clear-cutting—it just wasn't as visible. The land adjusted in ways that left it out of sync with the kinds of fires the region would experience. When a hiker in the Bear Wallow Wilderness left a campfire on May 30, 2011, the fire did what similar fires had not been able to do for a long time. It blew up and bolted across the length of the White Mountains, overrunning 538,000 acres as it went. The legacy of the past powered that conflagration as surely as the June winds. What happened on the Apache National Forest has been repeated across the American West for nearly 25 years. Attempting to banish open flame has made fire unmanageable, which may be synecdoche for the application of our new, industrial firepower generally.

We now know that attempting to abolish fire from natural sites for which it is indigenous is a mistake as profound as mindlessly extirpating wolves. Yet letting wildfire ramble amid the global metastases of the Anthropocene is an act of faith-based ecology. If all we want is the wild, we will get it. If we expect a usable mix of ecological goods and services, we will have to add our hand to nature's. We created an ecological insurgency, and only controlling the countryside can quell it. Yet to intervene may violate the norms of the wild.

The classic preservationist solution is to leave the landscape to sort itself out, even if this takes decades and the outcome is unlike anything experienced before. Nature has deep powers of recuperation; it has been coping with fire since the first plants appeared on land 420 million years ago. But when wildfires off the evolutionary charts are burning areas 100 times the size of the smallest legal wilderness, there may not be much resilience left. Our nature reserves, even the largest, will burn with properties probably not seen before and on a scale not previously experienced. Paradoxically, their recovery may depend on the character of their surrounding landscapes; these we can tweak into working landscapes to advance biotic goals. Such a conception inverts our traditional notions because it means that the preserved core may not be the refugia for its surroundings but that those surroundings may become the source of recovery for the nominal core.

Aldo Leopold's parable was written at a time when the state had to intervene to halt the ravages of global capital and folk migrations. Its primary means were to set lands aside as reserves and to shield them from the practices that wrought wreckage beyond those borders. That strategy kept out landclearing, settlers, and abusive burning. But the internal management of those lands has made them prone to flare-ups and crashes and the losses of what they were established to conserve. In the 21st century the state will likely need to intervene within those reserved lands to prevent the indirect ravages of the Anthropocene.

From the rimrock overlooking the east fork of the Black River we can see how the Wallow fire integrated everything done and undone over a century with the silent stresses encoded in the name Anthropocene. There is no way to keep out climate change, unhealthy biotas, invasive species, beetle and bud-worm swarms the size of states, and the relocation of carbon from the Paleozoic to the Colorado Plateau, there to burn again. To do nothing is to risk losing everything save the notion of the wild itself. To do something will guarantee errors. It was easy to identify the causes of land-scalping. It's tricky to track the tremors of the Anthropocene and to know too little from too much safety. The choice is not as clear as whether to pull the trigger or not. And what we might see in the red eyes of a dying fire regime is an exercise in pyromancy. What is clear is that we will likely have lots of occasions to look.

Top-Down Ecology

A LONG NARROW road winds steeply up into thickly wooded backcountry to an exclusive enclave of costly structures, all well beyond the periphery of settlement. It's the formula for the worst-case scenario of the wildland-urban interface, except that this is no subprime landscape stuffed with trophy homes. It's a telescope complex atop Mount Graham, and on the sky islands of Arizona the scene is repeated four times. Call it the wildland-science interface.[1]

Fire management accepts as axiomatic that it is science-based or at least science-informed and that good science is the antidote to the toxins of politics, land development, and a Smokey-blinkered populace that doesn't understand the natural ecology and inevitability of fire. Science is better than experience or history, and more science is better still. Science, preferably natural science, since even social science is tainted with the implied values of its human doers, is the solution. At Mount Graham, however, science is the problem. And the challenge is not simply that "science" here underwrites its own version of the WUI and opens paved roads to remote sites that complicate fire management and compromise biodiversity. The real challenge is the assumption that science stands apart from the scene it describes and from its Olympian heights can peer objectively outward and advise wisely.

The Mount Graham International Observatory (MGIO) suggests instead that science's lofty perch is not removed from land management and that science, too, has its self-interests that can influence what it sees, does, and says.

Science, in brief, is not an ungrounded platform for viewing the universe of fire and recording its observations. It is sited, and that siting determines what it sees, and decisions over such sites make science and its caste of practitioners as motivated by their own values and ambitions as loggers, ranchers, real estate developers, and ATV recreationists. Science has its own dynamic apart from nature, its own presence on the land, and its own politics. The 1.83-meter primary mirror of the Vatican Advanced Technology Telescope, while nominally looking out, is also a reflecting lens that looks back on its viewers.

•   •   •

As their name hints, the sky islands are ideal for astronomical observatories. They sit atop high mountains amid a dry climate surrounded by dark deserts (the exception is Tucson, but the city has adopted light abatement measures). The costs of constructing and operating such facilities favor clustering, and the region is dense with telescopes. Mount Lemmon in the Santa Catalinas holds one for the Steward Observatory, Mount Hopkins in the Santa Ritas hosts the Whipple Observatory's Multiple Mirror Telescope for the Smithsonian Institution Astrophysical Observatory, Kitt Peak in the Baboquivaris has a compound of 22 instruments including the famous National Solar Observatory, and in the Pinaleños three telescopes sit in a concrete aerie atop Mount Graham. In recent years wildfires have threatened them all.

The fires have come almost annually. In 2002 and 2003 the Bulloch and Aspen fires together burned 85 percent of the Santa Catalinas. In 2004 the Nuttall fire complex burned 29,000 acres of Mount Graham. In 2005 the Florida Peak fire in the Santa Ritas threatened both the Multiple Mirror Telescope and cabin inholdings in Madera Canyon, and forced evacuations. In 2007 the Alambre fire moved up the slopes of Kitt Peak before being contained at 7,000 acres and over $2 million in suppression costs. But such scenes are hardly news: the same dynamic is playing out across the country, and for that matter, throughout the industrial world, as a revanchist vegetation meets an out-migration of urbanites. Matter and antimatter—the astrophysicists needn't peer into nebulae at the fringes of an expanding universe to detect such explosions, they need only look around them.

But the deeper collision is occurring within the domain of cultural values, not subatomic particles. The observatories break up public wildlands into incommensurable blocks: they are in this respect no different from a private inholding or a clear-cut. Four sky islands have roads that extend to their summits; all lead to observatories. So long as Science seemed a greater good, as incontestable in its claims to public land as to public money, there were few objections. Certainly there were no doubts from the scientific or university communities. They were the good guys, far removed from grubby commodity producers and selfish summer homeowners. Their motives were unimpeachable. They were studying the heavens. Theirs were the highest values of civilization.

Then the science-industrial complex met the Wilderness Act, the American Indian Religious Freedom Act, and the Endangered Species Act. They spiraled together with particular force at Mount Graham when in 1984 the University of Arizona (UofA), heading a consortium, petitioned to create a cluster of seven telescopes, one of them an enormous six-dish, rail-mounted interferometer, at the summit. That catalyzed an opposition. An Apache Survival Coalition declared the peak a sacred mountain. Advocates for roadless areas wanted access limited rather than enlarged. And enthusiasts for wilderness and biodiversity noted that the mountain was a Pleistocene relic of Englemann spruce and cork-bark fir with 17 protected species, including a unique subspecies of red squirrel that inhabited the summit, and could go nowhere else; expanding the facility over two peaks would diminish its required habitat and perhaps introduce other disturbances. Under terms of the Endangered Species Act environmental groups protested and eventually brought suit.

The controversy—"scopes vs. squirrels"—became more bitter as the years passed, not lessened by the inability of the science community to admit that they were in fact upsetting a biotic order. Compared to the 156-billion-light-year width of a Hubble universe, Earth is less than a flyspeck and the addition of a few acres of telescopes on the Pinaleños tiny beyond infinitesimal. But compared to the habitat of the red squirrel and public claims on a patch of land that could not enlarge or go elsewhere, it was a significant, probably irreversible disturbance. The University of Arizona and the astronomical community refused to accept that fact or to place themselves within the scene. The Mount Graham International Observatory was only a platform for viewing. MGIO was not itself within the panorama viewed.

The controversy dragged on for years, a bitter war of bureaucratic attrition. On-site protests, fudged reports on the biological status of the squirrel in particular, intervention by the state's politicians, backroom deals, legal suits, court injunctions, appeals—from the time the UofA proposed a complex akin to that on Kitt Peak until three telescopes actually arose on Emerald Peak and an adjacent knob, a decade of rancor passed. Throughout, the tendency was to interpret the feud along familiar tropes such as jobs vs. environment, or the perversion of biological science by politics, greed, and hubris. For environmentalists, Mount Graham joined Glen Canyon as a martyred landscape. Yet the contest might as equally be viewed as one between sciences, and between science and wildland management, and between institutions of science. One science, astronomy, and a nominally science-supporting institution, the UofA, turned to politics to overturn the claims of another science and its nongovernmental auxiliary. Astronomy meant Big Science, while conservation biology had only acquired a name in 1978. Deep sky met deep biology, and sky won.[2]

• • •

In retrospect, it should have been obvious that Mount Graham would become a point of convergence for controversy, as much a focal point for gathering environmental themes as the lenses of the observatory were for light and microwave radiation in the night sky. That applies not only to its ecological status but to its human history.

The mountain is like a living natural-history cabinet of southwestern species. The Chihuahuan Desert laps against its eastern flank; the Sonoran Desert, its western. The floodplain of the Gila River runs along its north, and high desert along its south. So steep is the mountain that life zones become wafer thin, stacked like poker chips and tucked away into niches amid the deeply crenulated flanks. From Fort Grant on its southern shoulder to the peak of Mount Graham 6 horizontal miles rise 6,000 feet, or better than 1 foot of rise for each 6 feet of run. Forest types appear like thin sections from rocks; only along the rough-rolling summit is there breadth enough to create the semblance of a forest, and those 3,000-plus acres of Engelmann spruce and corkbark fir are a relic biota, a Pleistocene refugee as distinct as the California condor, and just as endangered.

The mountain's human history has been as mixed. Major mines bore into the mountains to the north. Ranching overran the lowlands as soon as mines developed and the indigenous tribes were pacified. Mormon settlers colonized the Gila River floodplain for irrigation agriculture, establishing regional entrepôts like Safford and Thatcher. Military posts dot the region; Fort Grant was part of General George Crook's famous campaign to contain the Apaches, blocking potential escape routes from the San Carlos Reservation to Mexico. The

Mormons and the military, from the north and south respectively, converged on the summit. Settlers developed flume-transport logging that ate away at the northern canyons. The army established a hill station, complete with a hospital and heliograph lookout, along the summit, with a wagon road to supply it. The settlers deposited a small cluster of cabins at Columbine, while Fort Grant evolved into a state prison, which continues to supply labor for forest-related projects. In 1902 Mount Graham became a forest reserve; and after the U.S. Forest Service acquired control over the reserves in 1905, a national forest in 1907. The federal government went from suppressing Apaches to suppressing fire; Heliograph Peak became a fire lookout; fires all but disappeared. Although formally vanquished, the Apaches continued to identify sacred springs like the Bear Wallow cienega on the top, and ritually revisit them. Then the University of Arizona and Big-Science astronomy staked a claim.

By that time Mount Graham, despite its extraordinary ruggedness, was becoming an ecological shambles. Logging, grazing, fire control, the introduction of the Abert's squirrel by the Arizona Game and Fish Department (as a putative meat source), all had broken the biological integrity of the mountain refugia and compromised its resilience. In addition to its summit woods, as isolated as though they were on Selkirk Island, the mountain had the northern goshawk, the Mexican spotted owl, and the red squirrel, all threatened or endangered. But underwriting everything was a thickening bloat of combustibles, both choking the old biome and stoking the potential for large fires. The scene worsened when spruce beetles (native) and spruce aphids (exotic) began stripping spruce as part of the living dynamic. The tiles of the old mosaic began falling away.

• • •

Some of this was obvious when the UofA declared its intentions. The potential biological (and cultural) competition was clear from the onset. What all sides failed to consider, however, was fire. It claimed no more than a nominal paragraph in draft assessments, as bureaucratically trivial as a burrowing mammal.

One reason is that attention was riveted on the summit and its residual woods, widely regarded as an asbestos forest. It had never burned in the memory of American settlement. Later studies demonstrated that the peaks had not burned for 300 years. (In a gesture of historical irony, the last major fire on the heights of Mount Graham had occurred in 1685, two years before Isaac Newton published the *Principia Mathematica*, the foundational opus for modern astronomy.) Lightning kindled no more than 6–12 fires a year, most of which quickly self-extinguished. People had long ceased to set burns deliberately. Fire seemed a relic from the past, like the heliographs now replaced by microwave repeaters.

The critical concern with fire, moreover, did not reside at the summit but along the lower elevations. Few ignitions would start at the top: there, lightning would typically be accompanied by copious rain, and spruce and fir lack the long needles that make an ideal bed for surface burning. Instead, ignitions would blast up the slopes from lower elevations; and part of what had spared the summits from repeated fires was that those montane woodlands had absorbed routine burns without blowing up and hurling flames to the crestline. That middle landscape was the one most in upheaval: it was fast morphing from a predominantly open Douglas fir woods to a mixed forest choking with assorted conifers, all as congested as a squirrel midden and as combustible as a crate of excelsior. Much of the change had occurred over the past 60 years. Now, if drought drained those fuels, fires could burn unheedingly across the old borders. The interface between woods had blurred and the interface between woods and scopes had sharpened.

Certainly, this was the reasoning of fire behavior science, which now also found itself on the summit, overlooking a fireshed much as the Large Binocular Telescope does the Milky Way. Here it appears to challenge the other sciences for space, and in fact might even seem to synthesize them, as it studies the bioburning of the landscape by the methods of physics. The logic of fire behavior would appear to favor its claims. Fires burn most fiercely where they have more to burn, which at Mount Graham is also where ignition is most common. Fires burn most savagely upslope, and both MGIO and the squirrel's habitat are at the top of the mountain. Add in a warming climate, in which the desert seas will rise, and fire may remake the landscape as thoroughly as glacial ice in the Pleistocene. In time, without aggressive action—not just suppression but preventative intervention—fire might well claim it all, like a sun going into supernova and taking its planets with it.

Then life imitated science. In May 1996 the Clark Peak fire, kindled from an abandoned campfire, burned 6,500 acres, spreading into known squirrel middens, giving the MGIO a thorough scare, and inspiring a fire inspection by a scientist who specialized in the WUI. In June 2004 lightning sparked two fires, the Gibson and the Nuttall, that together drove through nearly 29,000 acres on the mountain's northeast slopes and nearly converged exactly at the MGIO.

All three fires emerged from the mountain's middle zone. In the end, both squirrels and scopes survived. The squirrels took the bigger hit since the beetles and aphids had struck hard even before the burns. Afterward, the MGIO hardened its structures to prevent embers from entering the interior and burning the scopes from the inside out, and it laid out perimeter protection in the form of a network of sprinklers. Besides, bugs and burns had now insulated it from the threat in ways not possible for the squirrels. The observatory did not live off the lost trees.

• • •

It would seem that fire science also exerts a claim to the summit. Yet since science consists of methods and ideas, not a physical entity, how might it demand space akin to that of the Heinrich Heitz Submillimeter Telescope or the Mount Graham red squirrel? In fact, it can because it affects fire's management on the ground. Here is another wildland-science interface, where science as a mode of inquiry confronts fire management's need to act, and of the two wildland borders, this is much the trickier.

There are those who say that the barrier should not—does not—exist, that fire management is simply the best application of the best knowledge science produces, and that more science will spark better practices. But there are also those—fire's curmudgeons—who point out that humanity handled free-burning fire, and probably handled it far longer and more successfully, before modern science than after it, and that more money poured into research has not produced savings in fire management's costs or a reduction in area burned by wildfire. Rather than improve performance, science as an intellectual enterprise sits within fire management as the MGIO does on the peaks of the Pinaleños. It not only interrupts the landscape of fire but demands a peak and shoves aside competing claims. What then should be the relationship between knowing and doing, and what is the proper place of science?

From its origins in forestry, fire protection, and later fire management, have insisted that they are science based. It has been their self-declared mission to counter centuries of misguided folklore and superstition, and they, and they alone, should instruct administrative practice. Yet the record of their accomplishments is at best mixed. Science proved excellent at reducing complex systems into simple matrices and then creating machines (or rules) to apply particular actions. Fire protection became an intellectual complement to silviculture:

it could successfully decrease burned area much as tree farmers could pump up cellulose production. But public wildlands are not pine plantations.

The outcome was generally disastrous. Forestry used its status as an academic science to counter folk wisdom, to condemn western underburning as Paiute forestry and cracker-cowboy range burning as wanton vandalism, and to dismiss counterclaims by prairie restorationists and wildlife biologists as the wistful fancies of niche hobbyists. It used science to condemn fire's presence on the land and denounce its practitioners. When experiments in the southern pines showed favorable results from burning, the forestry profession and its state-sponsored expression, the U.S. Forest Service, suppressed them with the same fervor with which they attacked flames. Eventually, the outsiders were able to replace the "bad science" with their "good science," but the environmental damage had been done.

The response? An ardent appeal for more and better science. If the agencies had had more of the right science, so the apologists claim, they would have applied those findings and could have resisted the perverting pressures of politics. Rather, the story seems to be that the agency did have the best science of its day and applied it with consequences later scientists rejected. Still, one could legitimately discriminate between science as a mode of inquiry and science as a body of positive knowledge, and suggest that its ever-inquiring character, its restless skepticism, is its most basic attribute. It is possible to present the breakdown as not the product of science but of its political misapplication. And one could observe, philosophically, that given a couple of decades the scientific community had shown itself capable of righting its errors. The misreading of fire was simply a longer example of fads like cold fusion and polywater, which the community ultimately self-corrected. In the long run science was right. Yet a land agency is not a research institution: it must act, it needs workable knowledge to perform its tasks, and the consequences of error cannot be overturned by the latest journal article. Besides, in the long run, as John Maynard Keynes famously observed, we are all dead.

The reality would seem to argue that local knowledge based on centuries, if not millennia, of practical experience coded into cultural mores was far superior to field- and lab-generated (and later, computer-simulated) data. It just didn't have the same cachet, and it could threaten to undercut the claims to privileged knowledge that led to money and power. Interestingly, the Clark's Peak, Gibson, and Nuttall fires did not obey the fundamental logic of fire behavior by which the most vigorous burning would occur upslope. The steepness of the massif apparently encourages very strong temperature gradients as, at the end of a day, the top cools rapidly while the shoulders remain heated and a violent sundowner wind results. The worst blowups were actually blowdowns. Abstraction met local circumstance, and the facts won.

The general response has been to do more of the same. The solution to the problems of science is more science of the same sort. That the agencies responsible for administering the land are also the ones sponsoring research means that there is no way to segregate the two. Science cannot exist apart from politics because politics pays for the science. It cannot merely observe and analyze from a neutral vantage point because those operations and the vantage point itself deform the scene being examined.

The reality, too, is that major reformations in fire management have come not from new scientific discoveries but from changes in cultural values. Upheavals in social understanding determined the paradigm shifts in fire science, not vice versa. Critical thinkers came to value fire because they saw it as part of wilderness, not because they chronicled its evidence in scarred trees and soil charcoal. Those cultural revolutions further allowed society to sift through the competing claims of the various sciences. The ideas and beliefs that surfaced chose which kind of research to support and which to put on the shelf. Still, perhaps damningly, the most influential publication of recent decades, Norman Maclean's *Young Men and Fire*, was not written by a fire scientist but by a professor of Renaissance literature at the University of Chicago. After the disastrous 1994 season, the book directly affected the adoption of a common federal fire policy and helped convince fire officers to fight fire differently, one consequence of which has been a willingness to trade burned acres for enhanced firefighter safety.

After nearly a century of evidence, it should be clear that fire science is not adequate to the task of full-spectrum fire management, and that it will never be adequate. Science, as science, simply can't answer the questions most needed to live on the land, which lie in the realm of cultural values, a moral universe impermeable to the lens of modern science. It can improve technology and advise about possible outcomes of decisions, it can overgrow with data, but it cannot decide, and its record is such that acting solely on its existing data will almost certainly lead to errors if not disasters. It should be one scholarship among many, and one epistemology among a throng that includes the impossible-to-codify-and-reduce-to-numbers experiential reality by which people actually live. Yet there it resides like the MGIO, demanding ever more space to do what it deems essential, insisting that it sits above criticism, willfully agnostic about the scientific-industrial complex that supports it.

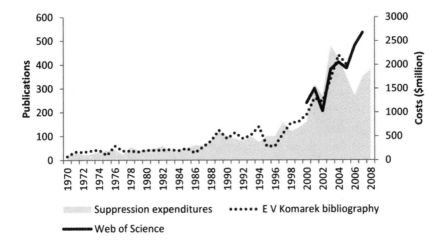

Suppression expenditures ••••• E V Komarek bibliography
━━━ Web of Science

The critics of fire suppression often point to graphs of increasing expenditures and swelling acres burned to make a case that more money fighting fire doesn't reduce either costs or burned area. Defenders will reply that worsening conditions—climate change, the WUI—are determining the fundamentals, and that these deep-driving circumstances are causing the megaburns that bring larger suppression costs. Yet, in the perverse way of correlations, critics could impishly hint that the rising expenditures are just as likely to be the cause of increased burning. The more we spend, the less control we get. A fire suppression–industrial complex is pushing up costs without regard to results on the ground.

This same logic can be applied to fire science. An uptick in fire research parallels the same upswings in firefighting costs and burned area. The USGS has joined the USFS as a funding agency, and the Joint Fire Science Program, established in 1998, has pumped significant monies into research. The number of scientific articles published shows an exponential rise: in the early 1960s some 13 papers per year were published, and in the early 2000s over 300. Partisans will argue that the growing crises, worsening circumstances, and emerging megafires are the reason for more research funding, and that the proper solution is still more funding for still more studies. Yet this is exactly the logic that long governed suppression. One could just as easily argue that the enhanced investment in science has not made any difference on the ground, or even that an emphasis on fire sciences has diverted attention from the real "drivers" of fire's management. An objective measure of applied fire science—analyzing science as science would natural phenomena—would probably show mixed results much like that from fire suppression. The more we spend, the fewer practical outcomes we get. A fire research–industrial complex is pushing up costs without regard to results on the ground.[3]

Apologists brush off such observations as they might an annoying deer fly. They know that authority goes to power, power goes to money, and the money goes to science. They might further demur that science only observes and analyzes with complete disinterest. In fact, science can deflect from other forms of inquiry and by counseling practitioners it actively alters the landscape it studies. Over the years it has measured a landscape shaped by decisions informed by past science. It has affected the Pinaleños as fully as the Mount Graham International Observatory. This places science, and its institutions, squarely on the summit; and like the squirrels it has nowhere else to go. If it affects the outcome, then it is part of the problem. If it has no effect, then why is it granted special status and funding?

A better explanation for increases in cost and burned area is that America has reclassified the purposes of its public lands, accepts that more burning is advantageous ecologically, and refuses to commit firefighters to go mano a mano with fire in remote, rugged landscapes. Fire officers back off, as federal policy has encouraged them to do. For 30–40 years the major federal land agencies have adopted goals to increase the amount of land burned under their care. The statistics suggest they are finally doing just that. Be careful what you wish for—and how you study it.

The stronger argument for supporting fire science is that fire management needs to engage its larger culture on terms other than merely as vernacular learning and folklore; the fire guild needs, somewhere, a sense of itself as more than backwoods mechanics and wildland sharecroppers. It needs to connect to high culture in order to truly engage its sustaining society. It needs sophisticated fire science for the same reason that a modern culture needs astronomical observatories. The difference is that observatories tell us little about how to manage Earth, while wildland fire science intends, as its announced ambition, to influence conduct, which is to say, to shape fire practices on the ground. But neither is without cost. Those observatories compete with other values—can literally shove them aside; they claim, and defiantly occupy, the high ground. So, too, fire science can push to the margins other scholarships and forms of knowledge.

Those scopes on the summits will not melt away: their internal imperative is to expand. To even appear to criticize Science is to invite charges of philistinism, politicization, and capitulation to faith-based superstition. A steely-eyed survey of fire science's actual achievements, however, would point to marvelous insights into nature and a much-flawed record of practical outcomes. And that, in the end, makes the wildland-science interface a far more troubling conundrum for the fire community than the better publicized wildland-urban interface. It's easier to defend those trashy trophy homes than to dismantle telescopes.

•   •   •

Yet, unexpectedly, the imperial model of science, in which science informs and management applies, is finding itself constrained. Nationally, countermoves are underway in the guise of traditional ecological knowledge and of adaptive management that blur the hard border between science and quotidian experience, in which science becomes an experiment in management and practice becomes a scientific experiment, with both needing to be constantly calibrated, compared, and adjusted. Granted some space the concept may return fire management to its ancient status as grounded in experiential knowledge. In principle it restricts science to standing as one form of information and political input, much as modern fire management identifies suppression as one option among many.

In the Pinaleños the general must always interact with the specific; and here, the alchemy by which principles meet particularities can yield unexpected outcomes. In this case the specifics are the reality of the Mount Graham International Observatory. While the University of Arizona and Big Science got their way—got exclusive rights to the site, built an edifice like a Borg cube that can be seen in reflected sunlight from the White Mountains to the Mexican border—they find themselves denied any larger claim. The road and facilities occupy 8.16 acres. That zone is fenced by stakes and a yellow acrylic rope, beyond which MGIO residents are not permitted to go. The grounds are patrolled by security officers and dogs; they are there to keep in as well as to keep out.

Ask anyone about Mount Graham, and you will be told the place is "political." Of course it's political: it should be. These are public lands and arenas for public values, and in a democracy politics is where competing values must be openly discussed and decided. What tainted the MGIO was that its politics was not open and honest. And what has compromised so much of fire science in the past is that it has confused its data with its values and has dismissed any other scholarship and any other competing values. Science there must be. But it has no claim to the whole of the summit.

Squaring the Triangle

ON APRIL 19, 2014, a small storm cell passed over the Nantanes Plateau, shedding virga and dry lightning. The next day a fire was sighted among grass and shrubs above Skunk Creek. Two days later smoke spindled up through the pine forest around Bloody Basin. When the fire and forestry staff of San Carlos gathered, the Skunk fire had scooted to 60 acres and the Basin fire had barely held beyond a snag and scorched ground.[1]

The group gathered around a table blanketed with two detailed topographic maps, exchanging confidences and opinions like a covey of physicians over a problem patient. Where, Duane asked, do we draw the box? Until recently such a question would never have been asked: response to a reported smoke would have been automatic. Both fires would already be out by now, or nearly so, as engines, hand crews, and maybe aircraft attacked it. That was still a live option. Now, Duane, Dee, Clark, Kelly, Bob, Dan, Bear, and later, Nate, poked and pointed and passed questions.

The San Carlos fire cadre viewed the reported smokes less as problems or threats than as complicated opportunities. Though larger, the Skunk held lesser interest. A high-pressure ridge had rolled through since the fire began, raising temperatures and stirring blustery winds. Were the fire primed to burn, it would have reached 1,000 acres instead of 60. It was, as its name implied, just skunkin' around and offered fewer options to manage. The Basin fire burned, if unsteadily, in country that badly needed burning. Whether this fire, at this time, was the right fire for that job occupied the conversation.

They discussed quietly, each offering thoughts in turn, pointing to the map, pausing often, a conversation of silences as much as of utterances, tracing terrain, searching for consensus, not yet ready to decide a course of action, trying to match the fire with landscape features not documented on maps and with purposes not coded into dispatch software, probing like the flames of the Basin fire as it desperately foraged for fuel. Where might they "draw the box?" They studied old roads, trails, rims and ridges, natural fuelbreaks of any kind. Any roads would have to be improved since lava flakes and basalt cobbles would shred engine tires. They would need to move "heavy iron" in the form of bulldozers and graders to do it, but that too was an opportunity because it would allow them to upgrade their infrastructure even as they reined in the fire. Bear recalled an abandoned trail. Dee lamented that the ideal solution might be the old military road that cut through the Nantac Rim. They traced a contour that, if joined by a handline, could tie two existing fragments together; for that they could use a shot crew, specifically the Geronimo Hotshots, San Carlos's own.

They proposed, in brief, to adapt policy options that allowed fire officers to confine and contain wildfires. They were neither direct-attacking the fires nor leaving them to burn, but loose-herding, boxing-in, and burning-out wildfire, and by doing so they were devising a surrogate for prescribed fire, much as pre-scribed fire had been promoted as a surrogate for natural fire. In this they were part of a trend throughout the West. The policy was latent in the 1995 federal common policy for wildland fire or for that matter with reforms that dated back to 1968 for the National Park Service and 1978 for the Forest Service. What made San Carlos interesting is that they were applying those permitted principles on a serious scale on the ground. Still, some tension was in the air. The best way to prevent a fire from going bad was to hit it when it started, and operations were instinctively geared to act, not to ponder.

They decided they needed more intel. Crews were dispatched to each fire to document, locate, take measurements, order a spot weather forecast, and assess the potential for spread. A grader was noted, parked not too far from the Basin fire. A bulldozer, now in the shop, would be dispatched before the morning was out. Another weather system, more powerful, was forecast to rumble through in another day, perhaps with rain, certainly with strong winds and a drop in temperature. It would determine if the Skunk could do more than scuttle through scrub, and if the Basin could burn through the forecast evening frosts. Today's intel and the coming front would decide. For now it was enough to confine and contain.

If the fires survived, they would move from statistics to operations. They would go from a threat to an opportunity, from sand gagging the gears to a lubricant. Their management became interesting in the same way that radio-tracking a bear is more interesting than killing it. Instead of automatically responding to chance ignitions, fire officers could manage by choice. That Easter Sunday weekend the resurrection of free-burning fire on the San Carlos was underway.

•   •   •

The San Carlos Apache Reservation sprawls, crudely gerrymandered, over 1.8 million terraced acres and many millennia of terraced time.[2]

Geologically, it presents a series of landscape steps crusted and filled with volcanic outflows. Each terrace replicates and lowers the one above, until the coherence of the rims and plateaus fragments into isolated peaks and valleys. The highest terrace is the Colorado Plateau in the northeast, edged by the Mogollon Rim; its lowest point, the sky islands and valleys of the Basin and Range to the southwest. The highest terraces are flat and filled with basalt. The lowest valleys are stony pediments, sinking into the floodplain of the Gila River.

Ecologically, the terracing appears as roughly contoured life zones. The Mogollon Rim boasts a robust ponderosa pine forest, grading into mixed conifer. The middle terrains are mixtures of woods and grasses; the basaltic terraces range from prairie to high-desert grasslands. The lowest landscapes are Sonoran desert, degraded into creosote. About a third is woodland; a quarter, grassland; a fifth, desert; some 13 percent, ponderosa; and the tiny remainder, human communities of some kind.

The heart of San Carlos is the Nantac Rim, a miniature of the Mogollon Rim, slashing through the middle of the reservation from northwest to southeast. Along it, ponderosa pine flourish, while the terraces that flank it are grasslands—Big Prairie to the north and Antelope and Ash flats to the south. Near its center lies Point of Pines, where a stringer of ponderosa reaches into Big Prairie. It rests at a kind of eco-librium midpoint for rock, biota, and human history.

The human history, too, appears layered, terrace by excavated terrace. In part, this simply reflects the source of the evidence from archaeological excavations, digging down through layered culture upon culture. Instead of cascading across space, eras pile one on top of the other. But people have actively made terraces as well. Ancient agriculturalists terraced hillsides to hold soil and water, 20th-century mining terraced whole mountains to strip off ore, and archaeologists reversing the process have pulled back layer after layer by terraced pits.

Yet, a great fact of the reservation is that it runs cross-grained to that texture of time and space. Drive from San Carlos, the tribal headquarters near where the San Carlos River joins the Gila, to Point of Pines, roughly at its center, and you pass through Sonoran lowlands up through the foothill flank of the Gila Mountains to the Antelope and Ash flatlands, rich with high-desert grasslands, and then through Barlow Pass over the pine-clad Nantac Rim and onto the sweeping plains of Big Prairie. Continue from Point of Pines to Malay Gap, and you rise through a lesser plain, clothed with juniper savanna, and up to the Mogollon Rim itself. Unless you follow the Gila River, to move around the reservation is to step up and down terraces of some sort.

So it is with San Carlos history: it moves in jumps and over barriers rather than along worn pathways or the meandering floodplains of mainstream narrative. Peoples come and go, sometimes abruptly, the strata of eras piled one on another. But as they cross the grain of place, they can pick up quirky and unexpected adaptations that can evolve into telling traits. That cross-texturing of land and history has created nooks and niches with unexpected outcomes that can hold larger, even national significance. This has happened more than once in the past. And it seems to be happening to its wildland fire program today.

• • •

The setting is classically southwestern. That means the scenery is broken with peaks, plateaus, valleys, and gorges, amid an arid to semiarid climate that leaves its rocky exoskeleton starkly visible. It means its history is full of sharp human entries and exits. And it means there is plenty of fire.

The broken, terraced terrain; the mixed, middle biomes between valley desert and mountain forest; the annual rhythm of spring drought followed by summer monsoon—all are ideal for sparking fires, frequently by dry thunderstorms. San Carlos averages roughly a hundred such fires annually, with some years showing much less and some far more, and it stands close to the epicenter for lightning fire in North America. If it were a pond and you tossed a stone at Nantac Rim, the resulting ripples would align roughly with the density isolines for naturally ignited forest fires. Like the landscape, its fires burn patchily. In the past, the land burned from the sheer numbers of starts, and when dry years followed several wet ones, from lingering fires that could creep and sweep over many weeks. Only central Florida approaches that intensity of natural burning.

The record of fire is continuous; the chronicle of people, less so; and how they interacted, almost unknown. The terraced history exposed by archaeology begins with an Archaic culture of big-game hunters and foragers. Then, corresponding roughly with the rise and fall of the Roman empire, a Mogollon culture appears and continues until the mid-15th century. The Mogollon exchanged not only goods but styles of pottery and housing, and no doubt other ideas, with the desert Hohokam to the west and the cliff-dwelling Anasazi to the north. The Anasazis began migrating into the region during the 10th century; by the 11th, as Europe commenced its cycle of crusading, they were dominant, and clusters of pit houses became suburbs to Anasazi complexes. Then, suddenly, they all—Anasazi and Mogollon both—vanished during the 15th century. When the westernmost bands of Apache began filtering through the landscape, sieving through mountains and mesas from the Great Plains, perhaps a century before the first Spanish entrada, they passed through a landscape of ghosts. The chroniclers of Francisco Vasquez de Coronado in 1540 called the land a *tierra desplobada*, a place abandoned, a land emptied of people. The small bands of Apache hunter-foragers still dribbling into the land seemed to melt into the scenery.

For a millenium, however, from 400 AD to 1450 AD, the human population had been sufficiently dense to carve agricultural terraces and erect stone pueblos. How those cultures affected the prevailing fire regime is unknown: the earliest fire-scarred trees come later. But deal with fire they had to, because it was all around them. When they left, whatever lines and fields of fire they established fell into ruin as fully as their kivas and apartment houses. Until the Apache established themselves, the land went feral, and the abundant sparks of nature reclaimed and remade the scene. Until Europeans arrived, or more significantly, until the modern era that followed Mexican independence and then the Mexican War that ceded the region to the United States, the characteristic fire regimes were those negotiated between the light-on-the-land Apache and the heavy rhythms of lightning.

Ethnographers and ethnobotanists are adamant that the Apaches did not burn widely: they didn't need to. What they wanted from the habitat—edible wild grasses, game animals, shrubs for baskets, many dependent on routine burning—they got freely from a land drenched with flame. They burned some horticultural plots for special crops, they burned patches (some large) to attract preferred fauna by greened-up pasture and fly-retarding smoke, they burned during raids to alarm enemies and to cover tracks, and they undoubtedly left fire as litter. But human numbers were small, fire numbers large. The Mogollon peoples had sculpted surface lava into stone villages; the Apache assembled shrubs and twigs into wickiups; and so with their fire regimes. There are plenty of peoples of comparable technology—many Aboriginal bands in Australia, the congery of tribes in California—who burned both meticulously and extensively. The existing evidence suggests the western Apache did not.[3]

Their primary contribution to the region's fire regimes was not to add ignition so much as to remove barriers to fire's free propagation. Specifically, they stalled the spread of European livestock. Instead of raising flocks and herds, they let the missions and Mexicans do it, and then raided them. It was only after the Apache wars sequestered the remaining tribes onto reservations in the late 19th century that livestock exploded and the regional fire load crashed. Across the Southwest the slow combustion of cattle in particular beat back the fast combustion of open burning. The break in fire history is as abrupt and distinctive as the abandoned dwellings of the Mogollon and Anasazi. Like the Apache, free-ranging fire retreated to reservations or isolated locales spared the general crush of cattle.[4]

San Carlos was first gazetted as a reservation in 1872 and became a collective for various Apache bands that had little in common save their language. The Yavapai, the Tontos, the Chiricahuas, and fragments of other bands of the western Apache were rounded up like mavericks and put into a common corral. The experience was wrenching. Most had no ties to their new estate. They had no cultural continuities that could bond them to the land. The old ways often had little value: they would have to rebuild a new culture, find new stories to pass on their inherited wisdom, relocate from their sacred mountains to the desert hills of San Carlos's lowland administrative post. The unsettled demography and scrambled culture meant that whatever fire practices they had known in the past would have to be reconstructed in a setting over which they did not control the basics of their existence.[5]

The gathered groups were fed ration beef, and then granted cattle to slaughter or raise. It made sense to grow the beef on site, so herds came to San Carlos itself, as throughout the Southwest, although most arrived through trespass from herders outside the border. Some order was instilled by establishing a leasing system for the Anglo ranchers, which continued until 1933 when the tribe began reclaiming control. San Carlos's flatlands were prime grazing land, and the herds were large; the resulting ranches were among the last driftwood deposited from the storm surge of the Texas cattle industry; and here, like nearly everywhere, abusive overgrazing was the norm. A ruinous cattle rush that had degraded landscape after landscape in the American West came to the desert grasslands, forest savannas, and high prairies of San Carlos. The fire regime that had prevailed for centuries, however modified by the Apache, was eaten and trampled away.[6]

At the same time notions that fire ought to be actively suppressed added to the pyric loss. The Bureau of Indian Affairs (BIA) developed forestry programs to promote timber industries, and foresters at San Carlos did what foresters everywhere did: they fought fire. Between intensive grazing and active firefighting, the lavish ignitions could no longer roam with the insouciance of black bears or the free flow of the wind. Increasingly, the land became, for fire, a tierra desplobada.

• • •

The BIA established a branch of forestry in 1910, but it was not until the reforms of the Indian Reorganization Act of 1934 that San Carlos felt its presence. The CCC added muscle, committed to fire control as much as to erosion control, building fire roads and towers as they did rock check dams. The tribe began reclaiming ranching leases and looked to forestry to enhance its economic development. In the 1950s commercial logging developed. The postwar era set San Carlos on the path that led, ultimately, to the April 23 briefing.[7]

Its foresters were conscious that they managed tribal lands "in trust" and accordingly transplanted the prevailing practices of their day to San Carlos. But even as modern forestry arrived, its norms were being challenged. When he came to Arizona in 1948 as regional forester for the BIA, Harold Weaver gave voice to the concern that fire exclusion was disrupting the land as fully as the overgrazing that accompanied it. He also discovered that parts of San Carlos had apparently been spared.

He found in Malay Gap, along the Mogollon Rim, at the far northeastern corner of San Carlos, a place that seemed relatively unscathed by hoof, axe, or removed flame. There, fire-scarred ponderosa suggested that surface fires had returned at least every seven years over several centuries and had bequeathed a forest so magnificent that "it is hard to see how she can be much improved on." The fire challenge at San Carlos was to propagate that old regimen of burning, not to find better ways to knock it out. Weaver urged controlled burning of the sort he had helped pioneer in the Colville and Warm Springs reservations of the Northwest and which Harry Kallander, north of the Black River at Fort Apache Reservation, was introducing. But just as simply removing cattle dramatically unburdened the land, so ceasing to suppress the fire that nature so lavishly strewed about the landscape, would help. In fact, lack of resources meant that places like Malay Gap could perpetuate something like the old ways. Major burns had washed over Malay in 1943 and 1946, just prior to Weaver's tour.[8]

Mostly, though, prescribed fire meant burning the slash generated by the emerging timber program. Instead of adopting the Weaver agenda, San Carlos committed to fire control, which paradoxically acted not as a drain on the tribal economy but as a stimulant. The San Carlos Apaches heavily staffed the Southwest Forest Fire Fighters program, fielding as many as 800 men a season. Those paying jobs were a significant source of otherwise scarce income; they did for the feeble money economy of the reservation what migrant farm labor and remittances did elsewhere. At the same time, San Carlos forestry built up its own suppression infrastructure. It acquired engines. It erected lookouts and founded an aerial reconnaissance operation. It built a modern radio network. It developed a spider web of fire roads. It had access to bulldozers and air tankers. It built a fire camp at Point of Pines, complete with a helitack crew. In 1991 it founded the Geronimo Hotshots, over which the tribe assumed control in 1996. Prescribed burning was mostly limited to activity fuels, which is to say, logging slash.

Even as reforms swirled in the convective plume that promised a national fire revolution to reinstate fire, a trend with which its natural and ethnographic history would seem to align perfectly, and for which Harold Weaver had become a prophet, San Carlos hardened its commitment to serious fire suppression. The 10,000-acre Black River wildfire of 1971, not a scheduled prescribed burn on Ash Flats, defined the thrust of the program. In 1972 Tall Timbers Research Station sponsored a field tour to San Carlos and Fort Apache reservations, with Weaver among its company, and noted that any attempt at prescribed burning was failing to keep pace with the wildfire threat; the failure to install a vigorous fire restoration program would only lead to more blowouts. Yet it was a hard choice. Fire suppression was the national norm, and if nothing else, firefighting meant money. Its crews were one of the few exports the tribe had. Big burns brought in big bucks, and while prescribed burning and managing wildland fires would still hire crews, one of their objectives was to reduce dollars spent.

So San Carlos missed the national fire revolution. Not until the revolution revived in the mid-1990s was there pushback against simple suppression. An understory burn around Point of Pines and rangeland burning (to dampen encroaching cholla) put controlled fire on the ground for the first time since Weaver. The prime mover, Bob Gray, recalled that the "beating I took from my own administration, the BIA, the wildlife biologists, etc. was brutal—all wanted the experiment to fail; my only allies were the elders who understood the importance of fire." It helped that the suppression program had become moribund, and in 1995 the tribe took control of the Geronimo Hotshots from the BIA, and then upgraded suppression standards and rewrote the fire plan to accommodate some cross-border burning with the Apache-Sitgreaves National Forest and to establish a large natural fire zone in the eastern lowlands below the Nantac Rim.[9]

The 2000 National Fire Plan funneled more funds toward equipment and fuels.[10] But it also required a plan. The plan would have to reconcile fire practices with land use, which is to say, the forest management plan (and environmental assessment, and Tribal Strategic Plan), but also with reformed national fire policy, first articulated in 1995, and then reaffirmed in 2001. One of the difficulties with those policies was that they encouraged and made possible an active program of fire restoration but did not mandate it. Instead of seizing on the possibilities proposed in the 1995 policy, San Carlos had taken over the Geronimo Hotshots. After 2001, however, the latent prospects of policy met the right personalities, the tribal council expressed interest in more traditional attitudes toward their natural estate, and the resulting wildland fire management plan of January 2003 announced a new era.[11]

The fire program picked up the pace of slash burns, but quickly appreciated that more was needed. If fire spread is a question of surfaces, fire management is an issue of edges. The more surface relative to volume a fuel particle has, the faster heat and moisture transfer, such that small fuels burn more readily than large ones. Similarly the more edge relative to area a burn plot has, the more complex its control. The fire program burned 600 acres at Baskin Tank in 2006, and 2,500 acres at Dove Tank; but even amid logging slash or chained juniper, approval could take several years and cost serious dollars. And they were only treating new fuels they created, not legacy landscapes. The San Carlos fire program had to enlarge the size of the sites beyond individual plots and quicken the tempo of treatments.

Most clustered along the Nantac Rim. Beginning with reconnaissances in 2005, and plans in 2007, they turned to the Hilltop region. Between 2008 and 2010, with help from aerial ignition, they burned off three big blocks that added up to 14,000 acres. In 2009 they confined and contained the 20,000-acre Bear Canyon fire, even as they attacked 45 new starts. In 2011 the Maggie fire added another 5,000 acres. In 2012 the Trail and Shorten fires brought in 8,000 acres more. Managing fires in this way reduced costs by an order of magnitude. Meanwhile, like the rest of the Southwest, bleached by a long-wave drought, wildfires blotched woodlands north and south of the rim on the order of 500–4,000 acres.

Then came 2013. Early on they set three prescribed burns, a few thousand acres each in woodland savanna, along with the complex 13,000-acre Pine Salt burn around Point of Pines. The wildfires arrived on schedule. An "annoying" fire burned through salt cedar near Bylas along the Gila River. Two fires broke out in the high country. The Fourmile fire burned toward the reservation's eastern border through a "scabby transition zone," not easily attacked nor worth risky suppression. Fire managers backed off to roads or barriers, burned out, watched, and otherwise confined and contained. The Creek fire on the north flank of the Nantac threatened the Dry Lake Lookout and a remote automated weather station, some commercial timber, and Point of Pines; but even in the height of fire season, there were options other than going toe to toe, and fire operations backed off, burned out selectively, used previous burns as cold trails, called for some air strikes near developments and canyons where the fire might bolt, and generally herded it to good effect. Though its fire behavior specialist, Bil Grauel, then handling the wildland fire decision support system for the burn, complained to Duane Chapman, fire management officer and former superintendent of the Geronimo Hotshots, that he "couldn't model the damn thing if he [Duane] was going to herd it all over the landscape." The Creek fire blackened 18,000 acres. That year San Carlos had the largest fires in the state. What Grauel said about the Fourmile fire epitomized the season: "Along with the fire, they decided that 14,000 acres was about right." That statement's odd phrasing, and the attitude it conveyed, could stand for the program.[12]

• • •

Fire management at San Carlos is under the Mogollon Rim and under the national radar. Like all programs it has its liabilities and its assets. How they balance says a lot about how the program actually looks on the ground.

The liabilities are numerous and obvious. Many, like patchy support, are shared by all fire organizations, and like metastasizing juniper woodlands, by those throughout the western United States. Some, like shortchanged funding relative to federal neighbors, belong to reservations and the oft-poisonous codependency between tribes and the Bureau of Indian Affairs. A few are specific to San Carlos. The tribe's instincts to turn inward into a kind of cultural as well as economic autarky. The long reliance on fire suppression as a seasonal revenue stream. The lack of Apache traditions for landscape-scale burning. The attitudes of neighboring landowners more cautious about free-ranging fire and far-reaching smoke. The collapse of the Southwest Forest Fire Fighters program as trainees failed physical and drug tests. The way social pathologies seem to channel into land degradation.

But the program's assets may be greater. Over the past 20 years, livestock numbers have shrunk dramatically, and today are perhaps 20 percent of what they were previously. That has freed up grass to carry surface fire. Tribal autonomy allows it a freedom to maneuver not available to the national forests to its east, west, and south. The lack of major cities, or even exurbs, means the interface is a minor concern. If Apache culture lacks traditions of burning, it also has a tolerance for allowing natural processes to work their own destinies, which can translate into flexibility in handling the fires nature starts. A small timber program means the tribal economy can absorb some burned stands as a price of building resilience into the land and preventing savage wildfires; reduced cattle stocking means it can accept burns that take away winter range. The land is so fire-sated naturally that reform does not depend on fire lighting; it's enough to modify fire fighting. Even isolation and lack of attention has its merits. If insularity means San Carlos doesn't get much outside support, it also means it doesn't get much outside scrutiny.

It means San Carlos lives for the present in a sweet spot. It can do things its neighbors can't. It can turn its relative poverty into a wealth of opportunity. It can replace ever-more-encumbered prescribed burning with hybrid fires in which burnouts and free-burns from natural ignitions fuse. In the past, decisions about fires were boxed in by the demand to suppress them as quickly as possible.

Now starts like the Skunk and Basin fires allow fire managers to draw their own boxes. That grace period won't last, but it grants San Carlos a time to stabilize its program so that, when something does go wrong, as it inevitably will, its fire program will survive.

• • •

On June 6, 1946, Emil Haury of the University of Arizona, one of the premier archaeologists of his day, was busy erecting a field camp at Point of Pines, what would become over the next 14 years a celebrated summer training site for students of southwestern archaeology. At noon he received a call from Paul Buss, the San Carlos forester, with a request to aid in fighting a fire at Malay Gap. "An unwritten law in the wilderness country," Haury sternly noted, "says that when the forest begins to burn, all able-bodied men must pitch in to help bring the fire under control, no matter what they are engaged in at the moment." Six members of the company made the trek to Malay, "about as remote as any place could be."[13]

Amid a "totally strange country, extremely rough, and heavily timbered," Haury and his students struggled to cut line during the late afternoon. One student found himself trapped in a "fire circle" and barely escaped with his life. That night, with aching muscles, they struggled to sleep while immersed in smoke "from which there was no escape." The "eeriness and gloominess of the situation, the crashing of burning trees, the sharp riflelike cracks of exploding rocks around us, and, the next day, faced by a roaring wall of fire racing up the hill to our temporary camp, were enough to sear the words *forest fire* deeply in any mind." Meanwhile, several score Apaches on the scene exhibited "no special concern." Without strong direction, the archaeologists "lost heart" and asked to be released. From Point of Pines they could watch the smoke billow up for some days, apprehensive and grateful that it was so far away. "I will say," Haury concluded, "that no baptism by fire, literally, was ever more exhausting and frightening."[14]

What happened that inaugural summer was repeated in the years that followed more often than Haury wished. He came to view fire as a serious inconvenience. In 1950, the group was called out "frequently" to lightning fires, which "sorely interrupted" their work schedule. In truth, fire became the secret catalyst for some of their major discoveries. They tracked the history of the built landscape through hearths. They traced the historic mingling of peoples through cremations. In 1950, that summer of fire, they followed gopher mounds to charcoal and carbonized corn, which led them to "the very thing we were looking for," evidence of a "conflagration." Charcoal is nature's great preservative. When they excavated a pueblo with 20-plus rooms burned, they had a mother lode of how the Mogollon culture lived, what they ate, and when they flourished, and in the form of scorched skeletons who they were. The next summer they attended fire school, had even more callouts for fires, and at one point had to devise evacuation plans. Fire good, fire bad—student archaeologists were learning the real lesson of Point of Pines.[15]

They uncovered some startling finds. In the 13th century, the Mogollon and Anasazi cultures had met in an awkward mingling and occasional fusion best exemplified by the emergence of a square kiva. The kiva was Anasazi; the squared corners, Mogollon. The square kiva came to stand for the prehistoric world of Point of Pines, and it might foreshadow the modern world of fire management there as well. Kivas come in circles; fire, in triangles. The Mogollon built out of rock, the moderns out of the triangle of ingredients that make fire—heat, oxygen, fuel; terrain, vegetation, weather; fire lit, fire fought, fire herded. Prehistoric San Carlos managed to square the circle of their times. Contemporary San Carlos is squaring the triangle of theirs.

• • •

As the Mogollons rendered an idea into stone, so San Carlos fire must transfigure its innovations into institutions. It has happened before. Twice, in fact, San Carlos has succeeded in transforming a mangled scene into a model system.

The first involved the wreckage of its heritage of free-range ranching. In the 1930s, as the tribe began reclaiming leases, it also became the site for the first national experiments in artificial insemination under John Lasley of the University of Missouri. The numbers of scruffy stock came down, the quality of the remaining Herefords went up. By the 1950s a model ranch, the R100, combined breed improvement with range management on Ash Flat under the direction of the University of Arizona. Meanwhile, Emil Haury helped turn generations of free-range pot-hunting into an academic discipline, and underwrote an archaeology suitable for a research university, again the UofA. Its peculiar isolation made San Carlos reservation ideal for both purposes.[16]

Now it may be fire's turn. Abusive practices—in this instance, fire exclusion—are being transmuted into an exemplary exercise in fire restoration. The old

Point of Pines fire camp, once a high scorch mark in the era of fire fighting, could be remade into a training ground for fire lighting. The Prescribed Fire Training Center in Florida has as its motto Every Day Is a Burn Day. Active prescribed burning is far trickier in the West, but San Carlos doesn't have to wait for the cumbersome protocols of fire by prescription. The land is sated with natural ignitions that only need to be brought under a system of management. San Carlos doesn't have to rely on fire lighting to reinstate a more resilient fire regime. It need only modify its acquired habits of fire fighting.

And that is what the fire staff at San Carlos did with Basin and Skunk. They focused first on the more interesting Basin fire. They drew the box, and the contours of their burnout looked for a while like a folding protein as it absorbed old prescribed fires and patches of wildfire before stopping at 6,018 acres. When weather moderated about a week after Easter, the staff ignited the 1,228-acre Point of Pines prescribed burn to clean out thinning around the camp before rotating those crews into the 524-acre Bee Flat burn to restore juniper savanna. A wildfire started at Willow and was suppressed at 21 acres. Another started in grassy scrub at Rimrock and was boxed in at 1,826 acres. The Barlow fire broke out along the main road through the Nantac, burned hot, and was fought hard to a tough 1,483 acres. And the Skunk began to move.[17]

It was 30 acres at the end of April and 80 on May 11. Then a mild cold front passed over and strong northers pushed the fire to 1,822 acres and put flame on the grassy plains to the south of the Nantac Rim. Here it found a fuse it could burn along while prevailing southwesterlies could drive it onto the rim. The fire was filling in a long blank spot in the need-to-burn map of San Carlos. From the 11th to the 21st the Skunk fire rose and fell, making daily runs as high as 6,336, 9,248, and 4,254 acres. The San Carlos staff ran along beside it, using roads and burnouts like drovers holding a stampede. They put a crew in to prep a repeater station before the flames arrived. The fire front continued to spread, unusually for the Nantac, to the northwest. San Carlos called for a Type III short team to help with logistics, although the newcomers had to be constantly educated into what San Carlos wanted (and didn't want, an expensive air show). They thought Kidde Creek might hold the progression; it didn't. They continued to burn out along 1500, the main road along the rim. They downsized to a Type IV crew, used some helidrops to help hold the burnouts and called in some retardant from small air tankers. On the 27th the fire rushed over 4,879 acres before pausing. At Rocky Gulch they pinched off the front and began burning out the interior with aerial ignition. That accounted for the last big burn—14,087 acres on June 2. The fire perimeter now ranged over 92 miles, roughly 22 miles long and 8 miles wide. When the smoke settled, the Skunk fire had blackened 73,622 acres and tied in with the 2009 Bear Canyon and 2004 Upshaw fires to the northwest and the 2013 Creek fire to the southeast. The summer fire season—with its normal dry-lightning fire busts—was still weeks away. Staff were already wondering how to handle the reburns that would be essential to turn a spate of fires into a functioning fire regime.[18]

Meanwhile, lightning had kindled the Black River Tank fire along the border with Fort Apache reservation. Fort Apache tried to emulate the San Carlos strategy, but instructions were confused, and an air attack operation saturation bombed to hold the burn to an expensive 3,244 acres. What San Carlos had accomplished depended on the dynamics of its people, not just the dynamics of fire burning through pine, juniper, and grass.

Fire behavior is fire behavior and universal, but behavior toward fire is specific to cultures and not transferable with algorithms. Box and burn is not a simple tool, like a Neptune air tanker or a D6 caterpillar that can be dropped into any landscape. It is a negotiation between fire and fire managers. Like all things human it has to be learned, but unlike many it is not something easily taught.

A Refusal to Mourn the Death, by Fire, of a Crew in Yarnell

I shall not murder
The mankind of her going with a grave truth . . .

O N THE 30TH OF JUNE 2013 a fire blew over the Granite Mountain Hot-shots outside Yarnell, Arizona and left 19 dead. Three months later, on September 30, a formal investigation released its findings. The inquiry focused on the mechanics of fire behavior and how the Granite Mountain Interagency Hotshot Crew might have understood their "situational awareness," which is to say, how the crew recoded the information they were given with what they saw for themselves. Instead of ascribing blame, the investigative team sought to appreciate how the hotshots engaged in sensemaking in an effort to explain decisions that, to nearly all observers, made little sense. But the need for sensemaking extends also to the meaning of the fire for American culture at large.

For anyone conditioned to read landscape for fire behavior, Yarnell Hill is a Google of clues ready to be coded into the existing algorithms of fire behavior. The fundamentals point to fuels of mixed brush and grasses, parched by seasonal drought, to the terrain of Yarnell Hill, and to record temperatures, blustery winds, and the downdrafts ("outflow boundary") from passing thunderheads. There is nothing in the reconstruction of the fire's behavior that suggests it was anything other than a high-end variant of what happens almost annually.

What made a difference was that the collective will of a hotshot crew crossed that flaming front. The reaction intensity that matters is fire's interaction with American society. The behavior we want to explain is the crew's, and what their death signifies, and for that we must look outside the usual fire-behavior triangle and into that triangle of meaning framed by literature, philosophy, and religion. The fire came as a tear in the space-time continuum, opening into a void for meaning. There are fires that belong with science, fires that stride with history and politics, and fires that speak in the tongues of literature. Yarnell Hill is a literary fire. It's a fire for poets and novelists, and maybe the stray writer-philosopher.

In a profile of Howard Hughes, Joan Didion observed that he had to be understood as a "great literary character" like Jay Gatsby. So, after the reports are filed, the lawsuits settled, and the scientific interrogations published, we may well linger over the Yarnell Hill fire as a great literary moment, for which character, conflict, and plot serve as the fuel, terrain, and weather resolved into the blowup of tragedy.

The greatness of the fire does not lie in its physical behavior or scale. Hannah Arendt famously spoke of the "banality of evil." It is likely that the mechanics of the 8,000-acre Yarnell Hill burn will prove equally banal, not with active evil but with an unsettling emptiness. It is not what happened but what it means that mesmerizes the public imagination. Giving story to that sentiment will be the task of literature.

Nor blaspheme down the stations of the breath
With any further
Elegy of innocence and youth.

When it studies fire behavior, the fire community reaches for models; so, as it ponders the fire's meaning, it will need to search for narrative templates. The vital parameters are a wrathful gust of wind and a devouring fire, the immolation of 19 men, a lone survivor, a landscape of ambiguous purpose as a determined brotherhood chases flames by a nearly deserted town. These facts the community will try to reconcile with fire behavior and to the interpretive models that it uses to account for why fire suppression today is dangerous. The problem is that its prevailing models don't fit.

Expect that the community will turn first to Norman Maclean's Talmudic *Young Men and Fire*. Attention will focus, in particular, at the point where science met literature, the study of comparative behaviors, the fire's and the crew's, that Maclean's inquiry inspired from Richard Rothermel. At Mann Gulch the steep terrain quickened the fire even as it slowed the smokejumpers.

Where the curves of their differing rates of spread crossed, the crewmen fell. In Maclean's hands the mathematics became the narrative lines of a Greek tragedy in which fire and crew each did what they were destined to do. Bob Sallee and Walter Rumsey were high enough on the slope that they could just evade the flames. For the others, only Wag Dodge's escape fire—like a deus ex machina—interrupted the logic of fate.

Yarnell Hill will become another "race that couldn't be won." But the analogy stumbles. The smokejumpers at Mann Gulch had no choice: they were trapped in a closed basin, almost a chute, and would perish unless they could outrace or outsmart the flames. The hotshots at Yarnell Hill were safely on a ridge, in the black, and chose to race with the fire by plunging downhill into a box canyon thick with boulders and brush. Theirs was an act of volition denied the jumpers at Mann Gulch. At Mann Gulch the fatal numbers were coded into the scene at its origin. At Yarnell Hill the Granite Mountain Hotshots did the calculations and added the sums incorrectly.

The numbers tempt: they are hard facts, recorded in the landscape, not unknowables embedded in the nebulous "sensemaking" of mind and heart. So one narrative will turn to explanations for the fire's "extreme" behavior. The story will look to long-unburned fuels and especially to extended seasons, record temperatures, and climate change. The explosive Yarnell Hill fire will become another signature of the Anthropocene's new normal. The loss of an elite crew will be tallied as part of the cost of ignoring global warming.

It was hot, dry, and windy, but it's always like that in the early summer lead-in to the monsoon. Central Arizona has known higher temperatures, stronger winds, and deeper droughts. When thunderheads collapse, particularly in the season's first storms, those blasting downdrafts stir up dust storms if they overlie desert and firestorms if atop flame. A similar outburst drove the 1990 Dude Creek fire (eastward at Payson, a sister city to Prescott) through a crew and killed six. The flames at Yarnell Hill leaped and spotted through grass and brush—combustibles ideally suited to react quickly to wind. Conditions on June 30 were not beyond the region's environmental or evolutionary scale. We don't need climate change to account for the fire's behavior.

Already, the community is turning to that other great narrative template invoked to describe the contemporary scene, the geeky-named wildland-urban interface. The Granite Mountain Hotshots were putatively on the scene to defend Yarnell, an exurban enclave that may have been indefensible and in any event was mostly evacuated. It's unfair to demand that fire crews risk their lives for property. That burden belongs to the community. If they build houses where fires are, they have to live with fire.

But Yarnell frustrates this generic model. The town sprang up during an 1860s gold rush. It was platted 50 years before Arizona became a state. It survived by being repopulated, most recently by retirees. Whatever the firescape at the time of founding, it was undoubtedly scalped by the miners, who burned off the cover to expose outcrops, cleared any trees and shrubs for firewood, and brought in meat cattle that stripped away the grass. The existing scene is the jumble of recovered pieces. The town was not plucked down amid combustibles; the firescape grew up around the town. In fact, by the early 1950s the state of Arizona was actively promoting Yarnell as a retirement community, which is what it became.

The dispersed outliers were mostly indefensible, so there would have been little justification in trying to shield them, particularly with fire bearing down imminently. The Granite Mountain IHC was outfitted with hand tools for cutting fireline, not with hoses, pumps, and shielding for defending structures. Before they left the ridge, they were working a free-burning fire; they were assigned to establish an anchor, not protect structures. This was not an Esperanza fire in which flames washed over an engine crew positioned to defend a house. The Granite Mountain IHC did not perish, arms locked, standing between the flames and homes. They died in a box canyon into which they had voluntarily hiked.

The other template—everything has to come in threes—is that legacy agencies are unable to overcome their culture of suppression. They fight fires where and when they shouldn't because they know no other way to respond; the endless roster of shelter deployments, near misses, and fatality fires is the inevitable outcome. Even the shock of the South Canyon deaths, the prospects for civil and criminal penalties, and the emphatic edicts from above have failed to dislodge that culture. It invites risk-taking that leads remorselessly to Thirtymile Canyon, Cramer, and Yarnell Hill. The rules keep being broken—must be broken to satisfy the nature of the beast. Implicit, too, is a tint of gender bias; the Granite Mountain crewmen were all young men engaged in what could seem an extreme sport.

Yet again this particular fire turns such understandings into shades of gray. Granite Mountain IHC was not a Forest Service or Bureau of Land Management (BLM) crew: it was proudly, defiantly, the only nonfederal IHC on the national register. Undoubtedly it absorbed elements of traditional mores and camaraderie, but perhaps without the institutional checks that have been cultivated over the generations since 1994. It belonged to a city fire department: it absorbed at least as much culture from urban fire-service expectations as from wildland agencies. It was fighting a wildfire on Arizona state lands, under the auspices of the Arizona Department of Forestry, the youngest state forestry bureau in the United States, established the last year fires burned in Peeples Valley. There were other crews on Yarnell Hill, including hotshots. None of them put themselves into the path of the fire.

In the least valley of sackcloth to mourn
The majesty and burning of the child's death.

If the traditional templates don't apply, what might? If this is a literary fire, they will come from literature, and in truth possible models seem to leap from the scene. There are those that help arrange the particulars into a story, and those that invest that narrative with the kind of meaning that speaks to the extraordinary reach of interest that this fire has sparked. They move the tale from physics to metaphysics.

Look, first, to *Moby-Dick*. The pivotal—the Ahab—character is surely Eric Marsh, cofounder and crew superintendent, not because he is mad or malevolent, but because he is driven, and whatever future inquiries determine to be the chain of events, this is his literary role.[1] Wildland fire was his obsession: he helped transform a brush-clearing crew into an IHC, he established the Arizona Wildfire Academy in his living room, he modeled the crew after his own character. They were all young, male, and Anglo. They had something to prove: they were a proud "oddity," he had written some months earlier, a city-sponsored IHC among a federal-dominated workforce; a "mystery," to city coworkers; "crazy," to family and friends. They prided themselves in showing up "to a chaotic and challenging event, and immediately breaking it down into manageable objectives and presenting a solution." They did not just call themselves hotshots, they were "hotshots in everything that we do." They "loved" the life they had chosen. They "managed to do the impossible."[2]

On the afternoon of June 30 Marsh was a division superintendent, but he left the line to scout and then moved the crew away from a blackened ridge and back into the action. He showed an élan and initiative that in many circumstances of life we would applaud but what here looks like an obsession. They were safe. He took them to the flames. His drive became a fatal flaw and carried the others with him. The crew, even the Starbucks among them, follow, all caught up in the chase. So at last that great white whale of a fire turned on them, and left one survivor, a solitary witness to proclaim, after the Book of Job, "And I only am escaped alone to tell thee."

Still, tragedies abound; what made this of interest in Toronto and London as well as Phoenix is that the fire dramatized with sudden and graphic violence the question of what purpose if any informs our existence, whether our lives reflect the workings of a plan or of accident. Look, in this case, to Thornton Wilder's *The Bridge of San Luis Rey* to account for the awful coincidences and the quest for patterns in the void. "On Friday noon, June the twentieth, 1714, the finest bridge in all Peru broke and precipitated five travelers into the gulf below. . . . People wandered about in a trance-like state, muttering; they had the hallucination of seeing themselves falling into the gulf below."

Why this crew? at this time? in this way? Why, as residents of Yarnell asked, were some houses spared and others burned? Was there a hidden order, which is to say, a deeper providence in the tragedy, or was it just an arbitrary collision of actions with no more design than the scatter of summer cumulus? The Granite Mountain IHC had for its original logo a pair of flaming dice that always came up seven. This time the dice rolled snake eyes.

What the *Bridge of San Luis Rey* model also suggests is that the way to narrate the meaning is not directly under the gaze of an omniscient narrator arraying events within GIS grids and plotted along timelines, but through the quest for their significance. Norman Maclean made that pursuit personal: *Young Men and Fire* became the story for his own search for the story. Thornton Wilder refracted that inquiry through Brother Juniper, whose pursuit of and meditation on the gathered facts leads ultimately to an auto-da-fé, the burning of self and book at the stake.

Beyond metaphysics and theology lies the tangible grief of the survivors. They want significance ascribed to the sacrifice. They want their loved ones honored and valorized. This, too, is an old provenance of literature, and it offers a full gamut of consolations, from doggerel and sermons to narrated emotion fused with artful intellect. For this task ponder James Agee's *Let Us Now Praise Famous Men* to appreciate how to ennoble ordinary lives without sentimentalizing their fate. And at its most challenging, read Dylan Thomas's inextinguishable "A Refusal to Mourn the Death, by Fire, of a Child in London" for the cold, obdurate turning inside out of emotional trauma in the face of a tragedy of innocents.

For all its splendor and pathos, American fire has no novel to match its stature in the American scene. It has some powerful nonfiction, most notably Maclean's *Young Men and Fire*, which has many imitators but stands alone, itself a survivor and posthumously published witness. But the fusion of fire and art remains unmet. In Yarnell Hill, however, the American fire community has the themes and latent structures for a great work of literature.

That doesn't guarantee it will happen. But my guess is that the enduring voice of the tragedy will be the writer who recognizes that this is not just human-interest journalism, a gripping story of a disaster, or a political parable about misplaced national priorities, but someone who appreciates it as great literary character whose meaning must depend on the ambiguities of art to extract significance from the indecipherable. Not just someone who can see patterns in the flames and hear cries among the roar, but someone who can say with Dylan Thomas that

After the first death, there is no other.

INTERIOR WEST

A Worthy Adversary

"I ADMIRE CHEATGRASS."

That's not a common sentiment. It sounds even stranger coming from someone who has spent a career grappling with what he calls the "invader that won the West." But if you are a competitor at heart, if you value tenacity, adaptability, and patience, if you accept that the nimble will prevail over the sluggish, then cheatgrass is not merely an ecological survivor or a stray gallows-humor success story but an extraordinary rival. It came. It saw. It conquered. And it did what few conquerors do—it stayed. We judge heroes by their villains, we value contests by how hard fought they are. For a land manager in the Great Basin, cheatgrass can seem an ultimate challenge. A weed that can remake whole biotas, that can reshape the largest federal land bureaus and bend research organizations to its will, can also shape a life. So it happened with Mike Pellant.[1]

• • •

He was born and raised in central Kansas, far from the seat of cheat. Fifty years earlier cheatgrass had pioneered sites from Reno to the Wasatch Front. Twenty years before he arrived in Concordia in 1952, it had exploded over the northern half of the Great Basin, and G. D. Pickford was publishing the first serious scientific studies of grazing, fire, and cheatgrass at the U.S. Forest Service's Great Basin Experiment Station in Utah. By World War II locals accepted cheatgrass as rural Africans did malaria and sleeping sickness. It was an indelible part of the countryside. It could no more be eradicated than mosquitoes.

In 1949 Aldo Leopold's posthumously published *A Sand County Almanac* included an essay that made cheatgrass public enemy number one for nasty invasive species. "One simply woke up one fine spring to find the range dominated by a new weed." With what, in historical perspective, seemed breathtaking speed, it had leaped from the straw of thatched roofs in Europe to whole landscapes of the West. Worse, "it is impossible fully to protect cheat country from fire," and the shrinking native landscapes, now assaulted by fires, were "the key to wildlife survival in the whole region." Cheat was here to stay. It couldn't be plowed under, grazed out, poisoned, or burned out, all of which only fed its spread. So the infested regions had found ways to make "the invader useful." It couldn't be beaten. It wouldn't self-extinguish. Leopold discovered among residents a profound fatalism. Cheat was awful, but it was shrugged away as better than grazed out pastures and bare ground.

Yet the postwar era witnessed large-scale experiments in land use and science applied to the service of political economy. The BLM, formed by administrative fiat in 1946, sought to rehabilitate the western range that was its administrative homeland. That meant taking on the weed that had taken over so much of the landscape. It was the postwar era that saw the last great dams raised, that chained off pinyon-juniper woodlands, and that converted large swaths of sagebrush steppe to crested wheatgrass. That last scheme served a double purpose. It transformed sage-infested wastelands into productive range and it stymied the growth of cheat. Crested wheatgrass made good forage, and if properly grazed, it could replace—serve as an ecological surrogate for—the native perennials lost to the era of bad grazing. The problem was that the project was expensive and it destroyed native vegetation that a not-too-future generation would regard as intrinsically worthy, a future more inclined to value native birds than imported livestock. The wholesale conversions rose and fell while Mike passed through elementary school and high school. By the time he entered college, an environmental movement was finding value in the once-despised sage steppe, much to criticize in the commodity-driven management of the public domain, and suspicion of a science-commercial complex that had loosed toxins, clear-cut wildlands, and sprayed DDT everywhere. The efforts to extinguish cheatgrass looked like the misguided campaign to eradicate fire ants.

Mike's most enduring education came on his grandparents' farm, 50 miles from Salina, the nearest town. He spent four summers there, from the age of 12 to 16. His mother had come from a family of 10, and he joked that with all the siblings gone, his grandparents needed another source of "slave labor." But they were delightful summers. He learned to observe nature closely. He grew to love plants. He thrilled to the annual spectacle of useful greenery reclaiming plowed earth. His grandfather had a self-informed but serious land ethic, and Mike absorbed it. Meanwhile from his father, a junior high school science teacher, he acquired a curiosity about nature, an appreciation for science, and a predilection to problem solve nature's questions by scientific principles.

Then, for his final two years of high school, he spent his summers as a biological technician for the Army Corps of Engineers at the Wilson Reservoir, a subtle segue from a naturalist perspective to a scientific one. When he went to Fort Hays State University he majored in botany and continued to work in the mixed-grass prairie at Wilson Reservoir; and there he became involved with prescribed fire. He pursued a master's in botany at Fort Hays, with an emphasis on range science, and used that rich prairie and its burns for his thesis. He even experienced his first escape fire. By then he was ready to leave Kansas for a wider world.

He got more than he perhaps bargained for. His advisor had spoken of the BLM, which suited Mike's passion for range management. In 1976, with his MS thesis in final review, he was appointed range conservationist at Monticello, Utah. It was, he recalls, a moment of "culture shock." The Colorado Plateau, with its incised gorges and mountain laccoliths, was unlike anything he had seen, and small-town Mormon country unlike the rural communities he had known growing up. He wasn't in Kansas anymore.

In retrospect it turned out to be the "best thing" imaginable for his career. He had to deal with badly disturbed landscapes. For two years he had to inventory the damages and inspect up close—a countryside so unlike rolling prairie that it might have been on Mars. He learned to listen to and understand people whose background differed from his own. Over the next two years his survey transitioned from degraded rangelands to landscapes deranged by uranium mining. Call it extreme restoration.

Then it was time to move on and move up. He applied for a supervisory range job in Boise; the interview took place in the foothills, amid lands converted to cheatgrass. Here Mike experienced his second culture shock. It was, he thought, the "ugliest country" he had ever seen. But as cheat had taken over the sage steppe, so a new aesthetic gradually overtook him, and he came to appreciate what he saw. He naturalized to the place. Boise became home. And he came to respect cheat as a competitor.

It was fire country of course, much of it burning within sight of the National Interagency Fire Center (NIFC), officially inaugurated 11 years earlier. Inevitably he became involved with the BLM fire program, first as a resource advisor, eventually in suppression. It was what you did. And fire, he quickly learned, paid much of the freight. The overtime and hazard pay from fire season was a welcome, and expected, annuity—not quite on the scale of Wall Street bonuses, but a Boise echo. Whatever its liabilities, cheatgrass boosted spring forage for ranchers, and so, it seemed, it did also for the summer wages of firefighters.

Two views emerged. People concerned with land health worried about the cheat infestation and its knock-on effects. People hired for fire crews saw it as endless employment, a problem but not one that aggressive suppression couldn't hold. There were many fires, but few conflagrations. Then the fires got bigger and meaner. Mike recalls a conversation with an old rancher at Mountain Home, gleefully explaining how his father had "made this land with a match," letting cheat replace sage, before some years later admitting that the overall deterioration and loss of forage had reached a point that was driving him out of business. So, too, fire managers came to recognize that the fire scene had mutated into something monstrous. Wildfires began burning into the outskirts of Boise itself.

• • •

The new regime demanded more serious rehabilitation. Here Mike found his calling. Coaxing good green out of black and bare soil was what he loved. He became a technical expert in restoring grass and shrubs. But emergency response was only first aid. What the land needed was an ecosystem wellness program.

In 1985 he became coordinator for an Intermountain-wide experiment in planted fuelbreaks and postfire rehabilitation that went under the name greenstripping. The teen farmer from Kansas now oversaw row crops of crested wheatgrass and forage kochia intended to break the sweep of cheatgrass-powered fires in the Great Basin. The width of the strips varied by location from 30 to 400 feet, but Idaho's averaged 300 feet, often adjacent to roadways. By 1987 the program had created a handbook that identified greenstripping criteria and protocols and with the help of Idaho's congressional delegation got funding for pilot projects. Four years later results looked sufficiently promising to scale the program up into Utah, Oregon, and Nevada.[2]

Was it a brilliant structural solution to cheatgrass, or an act of growing desperation, a final roll of the dice in a game of gambler's ruin? Fuelbreaks work best when they are built into the design of a planting program, not when they are retrofitted onto deranged landscapes. The green forage drew livestock and wildlife to them, which, if unchecked, reduced their value and invited roadway accidents. The strips had to be extensive enough to prevent cheatgrass from doing an end run: they were only as strong as their weakest link. And like all fuelbreaks. greenstrips had to be maintained and would survive against the odds only by persistent funding. Congress, and the public, love to build roads; they are reluctant to maintain them. The same considerations applied to postfire rehab. Sowing crested wheatgrass could replace cheatgrass, but at the cost of substituting one monoculture for another, and one that had, again, to bow to a grazing regimen that was unpopular with ranchers but because it supported herding was unwelcome to many environmentalists.

This was no final solution, but at best a chock block behind the wheel of a semi that could keep the truck from rolling further down the hill. Greenstripping was a stabilization measure. Rehabilitation was not restoration. Cheatgrass remained on the land, and in its genomic versatility was expanding into saltbrush. It had a persistence and patience and nimbleness that could dance rings around Congress and bureaucracies and disregarded the ambitions of ranchers and environmentalists alike. It milled with other noxious weeds. It defied simple control by grazing. It had an even more implacable ally in wildfire. Like a software virus it had infected and rewritten the operating system for Great Basin ecosystems. It took on all challengers.

• • •

By the mid-1990s, after ecosystem management had become the mantra of the federal land agencies, Mike Pellant summarized the options for coping with cheatgrass. They were the stuff that only a tenacious optimist and competitor could read without sinking into a salt flat of despondency.

The first need was to reduce cheat's spread. This meant keeping grazing from destroying the perennial native grasses where cheatgrass was present but not (yet) dominant. It meant interrupting the positive-feedback loop that made fire and cheatgrass so lethal. Where cheatgrass was already dominant, it meant replacing it with a more manageable surrogate, which is to say controlling it before planting by deep disking, spritely timed burning, seasonally targeted grazing, and herbiciding, all of which inflicted their own collateral damages. Then it meant reseeding. What had become the traditional plant of choice, crested wheatgrass, could compete if planted in cheatgrass-cleared sites and could support cattle, but it did nothing to help the indigenous vegetation and was tarred with the memory of mass conversions in the 1950s and 1960s. All this was a stop-the-bleeding first aid. The ultimate goal had to be restoring something like the presettlement biota, which among other impediments demanded sufficient native seed, which didn't exist. The solutions, Pellant noted, "are few."[3]

What had changed since the alarms of the 1930s was that cheatgrass had claimed yet more lands (and more kinds of lands) and that scientific interest and management experimentation had expanded. The first symposium dedicated to cheatgrass had been held in Vale, Oregon, in 1965. Some 89 participants heard 20 presentations. When the community next gathered, in Boise in 1992, 340 participants heard 140 presentations. Since then symposia have come on a cycle not unlike cheat fires.

So, too, the cheatgrass problem on the land had magnified by an order of magnitude. Cheatgrass had gone from an unfortunate exotic that had become vital as spring range to an existential threat to the entire ecosystem. What had changed was the escalating character of wildfire. The summer the Boise symposium met the Foothills fire had blasted over 257,000 acres around Mountain Home, the largest wildfire in Idaho since 1910, its plumes tauntingly visible from NIFC. Those fires made urgent what had been a remote and for most of the country a merely symbolic nuisance and ecological parable.[4]

The fires grew. In 1999 they exploded over nearly 1.8 million acres, the Great Basin's answer to the Big Blowup. The 1986 Dorsey Butte fire that rampaged over 17,000 acres of the Snake River Birds of Prey National Conservation Area had stunned observers. The complex of burns that massed into the 1999 season increased the scale by three orders of magnitude. The Dun Glenn Complex ran to 361,000 acres; the Sadler Complex to 209,500; the Corridor, 171,442; the Battle Mountain, 156,958. The Mule Butte fire blackened 138,915 acres; the Slumbering Hills, 103,641; the Jungo, 83,939; the Denio, 77,244. Even the smallest of the big fires clocked in at 50,000 acres.[5]

Twenty years after the Boise symposium (and half a dozen successor symposia) cheatgrass threatened to unhinge not only the biota of the Great Basin but the BLM and through it the national fire establishment. Cheatgrass had become a biotic equivalent to the WUI: it was sucking in resources that should have gone to general land management.

• • •

Yet after 1999 there were those who refused to walk away from the challenge. There were programs for postfire rehab, for weed control, and after the 2000 National Fire Plan for fuels treatment. But emergency fire rehabilitation only becomes active after fires. Weed funds only apply after the weeds have become serious. Fuels management has no metric for invasive grasses. Suppression could only hit the ground when flames did. All are fundamentally reactive. None address the comprehensive and, by national standards, alien character of the Great Basin. What was needed was a holistic strategy that could eradicate, rehabilitate, and restore entire ecosystems.

What resulted from the 1999 conflagrations were two reports, "Out of the Ashes" and "Healing the Land," that served as blueprints for the BLM to create a Great Basin Restoration Initiative (GBRI). The existing approaches each had their own mission and monies. The GBRI had to integrate the old approaches and find new ones, all under a novel conception. The vision of an appropriate response moved from greenstrips to whole landscapes. Mike Pellant moved with it. He had served on the post-1999 season review committees, upgraded his standing as emergency stabilization and rehabilitation coordinator for Idaho to the national office, proselytized on behalf of the GBRI, instructed at the BLM National Training Center, and ultimately became coordinator for the GBRI.

That required the skills of a ringmaster: there were programs to catalyze the right science, programs to build up native seed banks, programs for hands-on rehab and restoration, programs to design suitable equipment, and the tireless massaging of institutional arrangements. The USGS, the U.S. Forest Service, the Natural Resource Conservation Service, the Western Association of Fish and Wildlife Agencies, Joint Fire Science Program, regional universities, and of course the BLM all had commitments in research and practice. It was the kind of task Mike relished. In a way his entire career had escalated along with the cheatgrass-fire problem. He kept pace, however much the country lagged.

By now events were far outstripping the capacity to respond. Invasive grasses had leaped, as an issue, far beyond the borders of the Great Basin and were unmooring even the strategies that had emerged from the fire revolution. The

653,100-acre Murphy Complex (one of three adjacent complexes that totaled three million acres) kept the Great Basin in the national cauldron that was the 2007 season. Three years later, on the centennial of the Big Blowup, the Fish and Wildlife Service determined that the cheat-fire dynamo was "a significant threat to the greater sage-grouse" and considered whether "protections under the Endangered Species Act are warranted." That promised a political firestorm if the sage grouse did, in fact, get listed.

The GBRI suddenly found itself close to the center of the national fire establishment. The only hope of saving the sage grouse (and sparing agencies from a listing) was to save what habitat remained intact and roll back what was lost. Out of what had become an ecological equivalent of plowed ground it was necessary to grow new and improved ecosystems. How to do this administratively, politically, and technically, amid the still-festering injuries of settlement and the fast-marching insults of the Anthropocene, was a challenge only an inveterate competitor—a glass-half-full kind of guy—could savor.

On January 5, 2015, Secretary of the Interior Sally Jewell issued Secretarial Order 3336, *Rangeland Fire Prevention, Management and Restoration*. The Rangeland Fire Task Force charged with responding returned its final report in May. *An Integrated Rangeland Fire Management Strategy* incorporated a century of fire management, 20 years of understanding of what ecosystem management meant, and a decade of experience in collaborative enterprises. It was an "all hands, all lands," risk-based, science-informed, landscape-scale program to check, and if possible roll back, cheatgrass.

It was not a task for people who like marching orders that tell them what to do and how to do it and who could imagine analogues of ticker tape parades to recognize their success. There would be no final victory. Cheatgrass would remain; fires would continue; there would be no armistice, no end point. But with luck and pluck the devastation could be arrested and some of it nursed back to health. That had been Mike Pellant's career. Soon it would be someone else's job.

• • •

In December 2013 Mike retired. It didn't stick. His job was too much who he was, and the job was not done. He returned for a three-year appointment, relishing the unique opportunities to coordinate so wide a program and to mentor future staff, which he preferred to do on the ground. When that tour ends, he hopes to continue consulting on select projects that most interest him, push for experiments in targeted grazing, and train future BLMers.

He leaves with some parting thoughts.

He had seen fires since his days at Wilson Reservoir. But those of recent years are to the old burns what buffle grass is to Bermuda. After the Murphy Complex, he spent half a day flying the burn in a helicopter, speeding over black as far as the eye could see. It was an epiphanic moment, the shock that we are entering "unchartered territory." The uncertainties are greater than we thought, the future more portentous than we conceive. As much as a cause of change, the monster fires are a symptom of it. They have become the "ultimate challenge" for the Great Basin.

They can be fought of course, but at the cost of a huge buildup of suppression resources, and with no prospect of final victory. They simply kindle too easily, move too fast, escalate too abruptly, and every burn spurs on the next. The emerging treatment of choice across the public lands of the West—managed wildfire—will not work in cheatgrass since any kind of fire can cycle back in a positive feedback. The only hope of controlling cheat-powered fires is to control the land, which is say, the cheatgrass itself. Fire and cheat have come to resemble an M. C. Escher drawing in which one stairway rises to another, and another, and around until the series ends paradoxically at the place of origin.

Still, there are points of light amid the gloom. Targeted grazing, Mike believes, must be part of the solution; in a sense, prescribed slow combustion might do here what prescribed fast combustion does elsewhere. There are prospects for biocontrol like the fungus known colorfully as the "black fingers of death." There are cases of extensive cheatgrass die-off, perhaps from soil microbes. The causes are not understood, but those patches, some large, are points for intervention, a kind of emergency rehabilitation not unlike replanting after a wildfire. They can add up.

Yet it isn't enough to dampen or even extinguish cheatgrass. The crisis developed because the native perennials were destroyed by bad grazing. The ecological niche still exists. Something will fill it. Where a flower blooms a weed cannot, and where the flower fades, weeds will rush in. The replacement might resemble the indigenous bunchgrasses and co-sustain sagebrush, or it might resemble cheatgrass and wipe out the sage. That leads Mike to his nightmare scenario.

Cheatgrass is not the toughest, nastiest, most loathsome weed lurking in the ecological shadows of the Great Basin. It has some value as forage, it stabilizes soil, it's adaptable; until the big fires broke out, we managed to live with it, if unhappily. The horror is that the day might come when some future Aldo Leopold visits the Great Basin and finds the cheat suppressed, a more demonic invasive thriving, and the scene so far gone and the despair among residents so profound that they look back longingly to the era of cheatgrass, not perhaps as a golden age, but as one at least of golden hillsides that shone brighter than the biotic lead that succeeded it.

Not all newcomers are toxic. Some, Mike Pellant among them, naturalize successfully, and make their habitat the better for their presence. But all carry with them an evolutionary past that shapes their disposition to plant or plunder. William James noted that a "man's vision" is the great thing about him. A part of what makes the long- and multiserving coordinator of the nation's efforts to cope with cheat is the vision cultivated by a youth on his granddad's farm eagerly awaiting, with a patience born of hope, the new greenery that will arise from bare ground.

Deep Fire

For millions,
for hundreds of millions of years
there were fires. Fire after fire.

—Gary Snyder, "Wildfire News"

I N THE YEAR 02000 CE a fire—believed to have been kindled by lightning just before midnight on July 25—started on the south side of Mount Lincoln in the Snake Range.[1] Steep terrain and tricky winds, plus extensive fires elsewhere in Nevada, argued to monitor rather than suppress the fire, which reached an acre in size the next day. It grew slowly for several days, then absorbed larger chunks. By August 3 it had spread to 1,250 acres. By the 23rd it had dropped off the western cliffs of Mount Lincoln and entered the valley below. There it met a cache of dry combustibles and simmered before boiling up the slope of Mount Washington, billowing skyward in a vast plume that dwarfed the peaks. It overran the St. Lawrence Mine, purchased the year before by the Long Now Foundation as a potential site for a 10,000-year clock. It burst over the summit where it flash-burned a grove of ancient bristlecone pine. Seven incident management teams wrestled in sequence with the burn, the last departing on September 7, with the fire 90 percent contained.[2]

The clock and bristlecone move the Phillips Ranch fire from the realm of anecdote to apologue. It's impossible to look on those scorched sites and not meditate on fire and deep history.[3]

· · ·

The Snake Range in eastern Nevada is a good place to look. It contains four of the five tallest peaks in Nevada. But it is not the height of the peaks that matters:

it's their depth in time. That's true for its biota as well as its rock, and it's true for its peoples. It's a place Stewart Brand has called "timeless," but only because most humans think in terms of hours and weeks, and their digital machines in nanoseconds. We're a species for whom a lifespan of a century is an occasion for wonder. Even a millennium seems beyond our temporal reckoning.[4]

The larger narrative over the last 10 millennia is slippery because there are so many moving parts and they often move with extraordinary power and speed. The climate wavered and warmed, then cooled slightly. Glaciers shriveled to token cirques. Immense lakes—the Great Basin equivalent to subcontinental ice sheets, some 28 million acres in all—drained or evaporated. Half of ancient Lake Bonneville, originally the size of Lake Superior, mostly emptied over Red Rock Pass in the space of days. The flora scrambled across the new landscapes in the biotic equivalent of the mining rushes of the 19th century. Some 35 genera of mammals (mostly large: think mammoths) disappeared. The biota not only moved around and across basins but up and about the ranges that flank them.[5]

No less remarkable, through nearly all these wild upheavals, there were people present. The human saga is a story of movement because the Great Basin is a place not notably kind to fixed habitations and settled peoples. The earliest human narrative may date back 14,000 years. Clovis people arrived 1,000 years later. Others—how many is unknown—followed. The Fremont peoples probably migrated in from the Colorado Plateau as early as 00400 CE before they decamped wholesale by 01400. By then the ancestors of the present-day Shoshone, Goshute, and Paiute had reached the Great Basin.

Emblems of their appearance had preceded them, but Europeans arrived tangibly in the late 18th century CE. The pace quickened as Americans filtered into, then swarmed over, the Basin. The mountains then witnessed a historic cavalcade of peoples, moving in, passing through, departing, and more recently preserving. The markers of time have shortened, not only because more records exist but because the tempo of change has quickened, no longer keyed to the ponderous rhythms of climate but to the teleconnections of distant economies, beliefs, and institutions.

Most of the Snake Range was gazetted into the public domain in 01891 CE as a forest reserve, later reorganized into the Humboldt National Forest. The lower reaches of Mount Wheeler contain an elaborate subterranean cave system, Lehman Caves, named after Absalom Lehman, the rancher who discovered them and became their promoter, opening paying tours in 01885 CE. The caves became a national monument in 01923 under the jurisdiction of the

Forest Service. Ten years later responsibility was handed over to the National Park Service. In 01986 the monument and some additional lands, 77,000 acres in all, became Great Basin National Park.[6]

Throughout, there had to be fire—nature would see to that. Fires must have burned amid the forests that lapped between glaciers and pluvials, and then across grass, shrub, and woodlands that claimed lands vacated by ice and lake. Surely, there must have been some basal rhythms to burning, though the biota might still be quivering from all the massive shocks of the Pleistocene. Like the land under ice and water, which continues to flex upward in rebound once that burden was lifted, so the biota may yet be responding to the abrupt changes in that past climate, even as it must react to the sudden changes now underway. How many of the fires recorded over the past century are in some way legacy burns? How many are early-adopter harbingers?

But people also carried fire and put it on the land both purposefully and as an inevitable spoor of their presence; anthropogenic fire may be humanity's first camp follower. Migrating peoples added hunting, warring, trapping, mining, farming, and herding, and they introduced species, both deliberately and accidentally, including disease-spawning microbes. They rearranged the places and tempos of burning. The old rhythms must have found themselves syncopated, collated, isolated, quickened, dampened, and generally unhinged if not scrambled. A detailed study of fire scars and stand ages mumbled that "a distinct within-fire-shed contrast in fire frequency was difficult to explain without invoking the possibility of spatially-variable human-caused ignitions." In other words, people interacted, with fire, to reshape the natural scene. The study noted that a major inflection point—what for fire might be the equivalent of the onset of a glacial epoch—occurred in the 19th century CE.[7]

The fires that reached the ledgers of the 20th century CE are not likely those that characterized the presettlement era, though just when the first tendrils of European "settlement" arrived is a fraught topic. Trade goods, steel knives, iron pots, horses, and communicable illnesses like cholera, measles, and smallpox could precede and remake societies, and so remake landscapes, decades or even centuries before white Europeans officially recorded first contact. "Presettlement" makes an unstable baseline, related by unreliable narrators.

The fire history of the Snake Range is typical of that for the Great Basin's sky islands. An analysis of fire-scarred trees carries the story back to 01538 CE, two years before Francisco Coronado's expedition through the Southwest and 229 years before the Escalante-Dominguez expedition blazed the Old Spanish

Trail down the eastern Great Basin. Most scars, however, track to the mid-19th century CE. The majority occur in limber and ponderosa pine, the rest in Douglas fir, white fir, and pinyon pine. Low-severity fires burned on average every 11 years in ponderosa (2-to-22-year range), and 19 years in mixed-conifer (1–62 years). In pinyon-juniper, fire return intervals vary by the degree of grassy understory, but a conservative estimate is that fire occurred every 15–20 years.[8]

We know the most recent fires from formal records kept by the Forest Service and National Park Service. A large fire (200–300 acres) broke out at the time the Forest Service assumed control, then a few arson fires blackened 150–200 acres, then almost nothing. Lightning became the only ignition source, kindling an average of three small fires a year. The Park Service acquired control just as the West generally began to swing into drought and an era of larger fires. Most of its notable fires, however, began on private or bordering lands and burned into the park such as the Phillips Ranch fire of 02000 and the 164-acre Border fire in 02006. In August 02016 lightning started the Strawberry fire that blew up into 4,700 acres on the north foothills of Mount Wheeler and became notorious when a falling tree killed a Lolo hotshot. But over a century of detailed records roughly 5 percent of the park has burned—not much even for a mountain range in the Great Basin.[9]

Fires take their character from their contexts. The generally complex assemblages of forests on the Snake Range make for equally complex patterns of burning. The Snake Range is a sky island large enough to accommodate many refugia. Likely, it's a palimpsest of natural history, and if so, that applies to its fire history as well. How far today's regimes deviate from presettlement conditions is unclear, and perhaps unknowable. Fire's regimes must have been as varied and no more settled than the riotous movement of flora, fauna, and people. They remain unsettled today.

·   ·   ·

Bristlecones are the oldest known trees on Earth. The youngest old bristlecones match the age of the oldest sequoias. Not all bristlecones are ancient, but all of Earth's longest-lived conifers are bristlecones. The oldest, named Prometheus, dated 4,862 years—2,100 years older than Hesiod's eighth century BCE account of the mythic Prometheus in the *Theogony*.[10]

The oldest bristlecone groves in the Snake Range—one on Mount Wheeler, and a larger one on Mount Washington—date to almost 5,000 years; the oldest tree, to 5,600 years. They have flourished within two frames. One is climate—the warm altithermal that ended about 6,500 years ago, and the more recent warming of the past century. The other is geology. In the Great Basin plants grow in a broad zone between basin playa and range tree line, or between the salt left by Pleistocene pluvials and the ice left by glaciers. Bristlecones grow near the upper limit, near the cold. Mostly they grow in mixed association with other trees. They please the mind with their age; they please the eye with their sculptural forms, like flames petrified into wood. Their fire ecology is barely understood. The fire history of the Snake Range is not much better known.

The patriarchal bristlecones survive because they live in an ecological sanctuary so hostile to life that the rest of the biota including predatory beetles and blister rust, competitors and predators, and rival trees have been filtered out. They have little ground litter, and grow well apart, so there is scant occasion for fire. Bristlecones will thrive nicely where limber pine and Engelmann spruce do, but bristlecones in those habitats are unlikely to live to old age, which for Great Basin bristlecones is about a thousand years, much less achieve patriarchal status. They find sites where nothing else can survive. But also critical to their longevity, they seem not to senesce. They don't biologically age. Instead, the ancients die from cumulative physical insults, which for many ultimately means that erosion exposes and kills their roots. If their habitat is framed by salt playa and stone summit, so their life span is also set geologically not biologically.

Fires burn subalpine forests, and bristlecone with them. The trees can survive light-severity surface fires, but not much more, and so resemble a high elevation version of pinyon pine. The contrast with sequoias is striking. Old sequoias thrive in robust habitats; old bristlecones, in harsh, unforgiving ones. Redwoods live long because they accommodate fire. Bristlecones live long because they avoid it. Since the legacy groves burn rarely, their fire history is little known. But they do burn.

The Phillips Ranch fire stunned observers because it burned a large swathe through a grove of bristlecones. Such burns couldn't be common, or else the grove would not be old. Part of the explanation may, again, be the association with other species. For such fires not to have happened, fires could not have erupted up Mount Washington as they did in July–August 02000 CE for several thousand years. The bristlecone didn't change: its surroundings did. More savage fires were possible, most likely because regimes around it had been perturbed over the past century. A century is not long for a patriarchal bristlecone. But it's plenty of time to alter fire regimes downslope.

Equally astonishing, bristlecone appears to have regenerated preferentially in the burned area. How? Why? It's not known, but speculation suggests that once-occupied sites became available, and that the Clark's nutcracker was critical in distributing seeds. Perhaps, too, there was some genetic memory still coded from the times when bristlecone and limber pine dominated the flora of the Basin and had to accommodate fire regimes very different from those on Mount Washington and the quartzite summit of Mount Wheeler. Until ancient groves burned, research wasn't possible. Now, however macabrely, it is.

Still, the oldest bristlecones span half the length of the Holocene. They seemingly defy ecological time. They can't defy their geologic boundaries.

• • •

Fire raging forest or jungle,
giant lizards dashing away
big necks from the sea
looking out at the land in surprise—
fire after fire. Lightning strikes by the thousands, just like today.

The proposed Long Now clock would double the age of the oldest bristlecone. Ten thousand years back spans the Holocene, ten thousand ahead would presumably absorb the Anthropocene, or if the human hand shaping today's Earth proves too ephemeral, whatever geologic epoch will subsume it as the deep rhythms of the Snake Range would an afternoon's thunderstorm.

In 01999 CE the Long Now Foundation purchased 180 acres, including the St. Lawrence Mine, on the flank of Mount Washington with the intention of constructing a clock that would run for 10,000 years. It would serve as a tangible symbol to encourage thinking about the long term. Like ancient bristlecones it would survive because it would occupy a sheltered site, shielded from natural and human disturbances (fire among them). Within a year fire blew over the mountain.[11]

Ten thousand years is nothing for fire, which has thrived on Earth for longer than 42,000 times the length of the Holocene. Fire predates the genus *Homo* in all its avatars. It predates cheatgrass, sagebrush, and the Eocene origin of bristlecones. It predates the genus *Pinus*. It predates grasses and forbs. It predates beetles. It predates the evolutionary history of everything presently living on the Snake Range. It has outlived five glacial ages and five mass extinctions. It has outlived the creation of the range itself. It predates almost all of the rocks that make up the range.

It burned through the Pleistocene, while glaciers came and went, and talus, lacustrine gravels, and alluvium slid down the slopes. It burned through the late Tertiary conglomerates that eroded out and redeposited old Paleozoic rocks. It burned while lakes laid new deposits, and feldspar-rich and biotite-rich volcanics, rhyolites, and granites formed. It burned while the granites of the Cretaceous and Jurassic congealed. The Snake Range lacks any Triassic or Permian rocks, but fires scored trunks in Arizona's Petrified Forest that date from that era. The Permian was a time rich in oxygen, vegetation, and fires, probably the most burned era in Earth time. When the Ely limestone was laid down, earthly fires moved into uplands. When the Chainman, Joana, and Pilot shales were forming, fires burned in mires and the tropics. When the Simonson and Sevy dolomites began, fires first appeared, as plants, then forests, formed.

Yet, like bristlecone pine, fire is ultimately bounded by the deep-time evolution of life, which is to say, by geology. When the rocks that layer the lower Snake Range were quartzite, limestone, and shale, there were no fires. When the Osceola Argillite, Shingle Creek shale, and McCoy Group quartzites formed, there was only a thickening of oxygen in the atmosphere, a product of life in the oceans, the only life that existed. The geologic record goes back further, beyond even that border, but Earth at this stage in its history is indistinguishable from other dead worlds. Without life there cannot be oxygen, and without life on land, there is no fuel.

Fires. So many fires. Fires in glacial eras, fires in high oxygen eras, fires in eras of sparse fuels. Fires in ferns, fires in bryophytes, and fires in the first angiosperms. Fires when diplodocus browsed. Fires when ancestral mammals appeared. Fires in swamps that became the coal beds of the Pennsylvanian. Fires in prelapsarian savannas. Fires that burned as soon as life colonized land and have never stopped. Fires that are reburning the fossil charcoal of Pennsylvanian coal. Fires that will burn until terrestrial life vanishes, until planetary history erodes away its roots, like the dragons and serpents gnawing at the three roots of the Yggdrasil. So many fires. So many fire regimes. So many possible futures. So many choices for the Earth's keystone species for fire to make.

• • •

What is the proper fire regime for the Snake Range?

With its life-zone stratification and terrain-sculpted nooks, the Snake Range holds landscapes and niches, all with distinctive regimes, that stack vertically in short order what would otherwise span across hundreds or thousands of miles of geography. But that shortening of space is also a shortening of time. The mountain replicates the regimes of centuries, millennia, even epochs. The current regimen, which seems so out of whack, is a distortion of less than a century. Today's blowups are whispers compared to the fires that no doubt blew over what the Pleistocene's ice sheets and pluvial lakes didn't submerge. Today's most expansive megafire is barely a tick in the pyrochronometer of the mountains' deep time.

But we don't live in deep time—can barely comprehend its scope. We live in a present that spans a handful of years and can anticipate a future of perhaps a few decades. Amid the forecast upheavals of the Anthropocene, today's fires are a puff of snowflakes from a coming blizzard. Even the Long Now's admirable appeal to long-term thinking projects across 10 millennia, probably only 5 percent of humanity's past, and 0.004 percent of the Pleistocene. Compared to the antiquity of fire, that number becomes vanishingly small, an infinitesimal of pyrohistory.

Still, the fires of today are the fires we deal with. Most fire managers have working horizons of three years—this year, last year, next year. They must cope with the fires they face, not those of the Miocene or of a predicted post-Anthropocene epoch. "Think global, act local" also has its historical analogue: Think long term, act now. Fire rarely allows us the luxury of simply contemplating.

• • •

The myth of Prometheus tells how fire came to people, and so to Earth. Curiously, perhaps, there is not one myth but many. Hesiod tells one version. Plato tells one. Others tweaked the basics to account for some particular feature or other, or more correctly to illustrate some ethical or political point. They are less retold myths than reshaped parables. The story—or what by now may be the myth—of the Prometheus bristlecone pine also has many variants, but they tell how we should relate to that Earth, which inevitably will speak to how we should handle our unique firepower.

The basic elements are these: In the summer of 1964 Donald R. Currey, a graduate student from the University of North Carolina, wanted to use tree rings to estimate the age of glacial retreat on Mount Wheeler. He took cores with a Swedish increment borer. Bristlecones were understood to be old—this based primarily on studies from the White Mountains in western Nevada. But he stuck a borer—there may have been two—in the tree he labeled WPN-114, but which locals had earlier named Prometheus. He asked for permission to cut the tree, which would allow him to recover the borer, better date the rings in the complex tangle of trunks, and complete the science. The request went up the administrative channels of the Forest Service, then the responsible agency. The authorities agreed. A crew trekked up with saws, and when one member refused to cut, returned the next day, August 7, and cut several slabs. When the rings were finally counted, it was discovered that Prometheus had lived 4,872 years, and a recount suggested a number closer to 5,000, maybe 5,200. Prometheus was by far the then-oldest known (once) living thing. Protest over what many deemed an inexcusable act of vandalism boiled over.

For meaning, chronologies (which is what tree rings are) have to become narratives, which is to say, they have to acquire a theme to organize events over a designated span of time. No narrative is inevitable: each narrator can tell the story according to his own perspective and purposes. If the narrative is history, the narrator can't make up anything and can't leave out something whose omission would alter the outcome but otherwise is free to shape structure and meaning. If fiction, parts can be tweaked, added and omitted, for thematic clarity or aesthetic effect. If myth, fable, or allegory, the elements can assume a more abstract quality, in which nominal facts become symbols. What matters is that the resulting text is true to its genre and resonates with its readers. What complicates matters is that often even the supposed "facts" can be suspect. Facts, like fires, take their character from their context.

The Prometheus story has, like its referential myth, many versions. Michael Cohen has identified five "predominant" ones, which like root sprouts compete to become the primary trunk. Currey's version appeared in *Ecology*. His justification for cutting resided not only with the stuck borer (which could not be replaced easily), but for better access to the data, which the complicated growth habit of the tree made cumbersome, along with the need to finish his project. He was, after all, doing science. He concludes that "possibly no other living species presents such accessible long-term evidence relating to its biogeographic history and to the environmental histories of its sites," not only showing no remorse but appearing to invite others to do the same. Data is where you find it. Darwin Lambert, an advocate for transferring jurisdiction to the NPS, saw the Prometheus bristlecone as a martyr. Keith Trexler, an on-site witness and chief naturalist at the monument, interpreted the incident as reflecting the different values of two agencies. Galen Rowell, writing for the *Sierra Club Bulletin*, imagined a collusion between a commodity-blinded Forest Service and an instrumental science—the incident occurred, revealingly, the same year Congress passed the Wilderness Act. Charles Hitch, president emeritus of the University of California, also saw a clash of values, but argued for the free inquiry of scientists, even when errors can from time to time occur.[12]

The recounting of the events around the felling—the chronology, if you will—is secondary to the investment of meaning given them. In that sense the saga of the Prometheus tree remains a cautionary tale. We can hardly understand those events without their accompanying story; and this is no less true for fire history. How, then, can we tell the story of fire through deep time and know its place in our time? What is the purpose of fire history at all?

• • •

The value of placing fires in deep time is that it reminds us that fires change with their circumstances, that our understanding of fire also changes with circumstances, that only a few of those circumstances are under human control and people are unlikely to agree on them, that today's solutions can become tomorrow's problems, that there is no final resolution to fire. Fire will go on, with or without us.

Like the story of the Prometheus bristlecone, the saga of fire in the Snake Range, or for that matter fire on Earth, will have many narratives. More data will alter those stories, but data is just data, no matter how big or digital. Facts don't speak for themselves, are not even understood as facts without some context. Meaning requires a frame, whether as narrative, aesthetic sensibility, or ethical conviction. But even for natural phenomena, even for something on the scale of climate, those frames are not set simply by nature. Through their firepower, people are moving them. This is not to say that those processes are under human control, just that they are susceptible to human disruption. Worse, the same agents destabilizing them are the ones telling the stories to explain what is happening and what it means. At this point understanding starts to resemble a Möbius strip.

Our formal understanding of wildland fire is pitifully small. The first American scientific paper on fire ecology was published in 1910. The first Forest Service lab opened in 1959, five years before the Prometheus bristlecone was felled in the name of science, and 41 years before the Phillips Ranch fire flashed into the Mount Washington grove. Moreover, what modern science has studied are much-disturbed landscapes over short intervals of time. Most investigations last the one or two years allotted to graduate students. In truth, humans know a lot about fire—it's part of our species heritage, we've lived successfully with it for 200 millennia. But most of that traditional knowledge was lost in the chaos of American settlement and the various suppressions characteristic of fossil fuel–powered industrialization. What we have today are fragments, like the shards unearthed at a Fremont pit house. We hardly know the fire history of the Snake Range, and that for only a century. We don't really know much. We won't catch up. We won't get ahead. And the world we study is changing rapidly. We'll never know enough.

What caused the Phillips Ranch fire to burn into bristlecone? Was it just a chance event, a 5,000:1 bet that beat the house odds? Or was it the result of land-use changes, including recovered woods after mining and fire exclusion over the past century? Or did it happen because, thanks to humanity's combustion of fossil fuels, the climate inflected in ways that added punch to the fire, like an atlatl leveraging a spear? Those geologic borders that once hard framed the Prometheus bristlecone have become malleable under the fussy hands of humanity; and in the end, people even felled the tree, and then offered explanations for what they had done. Which narrative we choose matters because we will base future decisions on that understanding, and that choice will enable or shut down other choices to come. It's easier to break than to build. It's difficult to create what we wish for but simple to disrupt the present or derail the future.

Those most passionate about global change, and the Four Horsemen of the Anthropocene, argue for a no-analogue future in which the past provides no prescriptions. Even the Long Now clock will record the time to come, not the time passed. In truth, a knowledge of fire history won't tell us what to do now. Its deepest purpose is different. Technology can enable but not advise; science can advise but not choose. The value in an appreciation of deep time is that it can counsel us how to make better choices by means of narratives that speak to virtues such as prudence, humility, compassion, courage, and grace under pressure, character attributes that might well be described as timeless. An appreciation

for deep fire history won't tell us with mathematical rigor how to make landscapes more resilient. It might help prevent us from cutting down the future in
order to retrieve today's increment borers.

> *I have to slow down my mind,*
> *slow down my mind*
> *Rome was built in a day.*

Strip Trip

T COULD STAND AS a dictionary definition of hiding in plain sight. At 7,900 square miles the Arizona Strip is nearly as large as Connecticut and Delaware combined. It contains nine wilderness areas, three national monuments, a national forest, an Indian reservation (Kaibab-Paiute), parts of two national recreation areas, and a patchwork of public and private lands. It abuts two of the most celebrated tourist sites in the country—the Grand Canyon and Las Vegas. Yet it might as well be on a Jovian moon.

Geologically, it contains the last of the High Plateaus running south from Utah. There are four of them, stepping down from east to west. The highest, most famous, most accessible, and least typical of the Strip is the shallow dome of the Kaibab, whose east and south flanks were sculpted into the Grand Canyon. To its west—what is understood as the real Strip—lie the broad, somnolent Kanab and the narrow, violent Uinkaret plateaus. The most western and forlorn is the Shivwits, wide and featureless, whose canyon gorge is a series of terraces, fit mostly for cattle, the occasional mine, and dead-end roads.

Here history and geography fuse. The Grand Canyon diverted or dammed settlement. The southward flow of colonizing from Mormon Deseret spilled to the more passable corridors to its sides, crossing the Colorado River to the east at Lee's Ferry near the Arizona-Utah border or streaming down the Old Spanish Trail along Wasatch Front to Pearce's Ferry on the west. The Nevada-Utah corner is further isolated by the Virgin River gorge. The Strip became a slow historical eddy filled with transhumant cattle (or later on the Kaibab, transhumant tourists) and a few postfrontier homesteaders.

What made a marginal landscape into the Arizona Strip were the unintended consequences of a political process that had severed the land from Utah and Nevada and given it to Arizona, which in turn found it split between two counties. Two-thirds of the Strip belong to Mohave County, with its seat in Kingman, and the remainder within Coconino County, seated in Flagstaff. Grand Canyon National Park and the Kaibab National Forest, each of which have large holdings north of the river, have their headquarters south of the Strip. The BLM administers its lands from St. George, Utah. The Strip is a political no-man's-land, the Empty Quarter of America's Empty Quarter. For a long time it was best known for the polygamous communities that persisted in remote indifference from any reach of law. When a group of brownshirts in cowboy hats staged a Sagebrush putsch at Malheur National Wildlife Refuge for 40 days in 2016, their leaders, Ammon Bundy and LaVoy Finicum, hailed from the Strip. The Strip is so isolated that it became a site for reintroducing California condors.

For the American fire community the Strip distills the lesson that significance is more than a matter of size or celebrity. Much as the remoteness of the Mogollon Mountains invited the first primitive area, the model for wilderness, so the isolation of the Strip prompted a national experiment in landscape-scale fire restoration. For a while the core woodlands of the Arizona Strip, the Area 51 of fire management, became the epicenter of a national debate about rehabilitating fire-famished forests. For a decade it was as much a ground zero for field testing the tools of wildland fire management as the Nevada Test Site was for nukes.

•   •   •

That happened because the Strip was a cameo of fire history in the American West, though one with historical and ecological quirks of its own, and one that came with layers of isolation that, paradoxically, could entice a program of active experimentation.

The lower shelf-lands, broad and elevated above the norm, were grassy and shrubby, grading in higher reaches into pinyon and juniper woodlands. The critical site was the Trumbull Range, a chain of volcanic cones, piles, and peaks that flanked the eastern fault line of the Uinkaret Plateau. In Paiute *Uinkaret* means "place of pines." These were ponderosas that grew in classic southwest fashion—huge, clumpy, a score or two per acre amid grasses. Even before the pine arrived, roughly in sync with the modern climate 6,000–8,000 years ago, people lived here. Likely they moved seasonally as water and favorable habitats became available. They dug irrigation canals and farmed Toroweap Valley before abandoning the enterprise in the 13th century. Lightning, torch, grass, brush, and forest coevolved.

By the time American settlers arrived in the mid-19th century, Trumbull's fire history, as recorded in scarred boles, spoke of routine fire every 4.5 years and widespread fires every 9.5 years. These are minimum values, but they correspond nicely with chronicles throughout the Greater Southwest. Then that pyric saga ended. The last outbreak occurred in 1863.

Colonizing Americans had arrived, along with sheep and cattle, axes and sawmills. While they often followed similar seasonal rhythms, they introduced new elements and had novel outcomes. Instead of harvesting pinyon nuts, they harvested whole ponderosa pines, and instead of hunting deer and rabbits, they loosed livestock. Instead of adjusting settlement patterns to the cadences of climate, they doubled down to compensate. The result was an ecological shock-wave. By 1870 landscape burning ceased to be a routine feature of the biota. Save for the occasional outbreak of a few dozen acres, fires disappeared. Instead of stoking flames, combustibles fed livestock and steam engines.

What followed happened all over the arid West, though it was told here with a local twang. The luscious range crashed, eaten out and trampled. By 1880 Captain Clarence Dutton, then detached for duty with the U.S. Geological Survey, viewed the panorama from Pipe Spring, ranching's point of entry. "Ten years ago the desert spaces outspreading to the southward were covered with abundant grasses, affording rich pasturage to horses and cattle. Today hardly a blade of grass is to be found within ten miles of the spring, unless upon the crags and mesas of the Vermilion Cliffs behind it. The horses and cattle have disappeared, and the bones of many of the latter are bleached upon the plains in front of it." The close cropping might have been enough by itself, "as has been the case very generally throughout Utah and Nevada," he argued, but a decade-long drought had also settled on the scene like a down blanket, smothering everything beneath it, and the double blows were too much for the native flora to survive.[1]

The assault paused, then continued. Cattle, horses, and sheep from Utah; then Texas longhorns; then tramp flocks; decade after decade, another grazer appeared, taking what water and forage had survived, like waves of prospectors sieving through the slag heaps of earlier booms. In May 1933 a party from Grand

Canyon National Park traveled to Toroweap. They found that the landscape was "badly over-grazed, erosion is severe, and loco weed and other undesirable indicators are present." The next year the Taylor Grazing Act was passed, closing the public domain to further entry. An estimated 100,000 cattle were still on what remained of the range, along with an unknown number of sheep and feral horses. The Grazing Service created by the act brought some order to the pandemonium and reduced livestock numbers. In 1946 the Grazing Service was corralled with other Department of the Interior strays into a Bureau of Land Management. Today, 117 ranchers graze 15,000 cattle under permit.[2]

In 1879 a dairy operation located at Trumbull. But herders generally were moving their flocks to the higher forests during the summer, and there they met loggers. The Trumbull Range timber was good—a small mill was in place in 1870. The problem was, as always, isolation, the long haul to market. That objection vanished when Mormons began construction of the St. George Temple, found the Trumbull timber suitable, and built a Temple Trail to haul the wood by ox team. A steam sawmill speeded the process. When the Temple was dedicated in 1877, the Mormons sold out. The felling and milling went on. What was happening in the range was occurring in the woods as well. If the havoc was less severe, it was because the U.S. Forest Service imposed some institutional order after 1905.[3]

This was the third leg of the land-history stool, federal management, which added formal fire protection at least in principle. It meant that fires big enough to be seen would be fought, but more, perhaps, it meant that the ancient legacy of human burning would cease. The Strip's peculiar isolation, however, made its lands awkward to actively manage. Whatever the agency or administering unit, the Trumbull Range in particular remained an outlier, handed around an extended family of federal land bureaus like an unwanted orphan. The agencies seemed to change tenure as often as ranchers swapped spreads.

The Trumbull Range and Mount Dellenbaugh (an outlier of an outlier, located on the Shivwits) were incorporated into the Dixie Forest Reserve in 1904. The Wilson Administration dropped the Parashant (Dellenbaugh) division from the national forest system. In 1924 the Mount Trumbull holdings were transferred to the Kaibab National Forest. Ten years later Grand Canyon National Monument was proclaimed for the lower reaches of the range and the Toroweap Valley where it empties into the Grand Canyon proper. In 1974 the Kaibab transferred the Trumbull unit to the Bureau of Land Management. Mount Trumbull and Mount Logan wilderness areas were gazetted in 1985. In

2000 the expansive Parashant-Grand Canyon National Monument swept up the mountains into its domain, acting as a kind of Taylor Grazing Act to support an environmental ethos appropriate for a service and amenities economy.

The response of the woods was not, like the early range, to shrivel into dust but to morph into a thickening scrub. The forest overgrew, in the manner typical of the region. Instead of open glades, brush and dog-hair thickets flourished, and instead of the classic colonnades of old-growth yellow pine, forests became the woody "jungles" that tracked the failing health of Southwest ponderosa. Average tree density went from 22 stems per acre in 1870 to 349 by 2000. The exception was the rumpled summit of Mount Trumbull itself, which had resisted the logging technology of the early days and still held significant numbers of old-growth yellow pine.[4]

Meanwhile sparse local population increasingly slipped off to the towns and ran cattle as a side business, though in full costume. At times the Strip resembled a reenactment event, recycled from the 19th century.

• • •

Fire as an integral part of the Strip was gone. Lightning still kindled snags, and a few fires reached a few tens of acres, but they appeared as someone might stumble upon a basalt arrowhead or the wall of an Ancestral Puebloan ruin. They were relics of a former era. The ancient yellow pine ceased to record any further scars. When two "severe" wildfires broke out in 1996, one was 20 acres and other 150. For a long time the absence of routine fire had gone unnoticed—fire's removal had been policy, after all. The consequences of fire exclusion elsewhere in the West sparked concern, however, and then alarm, and finally inspired a famous experiment under the direction of Northern Arizona University professor Wally Covington on the Gus Pearson Research Natural Area outside Flagstaff, Arizona, that sought to explore what it would take to restore fire in the pines.

Quite a lot, it seemed, primarily by mass thinning of the small-diameter intruders, followed by prescribed burning, and then probably by seeding to leverage the native grasses. But before the Flagstaff model, as it came to be called, could leap from its tiny prototype plots on the Gus Pearson, it needed a major field trial. It needed experiments not only to sharpen the science but to sculpt the protocols for operations: it needed a full-spectrum enterprise in how science and management might coevolve the theory and practice of ecological restoration on a significant scale. In 1995 Secretary of the Interior Bruce Babbitt, who had grown up on a Flagstaff ranch, and Arizona's senators, particularly Jon Kyle, met with Covington, and all eyes turned to the Trumbull Range as an ideal locale. The catalytic 1994 fire season (which had savaged areas around Flagstaff) spiked the discussions with a sense of urgency. The upshot was the Mount Trumbull Ecosystem Restoration Project in 1996, sponsored by the BLM in coordination with the Ecological Restoration Institute at Northern Arizona University and the Arizona Game and Fish Department.

The Trumbull Project was big science for its place and time. It sponsored a full-spectrum ecological CAT scan and laid out systematic plots for field trials primarily in ponderosa pine, but also in pinyon-juniper and cheatgrass, all with monitoring and control sites for various combinations of treatments, all at a landscape scale. For a restoration point the project selected 1870, when the long rhythms of fire had abruptly ended. Comparative studies were initiated at Beaver Creek Biosphere Reserve, and Grand Canyon's North Rim, both in Arizona, and at the Sierra Tarahumara and the La Michilia Biosphere Reserve, Chichuahua and Durango, Mexico, respectively. The Arizona Game and Fish Department signed on as a collaborator and added faunal studies.[5]

An environmental assessment began immediately. The scientific inventorying, and the site preparations for treatments, occurred mostly between 1997 and 1998. By 1999 the first data were in and reports were issued. Second entry treatments were underway by 2001. These proceeded more slowly due to costs, changes in national administrations, and adjustments based on early results. From the onset the project's design had insisted on "adaptive management" in which the original prescriptions, and those shaped by the research, would be modified as field conditions warranted. Experiments were field trials; field trials were experiments; they both coevolved. On some topics surprises were technical: restored fires did not behave as forecast. On other topics the snags were cultural and political, conveying disagreements over means and ends. A final report was issued in 2006. But some research persists, and updates continue to appear.

Where had the rubber hit the road and gained traction, and where had the wheels just spun in place? The surprises were many, but the designers had expected to be surprised. *The number, size, and arrangement of "replacement trees" that would re-create 1870 conditions had to be reevaluated.* Critics wanted more trees. *Gambel oak thinnings were scrapped.* The copses were valuable wildlife habitat, and their contributions to fuels around old-growth pines could be compensated for in other ways. *Control plots were expanded to accommodate birds. Cuttings burned too hot,* but smashing the slash, which reduced fire intensity, threatened soil compaction. *Native grasses, shrubs, and forbs did not return after slashing and burning, which led to reseeding, which failed,* until the treated areas were dragged to cover the seeds. What had seemed transparently obvious often turned murky. The "challenges" were endless. Insufficient funding and staff. Reliance on third parties for cutting and commercial use. "Excessive mortality" of yellow pines and Gambel oak from treatments. Continued grazing. Lack of native seed sources. Labor-intensive and expensive National Environmental Policy Act analysis. Complexities about maintaining every habitat for wildlife while doing the treatments. And that old incubus: "Addressing concerns of a variety of constituents and avoiding litigation."[6]

But the most vexing was the failure of restored fire to do what it was expected to. Prescription windows were too narrow, the lopped-and-left slash could burn too intensely, the fires encouraged cheatgrass, and despite raking around their bases more yellow pines died than anticipated. Surely, the millennial drought of 1999–2000 aggravated conditions, but protecting the old growth was the practical and ethical justification for the program. The point of restoration was to make those biotas more resilient and accepting of fire, which in turn would catalyze other vital processes. More troubling, perhaps, was the fact that the landscape—and this was an experiment in landscape-scale treatments—was a shambles of the treated and the untreated, a crazy quilt of retouched, untouched, and untouchable fuels. Large as the project was, it was not large enough to produce the protective buffers needed. Still, the program adapted, prepared for a second entry, and planned for larger plots.[7]

To the prime movers the results were good enough and the need to expand dire. Wally Covington urged that the BLM extend the treatments into the Mount Trumbull Wilderness to save its old-growth ponderosas. Here, one great idea met another. Restoration confronted wilderness, and wilderness won. Restoration is flexible, adaptive, learning by doing. Wilderness is nonnegotiable. It isn't based on experimental science or empirical knowledge from practice. It celebrates the wild, which transcends museum relics like Trumbull's many-centuries-old yellow pines. Restoration is based on the belief that we can use our science to intervene constructively. Wilderness ideology considers our science another potential form of meddling. Besides, the Mount Trumbull Wilderness had not been unhinged by logging and still retained many of its native grasses. Let nature sort it out.

Instead, the lessons went into the Healthy Forests Restoration Act and other programs for hazardous fuel abatements, and eventually the Collaborative

Forest Landscape Restoration Program. They did not, paradoxically, expand into nonwilderness areas of the Trumbull Range. In 2000 much of the Strip, including the Trumbull Range not in Grand Canyon National Park, was put into a Parashant-Grand Canyon National Monument, which introduced still other values and points of friction. A phase three treatment plan was approved in 2008 and hovers in the wings. A Uinkaret environmental assessment, scheduled for completion in July 2016, targets 128,000 acres, of which 30,000 would be treated mechanically and 18,000 by prescribed fire. But little funding is anticipated even if operational plans based on the environmental assessment pass muster. What made the restoration experiment possible at Trumbull, the Strip's isolation, is also what makes continued restoration implausible.

Mostly the Trumbull forests will fall under national fire policy and practices, with local adjustments based on that remarkable era of the restoration experiment. Most probably something like managed wildfire will replace prescribed treatments. Between July 23 and August 10, 2012, lightning kindled three fires on Mount Trumbull. They were handled as a complex under a strategy of "resource benefit" previously known as wildland fire use. It was a wet monsoon, but the deep duff kept the fires smoldering. For a while crews raked around the bases of yellow pines to protect them from root scorch, and a few patches of pine torched, but then the crews backed off. The fires burned into December. Mortality was within historic ranges. Landscape patchiness returned. Bark beetles came, but so did the three-toed woodpecker, long absent from the unburned Strip.

What of the future? The Ecological Restoration Institute wants a 20-year review. It would be an excellent capstone to an extraordinary adventure in American fire and a confirmation of the belief that if restoration could happen anywhere, it should be able to happen here. Whether that science might inform a new era of practice and policy is less certain. The opportunity to apply such principles in big projects—even on the modest scale of Mount Trumbull—may have passed, or at least passed by the Strip. It may be the best lessons get absorbed into larger agendas. It may be that a bold concept came and went as others on the Strip have before it.

• • •

In 1880 with Kanab as a reference point, Captain Clarence Dutton, along with William Henry Holmes, the expedition's artist, made a series of treks to understand the geology of the High Plateaus and their complementary Grand

Canyon. From the Aquarius Plateau he pondered the limitless horizon and the unimaginable scale of erosion that had stripped away the Mesozoic into the Great Rock Staircase. At Zion he meditated on the fabulous sculpturing of cliffs. Then they went to Trumbull and Toroweap. They ended on the Kaibab at a narrowing peninsula Dutton named Point Sublime.

Trumbull and Toroweap offered two geologic lessons, one on volcanism, one on fluvial erosion. The two converged at the Canyon's rim, where the Uinkaret fault that traced the Toroweap Valley met a cinder cone and basalt dikes that spilled into a Canyon gorge that was itself the final incision of serial erosion. The tableau marked an extraordinary conjunction. "Apart from merely scenic effects," Dutton asserted, "it would be hard to find anywhere in the world a spot presenting so much material for the contemplation of the geologist." From a perch on Vulcan's Throne, Holmes drew one of the greatest scenes of Canyon art, *The Grand Cañon at the Foot of the Toroweap—Looking East*, while Dutton contemplated how to make sense out of the unprecedented scenery before him.[8]

His key point was that all the facts he had collected needed to be assembled if they were to achieve meaning. The whole—the "grand *ensemble* [italics in original]"—was what mattered. So for his monograph as a whole, Toroweap's dramatic features would become one theme of many synthesized at Point Sublime, and it was there, in the form of a triptych, that Holmes created the greatest single work of Canyon art, *The Panorama from Point Sublime*. Dutton's science has dated; Holmes' art has endured. So likewise the Trumbull science may expire, while the idea that inspired it may survive.

Such may well be the fate of the Mount Trumbull Restoration project. Like Toroweap for Dutton, it marks the intersection of large forces—in this case, of ecological science, politics, and wildland fire. It inspired a great work of artifice, perhaps unrivalled for its era. And it may end up not as a centerpiece but as a vital contributor to an ensemble of experiments as Americans struggle to live with fire. Like the politics that ultimately created it, it was an exercise in the art of the possible.

It's worth recalling that great innovations take time. Scanning the scene at Toroweap overlook, Dutton noted that "the human mind itself is of small capacity and receives its impressions slowly, by labored processes of comparison. So, too, at the brink of the chasm, there comes at first a feeling of disappointment; it does not seem so grand as we expected. At length we strive to make comparisons." So exalted were the hopes for the Trumbull project that a passing retrospective might leave a similar deposit of disappointment. We need time, we need comparisons. It took 37 years to translate Dutton's text and Holmes's art into national park status for the Canyon, and 52 years for Toroweap to acquire standing as a national monument. It may take decades before the full ripple and rebound effects of the Trumbull experiment make themselves known.[9]

For now, the Trumbull project may be best understood as part of suite, not sufficient by itself, but not derived from anyplace else. It's the Strip Trip of American fire. It was, in its way, a pragmatic vision quest to a difficult place that is hard to reach and perhaps even harder to know.

Mesa Negra

BETWEEN 2000 AND 2003, on the imposing cuesta overlooking the Mancos River, wildfires burned 39,178 acres, or 70 percent of what had burned over the past century. The 2000 Bircher fire alone scorched 23,607 acres. Those burns amounted to 75 percent of the total land under management. The fires menaced structures. They threatened prized cultural and biological assets. They prompted mass evacuations over a single, lengthy egress road. They defied normal suppression methods.

So far, so typical. These were the years in which a millennial drought blasted the greater Southwest, in which fires lit and fires fought escaped control, and in which the American fire community struggled to double down on efforts to restore fire and rekindle ecosystem health. A casual observer, even a knowledgeable pundit, might regard the great cuesta as synecdoche for the American West. The power of abstraction might toss this fire scene into the growing lump of lands with too much bad fire and too little good.

But a fire is a specific event in a particular place, and here the consequences of abstracting and amalgamating disguise what makes the massif trending into the Mancos significant. The grand cuesta is best known as Mesa Verde, and it holds one of the largest and most significant collections of Ancestral Puebloan ruins anywhere. Multifloor structures like Cliff Palace and Far View Tower are its WUI. Line construction must thread among lithic mounds, surface pit houses, and terraced fields—the analogues of threatened species and sensitive habitats. The combustibles powering most mesa conflagrations are pinyon-juniper woodlands, many of the landscape patches ancient—biotic relics as valued by ecologists as the stone walls and pot shards are by archaeologists. The usual treatments, thinning and prescribed burning, will likely damage old-growth trees and invite regime-changing invasives such as cheatgrass. The only management response is suppression, and suppression of a sort that must plan for the once-a-decade-or-two blowup.

In fire as in other matters, Mesa Verde harks back to an earlier era, and it reminds us that fire management varies place by place, time by time. Fire regimes, like civilizations, can rise and fall.[1]

• • •

Mesa Verde is not known as a fire park—not a place that routinely burns, that hosts a distinctive fire culture or fire species like giant sequoia or longleaf pine, that year by year commands national attention for its fire scene. The dominant vegetation is notorious for either not burning or blowing up. Most fires are one-day wonders, or as the saying goes, they either burn a tree or a mesa. Partly that reflects efficient fire suppression that has eliminated the middle class of out-breaks. The cuesta is, in fact, littered with the legacy of 10-to-100-acre patches of previous burns. The blowups of the modern era run 2,500–4,500 acres. At 23,607 the Bircher burn stands alone.[2]

It's hard to support a vital program, largely staffed by seasonals, when most fire personnel may never see but a small fire or two during their tenure. Human-caused fires are nil (access to the backcountry is prohibited); ignition depends on lightning's lottery. When the park was established in 1906, fire was not per-ceived as a problem. Fire records only begin in 1926. An effective infrastructure commenced with the Civilian Conservation Corps, whose buildings still house the park fire cache and fire management office. That year marked the first of the modern blowups, two fires that splashed over from the Ute Mountain Res-ervation border. In 1959 the Morefield fire burned 2,500 acres.

Then the 1972 Moccasin Mesa fire scoured out 2,680 acres while a pha-lanx of bulldozers cut a fireline six blades wide. Whatever damage the fire might have wrought was negligible compared to dozer lines that ran relentlessly through surface ruins. The Park Service moved to change fireline practice to avoid a repeat. (During the 1977 La Mesa fire in Bandelier National Monument, archaeologists accompanied line scouts in what has become standard protocol.)

But the fire itself fell within historic bounds. Despite occurring at the climax of a revolution in fire policy, the issue was never about fire's possible restoration—it wasn't wanted—but the style of suppression. In 1996 the Chapin 5 fire seemed to mark a phase change in the character of landscape burns. Then came the Bircher and Pony fires of 2000, the Long Mesa fire of 2002, the Balcony House Complex in 2003. The blowups were beginning to burn into one another. Suppression was beefed up with better aerial attack. If the fire scene thereafter calmed, it was partly from exhaustion.

These were the kind of high-publicity burns that set many parks and forests to reevaluate and repurpose their programs. Places with wilderness or undeveloped backcountry began to back off and allow fire more room. Places suffering from ecosystem degradation sought to restore fire by prescription burning. Places hamstrung, and hammered, by the presence of urban and exurban sprawl sought to shift the burden of protection from land management agencies to communities and homeowners. And a few places, most spectacularly Southern California, were simply, admittedly, unmanageable and had no future prospect of a sustainable fire program. They could only bulk up their suppression capabilities.

Mesa Verde was too small to tolerate free-burning fire, which in any event would ramble for a day or a few (the Bircher blowup lasted three days). It was hobbled in prescribed burning because surface burns damaged its fire-sensitive woodlands and invited invasives. It could not negotiate with county commissioners and homeowners because they had decamped in the 13th century. Any mechanical fuel treatments met with resistance: locals didn't like it because it exposed their residences to public view, archaeologists didn't like it because it exposed surface sites and might encourage visitors to walk through them, and environmentalists didn't like it because the disturbed sites could siphon in invasives and might damage old-growth pinyon and especially juniper. Still, some clearing and cleanup around developed areas and exhibits occurred during the 1990s. The treated areas survived the blowups while everything around them burned to white sticks and ash.[3]

Meanwhile, the firescape was worsening. The cuesta dipped southward and was gouged into north-south trending flutes and mesas, which aligned nicely with prevailing winds. In extreme times they became raceways and flumes to carry flame. And although a pinyon-juniper woodland might burn in stand-replacing immolation every 400 years or so, and the petran chaparral characteristic of the northern cuesta every 100 years, the lack of middle-range fires was allowing a more homogeneous cover to blanket the rims and canyons. The upshot is that the Mesa Verde fire program is left with minimal treatments to counter their maximum fires. They have to rely on rapid detection and initial attack. A helitack crew has the capability to land within a quarter mile of any start. A single-engine air tanker is stationed at Cortez. An air tanker base lies ready at Durango.[4]

Even so, minimum-impact suppression in a World Heritage Site might require an on-ground archaeologist to advise on tactics. So dense is the record of those earlier peoples, so thronged is the surface with lithics, structures, shards, and even refuse pits, that anything a fire crew might do will impact something of apparent cultural significance. Kick a stone and it might be part of a pit house. Scrape a shovel line through dirt and you will probably disturb a pottery shard, an arrowhead, or a bone scraper. It's as though, in a typical park, every plant and animal was a threatened or endangered species. Given Mesa Verde's core mission, one can understand why the park might wish the fire crew and all its apparatus away. But banishing a fire crew will not banish the fires. Catching snag fires before they can spread is the least intrusive approach.

Suppression, however, is only one of a series of fire-archaeology encounters. Cultural resource management, too, has its good fires and bad fires. As happened at Custer Battlefield in 1984, fires of this severity can both damage exposed relics and unveil new ones. After the Chapin 5 fire, survey teams discovered 372 previously unknown cultural sites—this in one of the most intensively studied archaeological sanctuaries in the world. It is estimated that another 2,000 sites may emerge from resurveys of the Bircher and Pony fires. Some damage to existing sites did occur, mostly by spalling, but fires had undoubtedly rolled over these sites many times before the park began to document them.[5]

• • •

Modern Mesa Verdeans put their fire into machines. Ancient ones put theirs into hearths. And there, in the layered ash, is where the record of landscape fire is recorded as the composition of species used for fuelwood shifts over time. At first it was pine and juniper, and then oak and cliffrose, and then whatever shrubs and combustibles could be found. Undoubtedly people used fire for hunting and foraging outside their main settlements, but as those settlements thickened, the reach of firewood harvesting, and then of scavenging, expanded. The land was being picked clean. Landscape fires thinned, and then shrank into the hearths of kivas and kilns. In the later, American Southwest, landscape flame disappeared down the gullets of sheep and cattle. In Anasazi times they vanished into domestic appliances.[6]

The scene recalls the story of early colonists in New England who were asked by the natives if they had left their old homes because they had exhausted their reserves of firewood. In swidden agriculture the village would be moved routinely, not only to let the fields go fallow for a few years (or decades) but because the fuelwood needed for domestic uses was exhausted. The great dwellings at Mesa Verde, however, could not be packed up like tepees or rebuilt like wickiups. They rose higher and wider, and they took more and more fire out of the land and put it inside their stone edifices.

There are ruins in Long Mesa, known as Fire Temple, where fire was believed continuously maintained, likely a public utility from where new fires could be kindled that may over time have become a sacred site as well. But the deeper story may be that it indirectly speaks to the narrative of landscape fire because those once-ranging fires that searched out fresh fuels were here domesticated and sustained by bringing fuels to them. Unlike landscape fires that burned sporadically and by seasons, this one burned continuously.

Only when that combustion pressure lifted, only after the population abruptly left, did free-burning flames begin to reclaim the cuesta. Eerily, it began when the abandoned villages were fired, apparently from the inside, whether as part of a ritual cleansing by those leaving or as fire-and-sword destruction by those who replaced them. At 700–800 years the oldest junipers known at Mesa Verde may very well date to the time of abandonment. More date from the 17th century when regional replacement populations crashed following the introduction of Old World diseases. The rest took root during the more developed era of heavy grazing that began in the late 19th and early 20th centuries. As people and livestock left, however, the woods not only aged but thickened, and fires burned out patches from a single tree to a few dozen acres. Then, before landscape fire could reestablish a quasi-natural regime, modern fire protection arrived and big burns defined the prevailing scene.[7]

• • •

The past is indeed a foreign country. But so, too often, is the present. Geologists are fond of asserting that the present is the key to the past. But for historians, and any strays from the fire community fascinated by Mesa Verde, it's the past that is key to the present because along with Spruce Tree House and Long

House, along with one of the great concentrations of archaeological wealth in North America, field schools that trained several generations of archaeologists, and contributions to America's understanding of its *longue durée* past, the Mesa Verdeans bequeathed today's park fire policy and protocols.

The people who abandoned Mesa Verde left two legacies. One is the built landscape that attracts tourists, archaeologists, and the National Park Service. The other is the recovered natural landscape, which led to the modern firescape. The two continue to interact, but moderns are not allowed to do with their surroundings what the Ancestral Puebloans did, which is to thin and clean up so thoroughly that surface fires became implausible and crown fires impossible. By living on the land, the Verdeans used it, and by using it they kept wildfires at bay. Modern Verdeans don't live on that land, don't use it except as a source of information, and can't manipulate or transfer its fire regimes. Despite their industrial firepower, they can only react.

What remains is a quirky WUI. Merritt Island National Wildlife Refuge, where fire managers have to cope with a fire-thirsty landscape and a fire-intolerant NASA, likes to call its setting a wildland-galactic interface. Mesa Verde has a wildland-antiquity interface. Most fire interfaces trace borders across space. Mesa Verde's intermixed landscape runs across time. This is a WUI that was laid down a millennium before. It is more protected than a legal wilderness. There is no chance to change the conditions that support the current scene. The surface woods and chaparral can't be removed, and the cultural sites can't be relocated or hardened. The effort to manage backcountry wildlands in a place dedicated to protecting cultural relics is awkward at best, and at worst ruinous. The fire program at Bandelier National Monument tried both, and in 2000 the Cerro Grande fire showed what can happen if conditions go south.

The only response—no one seriously proposes that it's a solution—is suppression. The built landscape must be shielded. But so, also, surprisingly, must the quasi-natural landscape. Researchers have argued that thinning and burning threaten the fire-sensitive pinyon-juniper woodlands, that prescribed fire will harm the trees and invite invasives, that old-growth forests deserve as much protection as cliff dwellings. The woods and the ruins are both relics. Neither has a place for open flame. Fire happens, but over such long expanses of time that it is not part of the routine creative destruction endemic to nature's economy.

What the Mesa Verde fire scene most resembles is Southern California where partisans of the indigenous vegetation and of old-growth chaparral join urban fire services in arguing for aggressive suppression. They know they cannot halt all fires, and the ones that erupt will be huge and costly. But so long as urban sprawl pushes against the land, there is little opportunity for a middle range of burning. So they argue for full-bore suppression while knowing that they will have to accept the occasional conflagration. In the South Coast that defining sprawl occurred in the postwar era. At Mesa Verde it was created in the 11th and 12th centuries. Fire officers in both locales project their anticipated fire behavior by using heavy brush models. The Mesa Verde cuesta and the Hollywood Hills make strange bedfellows.

The best preparedness strategy is those big burns. They weren't wanted, but having swept over mesa and canyon, they offer buffers against further blow-ups. With fuel cycles of perhaps hundreds of years, the plague of big burns has inoculated most of Mesa Verde for some time to come. The park has even incorporated them into its interpretive programs with signs and exhibits. They are, after all, the largest fraction of what its road-cruising visitors will see.

•   •   •

In the popular imagination the fire revolution of the 1960s was about restoring fire and living with fire. But the essence of the revolution's philosophy was to align fire management with land management. This proved tougher than antic-ipated because Americans frequently could not agree on what they wanted their public estate to be, or often even agree on how to talk about agreeing. But it was also based on an implicit fallacy that we could, after agreeing, craft fire to suit our ambitions. Nature, however, has its own reasons and fire its own logic. The inevitable upshot has been a series of compromises.

In Southern California it has meant substituting internal combustion for open burning, and where free-burning fire exists it comes as eruptive wildfire. In Mesa Verde it means minimum-impact suppression to prevent fires from leaping from struck snags to horizon-filling mesas, but otherwise accepting the inevitable blowups while protecting critical assets of the built landscape, although point protection is problematic when lithic points litter every square meter. It's as though a thousand years from now fire officers in the Los Angeles Basin, its human settlements abandoned for hundreds of years, had to protect the disheveled ruins of farm-terrace suburbs, celebrity pit houses, cliff-dwelling Disneylands, and car-junkyard refuse mounds against the fires that, from time to time, rush down the Transverse Range. Maybe Shakespeare was right. The past *is* prologue, even if it's in another millennium.

Fatal Fires, Hidden Histories

COLORADO IS NOT WHAT comes to mind when most observers think of wildland fire fatalities by burnovers. The giant is California—a universe unto itself. Then comes Idaho, but its figures are distorted by the Big Blowup; the era of record begins in 1937. Then comes a cluster of states that includes Montana, Arizona, and Colorado. What makes Colorado's ranking particularly intriguing is that the two major events, both involving hotshot crews, occurred in scrub oak, a fuel not normally considered volatile. But at Battlement Creek in 1976 and again at South Canyon in 1994, frost-killed Gambel oak powered an explosive run that trapped crews.

Each resulted in reforms. The Battlement Creek burnover led to an emphasis on personal protective equipment, especially fire shelters, which became mandatory. The South Canyon fire set into motion a series of corrections that made firefighter safety a primary goal of wildland fire management. It changed how fires would be fought. Then, more slowly, nudging the system the way a locomotive might crawl around a tight curve, it changed what fires would be fought.

SOUTH CANYON: A PROSE ODE

For a fire historian the treks can take the form of a pilgrimage. I have been to the Big Blowup and Blackwater, stood before the memorial at Bass River, looked along the ridge at Rattlesnake, into the ravines at Inaja and Honda Canyon, at the hillsides of Loop, Battlement Creek, Yarnell Hill, Mann Gulch, and now South Canyon. It's different being there. The slopes seem steeper, the scenes more compact, the setting more intense. Most of the burns seem surprisingly small. Most sites today are very quiet.

When I was there one bright August noon, the memorial plaque overlooking the South Canyon blowup was empty of other people. A light breeze skipped through grass. The sun passed over white-trunk juniper, dull as bone. A fly. A bee. Lizards. A landscape haiku: stone, a burned stump, weeds. There is no shade, there is no relief. It's a place at which to contemplate but not at which to be comfortable.

But what exactly to contemplate? The interpretive signs that line the trail speak to the job of firefighting. They speak only of doing, not thinking. Yet everyone dies. People die at their jobs. They die at home. They die in mines. They die in vehicle accidents. They die in floods, tornadoes, hurricanes. The wildland fire community has never been known for its meditative powers. It's a community that thinks with its hands. In 1994, when the South Canyon fire blew over a mixed crew on the slopes, 20 other firefighters also died that year. Why memorialize these 14? Why this way?

One answer of course is that this disaster took the lives of four women, which made it a novelty. Another answer is that those earlier crew disasters did get attention, though nothing like that granted to the South Canyon dead. Twenty-three years after the Big Blowup, the unclaimed fallen got a dedicated burial circle and commemorative stone at Woodlawn Cemetery in St. Maries, Idaho. After Blackwater, the U.S. Forest Service and American Forestry Association created a medal for forest firefighters killed on the job. At Mann Gulch concrete crosses were erected.

Generally, the dead were viewed by whatever lens caught the prevailing sentiments of the day. The fallen of 1910 were workers, hired by the government, who deserved a decent burial. The fallen at Blackwater were CCC boys, whose death reminded the country that they were engaged in a great experiment to rehabilitate a society ravaged by the Depression and a land wrecked by heedless settlement; regrettably, accidents happen, sometimes "without fault or failure." The fallen at Mann Gulch got a spread in *Life* magazine. Memorializing the Rattlesnake dead was tricky because they were temporary hires from the New Tribes Mission, a Bible college, and the fire had been set by a man looking to be hired as a cook for the suppression effort. The Inaja dead—most of them an inmate crew—were memorialized by Chief Forester Richard McArdle as having died "in the defense of the free world." The Loop fire fallen, the El Cariso hotshots, had worn berets and could be seen within the flawed buildup of the Vietnam war. The Battlement Creek dead were accident victims for whom better equipment might have prevented their loss, as though the wildland fire community needed the equivalent of airbags.[1]

In 1992 Norman Maclean changed how we would remember dead fire folk. He could appeal to deeper, even existential concerns, and allude to cultural talismans well outside the scope of the fire community. He changed the terms of how we understand firefights when something goes lethally wrong. *Young Men and Fire*—published two years before South Canyon—demanded that we recognize each of the fallen as an individual. It saw the fatal moment not as an accident, in which unlikely events randomly colluded, but as a tragedy, a clash of wills, as a crew attempting to control a fire met a fire that would not be controlled. What had been labeled "disaster" fires now became "tragedy" fires. In this way his meditation spoke to universal truths of the human condition.

Every fatal burnover since then has been interpreted through the prism of his text. All the old fires have been revisited and rememorialized in the vogue of South Canyon. On the occasion of its centennial even the Big Blowup had a new commemorative stone installed amid the burial circle. In fire, as in geology, it would seem that the present is indeed the key to the past. The Big Blowup was remembered on the job by the pulaski tool, a means to do the work more efficiently. South Canyon would be remembered with a change in practice, coded into norms and mores.

• • •

Maybe it's time to move on. If *Young Men and Fire* added a new narrative to American fire, it still told the story, once again, of a firefight, though one gone bad. It did not ask whether or not that firefight should have occurred. The jumpers at Mann Gulch stood for all of us. We all die. Mann Gulch itself offered literary possibilities that made a blowup fire that crashed through the crew an attractive symbol for the fate that awaits each of us. The traditional response to disaster fires has been to double down, finish the job, man up, make sure the fallen had not died in vain. The response to tragedy fires is to not forget.

The literature on fire seems so haunted by Maclean's masterpiece that it appears unable to move beyond recycling the story, often not even speaking to the Mann Gulch fire so much as to Maclean's telling of it. It's the literary equivalent of doubling down, of insisting that the dead be not forgotten by doing better, telling the story with new information and fresher insights, like finding a better combi tool or a leadership rule of thumb. For a community committed to doing, however, one can wonder not only if retelling can replace doing, but if the theme is an adequate narrative for what the fire community must now do.

Before, responses were always about doing the job better and safer, not whether or not it should be done at all. Blackwater led to reforms in the CCC and better crew organization. Inaja to the 10 Standard Fire Fighting Orders. Battlement Creek to personal protective gear and fire shelters. South Canyon to making firefighter safety an informing principle in plans. Not until the Yarnell Hill fire was the issue raised formally—and then by Arizona OSHA—that some fire settings are so toxic that there is no justification for having anyone on them at all.

The trail of plaques at South Canyon continues the older logic. It would ask too much that those reeling and mourning should dispense with this long tradition or with the closure and maybe catharsis it might bring or to deny the impulse to *do* that lies at the heart of the culture. So it would seem disrespectful to the point of moral callousness to suggest that a fire whose location was first misplaced, a fire left for three days before any action was taken, a fire that did not threaten vital resources or communities might not have needed an undercut line through dense, frost-killed Gambel oak.

Maybe the whole ethos is misguided. We made a wrong call back in 1910 and cannot return to those times to correct it. Instead, we renew those commitments so as to memorialize those who died trying to make it happen. Then, when the story goes south again, we search for catharsis. But if we need catharsis, maybe we're doing the wrong job, refurbishing the wrong story. Maybe it's time to pull back from the flaming front. Maybe we need to go indirect on the narrative.

That won't happen anytime soon. Instead we will honor the dead by learning from the compounded errors and ensuring that others don't die as they did. A narrative that speaks to fire as a coming of age story, that replaces fire fighting with fire managing, that can remember the fallen as men and women who did not die in a way of their choosing but did die doing what they chose, will have to wait for another time.

• • •

Not all places are glamorous, not all events sublime, not all lives heroic. We all die, or as Ernest Hemingway, a contemporary of Norman Maclean, put it, every true story ends in death. Some deaths and some stories seem particularly compelling.

Death by fire, in full-throated roar in wildlands, is one of those that can grip us by the lapels, get in our face, and shake us out of our routines. It appears to burn away the veil of normalcy, exposing a tear in the space-time continuum by which we understand the quotidian world. The usual tapestry of tropes, anecdotes, clichés, habits, and expectations is ripped apart. We need explanation. We need catharsis. We need closure. We need the rend mended.

That's what interpretive trails like the one at South Canyon do. By speaking to doing the job better, and by remembering the fallen with photos and bios etched into plaques, they close the hole. They suggest that the normal can return, just better equipped, better trained, better led, making sure that that vital weather forecast doesn't get mislaid or those spot fires in the ravine aren't overlooked, that the fire next time will not kill those sent to cope with it. The message is that we might be among the saved. The fate that awaits us all will not be in the flames of perdition.

BELOW THE FLAMING FRONT

Even celebrity events can have their hidden histories. There is no one way to understand the past. We know it through context, and as that setting shifts, so does our appreciation for what something from the past means. Mostly we rely on interpretive templates to contain free-burning intuition. With fatality fires, these are typically technical that focus on how fires happened (and were fought) or sentimental that say why we should care. But since these events are fires, they also have a context in the larger fire history of Earth. That holds for the two fatality fires for which contemporary Colorado is known.

The South Canyon fire burned outside Glenwood Springs. But another fire, origin unknown, has been continually burning since 1910 in the coal seams outside town. Mostly it's a case of out of sight, out of mind. It's known at the surface through subsidence patches, venting gas and ash, condensates, and red oxidized shales. If the mini-eruptions from South Cañon Number 1 become troublesome, crews fill in vents with dirt. In June 2002, however, when the West was hunkered down in drought and big burns, the surface venting kindled a fire that roared over 12,229 acres, incinerated 29 houses in West Glenwood Springs, and cost $2.5 million to finally suppress. Then August thunderstorms triggered mudslides. Burned slopes sloughed off soil and detritus that rushed into Mitchell Creek and blocked I-70.[2]

The Battlement Creek fire crossed another narrative of fossil fuels. The Mesa Verde formation is dense with gas, sandwiched between sandstones and shales. If it can be fracked, the gas will flow, but old-style fracturing was slow and costly, so the scheme bubbled up to use 40-kiloton nukes. Project Mandrel Rulison, one of three test sites, combined underground nuclear tests (Operation Mandrel) with Operation Plowshare, a broad enterprise to turn atomic energy to peaceful purposes, and commercial uses through the Austral Oil Company. (The idea probably belongs with other delusional proposals such as an atomic airplane.) The Mesa Verde formation lay some 8,400 feet below ground, thus shielding the surface from any radioactive fallout. The government contributed public land at Battlement Mesa and 10 percent of the costs.[3]

The experiment succeeded in liberating gas, but it was (of course) radioactive and deemed unsuitable for household use. Even the energy crises that engulfed the country during the 1970s could not argue for the program, although subsequent wells, with nonnuclear technology, now dapple the slopes down to the town of Parachute. From above the landscape looks like the high-tech version of a prairie dog town. The Department of Energy began cleanup operations from the explosion that continued into 1998. The State of Colorado established a buffer zone. A placard was erected in 1976.

That same year nature found ways to liberate more traditional combustibles. On July 11 lightning kindled a fire that flared into view the next day in Eames Orchard outside Parachute. By 5 p.m. the Grand Valley Volunteer Fire Department had controlled the fire at half an acre. On the 15th it apparently reignited, and this time with wind and cheatgrass to carry its spread upcanyon toward the tangle of gas wells, pipelines, and the newly plaqued Project Rulison site. These were still the early years of the fire revolution, in which interagency cooperation was promoted for fire suppression but the 10 a.m. rule remained as an inflexible principle. Still, there was added urgency to have crews and air tankers keep the fire away from those modern ancient fuels, even as those crews were transported and planes flown by distilled petroleum.

Of the three air tankers one, a B-26 piloted by Donald Goodman, crashed the next morning. The following day, during burnout operations along the ridgetop, three members of the Mormon Lake Hotshots were killed and a fourth badly burned. On the 18th rain fell. On the 19th the fire was declared controlled. On the 20th it was out. The final burn blackened 880 acres. The next year Carl Wilson published his famous study on common factors in fatality fires, which tend to be small, susceptible to sudden changes in relative humidity and wind, often in steep terrain, and happen during transitions. The Battlement Creek tragedy fits that profile nicely.[4]

And that is where the American wildland fire community has been content to have it reside. As with South Canyon it offers lessons in how to keep crews safe. Like the Eames Orchard reburn similar scenes keep recurring. The plaques, the memorial trails, the staff rides—these are the context for remembering the fire. Yet if you hover over Battlement Mesa today (or Google Earth it), you will see a landscape patchy with gas wells. Whatever its needs to manage wildland fire, the country has a deeper political commitment to keeping unimpeded the flow of combustion from fossil fuels. The South Cañon Number 1 coal mine continues to burn. In fact, some 35 coal mines in Colorado alone hold fire.

These fuels, and their associated fires, are also part of American fire history, or more broadly, of Earthly fire history. They speak to humanity as a fire creature, able to burn lithic landscapes as well as living ones. They remind us that our environmental power is fundamentally a fire power, that our industrial civilization is one founded on burning fossil fuels, and that these, too, can sometimes turn feral or lead to unhappy choices.

To date, we have no memorials toward our errors or tragic outcomes in nominally managing those burns. The alarm over global climate change, leveraged by combusting fossil fuels, is starting to alter that perspective, but it is not seen as a fire story. It is. Since the moment *Homo* first picked up a firestick, our decisions have been unsettling the planet. It's no longer good enough to pretend that fire history is a subset of climate history. Rather, climate history has become a subset of fire history. Our fires no longer interact only with air, and afterward, with water, but with earth.

Viewed this way the fatality fires in Colorado can merge with all the other fatalities caused by our addiction to fossil fuels. Maclean's meditation embedded fire within the human condition. But our fire technologies are also a foundational part of the human condition. There are times on a fireline when you have to choose, and those choices can have lethal consequences. The same is true for a society, or since all human societies are desperate for more power, and find it in fire, for humanity. Our species history is a fire history.

That's a much broader frame than the wildland fire community is accustomed to, or even wants. It's probably a narrative that only an intellectual could love. It's good to have—good to be reminded that our firepower saturates our civilization. But it's also good to remember those names on the plaques at South Canyon and Battlement Creek. The fire community can't solve the problems of global warming, urban sprawl, invasive species shunted around the globe by oil-powered cargo ships and airplanes, and the toxic residues left from mines and wells. It has to look after its own. But sometimes it's good to recall that we are members of a much larger fire community that spans the globe and a nontrivial fraction of Earth time and that also knows fire disaster and fire tragedy without ever donning Nomex or hefting a pulaski.

NORTHEAST

A Song of Ice and Fire—and ICE

THE TEXT BEGINS as a blank page, a vast sheet of white ice.

The ice resulted from the interaction of Earth and Sun, mediated by the distribution of lands and seas and the planetary wobbles and tilts, synthesized into Milankovitch cycles, that determined how sunlight struck them. The ice erased what had been written before, though never completely. It could mold, though not remove, some hard rock; it could widen and deepen, though not flatten or fill, some gorges; it could cover or expose the continental shelf, though not remake its contours. The ice was plural, not singular: it came and went repeatedly across Northeastern North America. Each time the ice receded, it left a geologic palimpsest, a hard parchment of terrain roughly organized into inland mountains, coastal plains, and median piedmont of rolling hills, along with scribblings and blotches of outwash sands. Here was first nature, the Northeast as a patch of planetary Earth.

But even as the ice broke, melted, and retreated, and the ocean rose, life appeared along the frontier, moving northward, a slow scramble to claim swathes and niches. People, too, were among that throng. Perhaps as early as 12,000 years ago, they prowled and probed along that border and helped the infilling behind it. They interacted with the biota, perhaps profoundly in the case of megafauna, until over the millennia, they expanded their interactions to reshape the mobile features of the varied scenes bequeathed by the ice. A world of life replaced the abiotic world of ice. With life came fire, and catalytic fire made possible more interactions and widened the reach of the human hand. Humanity's impact quickened over the past millennia, then accelerated and broadened over the past 40 centuries. People moved soil and sand, dammed and diverted streams, killed wildlife and felled forests, replaced a wild biota with a domesticated one, and they burned. People mediated and fire enabled. This was a world of second nature.

Then that, too, felt a new fundamental force not known on Earth before. Humanity, the planet's keystone species for fire, in its search for greater fire-power, changed its combustion habits. Mostly that burning has occurred in special chambers in one form or another of internal combustion engines. It began to burn lithic landscapes rather than living ones. The effects have cascaded through the Earth system until, in sum, they threaten to rival that of the ice. They have begun to challenge even the Milankovitch cycles that have powered the ebb and flow of ice throughout the Pleistocene. Most of that geologic era—roughly 80 percent—has been glacial. The last 10,000 years, the Holocene, have been an interglacial, an unusually long one, primed to flip back into ice. The Little Ice Age may in fact have been a tremor heralding the return of continental ice sheets.

It didn't happen. Instead, the Earth began to warm. Human meddling reached gargantuan proportions, attaining in the eyes of many observers the stature of a geologic epoch in its own right, an Anthropocene. The release of greenhouse gases from the burning of fossil fuels has, in particular, stalled the tidal shuffle between glacial and interglacial. Internal combustion engines—give them the acronym ICE—may be holding back the old rhythms. But like ice, fire now has assumed positive feedbacks. The Pleistocene ice ages are being replaced by an Anthropocene fire age, a Pyrocene. The upshot might well be called a third nature.[1]

• • •

Look more closely at that fire scene and its history.

As a fire province, it most resembles northern Europe, also an outcome of continental glaciation. There is a boreal portion—Fenno-Scandinavia, Maine—that has some natural fires. There are sandy zones that favor burning, largely along littorals, but in prominent patches elsewhere—the Landes and the Baltic lowlands, the coastal plains from New Jersey to Cape Cod to Maine. But lightning kindled more barns than snags. Fire-adapted species like pitch pine spread as fire promoted them, which depended on regular ignitions, which had to come from people. Like temperate Europe, the Northeast was an anomaly in a fire planet because there was little climatic basis for fire.

Mark Twain famously commented that if you didn't like New England's weather, wait 15 minutes, and people hold forth endlessly about the changing seasons. From a fire perspective, however, the climate is remarkably uniform because it lacks a rhythm of wetting and drying. Fire must squirm into seasonal cracks; spring after a dry winter and before greenup, autumn after dormancy and before snows, especially during sunny Indian summers. Flames interacted with blowdowns, with the once-a-century hurricane, with ice storms, with beetles, blister rust, and gypsy moths. Throughout it all, the forest has proved remarkably durable, but also equally in a state of continual churn. There is plenty of fire in regional history, though constrained, one of many disturbances. Again, as with temperate Europe, fire became prominent when it was put there by humans who slashed, drained, and kindled. The character of human settlement set the character of landscape fire.

What was fire like in precontact times? The most likely regimen would be that typical the world over. Aboriginal economies of hunting, fishing, and foraging display lines of fire—routes of travel burned deliberately and accidentally—and fields of fire, burned to improve hunting habitats, to help cultivate berries, to improve fields of vision. And here, as around the world, one should not overlook the effect of accidental or careless fire, what we might call fire littering.

How such patterns express themselves depends on the capacity of the landscape to carry fire once lit. The sandy coastal plains, rich in pyrophytes like pitch pine and oak, could (not surprisingly) burn easiest. The hummocky piedmont burned patchily by place and season. The absence of a grassy understory, the abundance of wetlands, the broken terrain, all would require repeated, site-specific burning that would leave a fire-mottled landscape. The mountains and coniferous northern forest burned even more selectively, leveraged by occasional lightning fires. In most years, fires would spread haltingly; in those years that opened wider windows for burning, those point ignitions could spread, sometimes explosively. What humans did in such landscapes was not underburn on the southeastern pattern, but leave spot fires that ensured that, when those exceptional conditions for big fires were right, there was always ignition present. There were always campfires and smudge fires unattended along rivers, along trap lines, around busy trails. It's the pattern that characterizes fire in Canada's and Eurasia's boreal forests, and it is hard to imagine why it would not apply to New England's.

But there was an agricultural fire economy, too. New England was the pointed tip of a frontier of maize cultivation that had entered hundreds (perhaps a thousand) years before European contact. The consequences for fire were both direct and indirect, as with swidden the world over. The direct effects were to expand the realm and to recode the pulses and patches of routine burning. Fire flared where it would not have under natural or aboriginal conditions; by burning to different purposes, it branded a different regimen on the land. Obviously, as with hunting and foraging sites, some places were better disposed than others. Places susceptible to swidden assumed the fire equivalent of being moth-eaten. Fires returned on the order of decades.[2]

The indirect effects came by establishing where people lived and how they interacted with the surrounding countryside. Near environs would be stripped of dead wood (gone to feed cooking and warming fires). Hunting and gathering would cluster in new ways, which would rearrange the mosaic of landscape burning. River bottoms, for example, would be fire-farmed; uplands, burned in patches to improve travel, gathering, trapping, and hunting. Of course there were limits inherent in the larger setting. Fire could not burn through extensive wetlands, could not propagate where surface fuels like mosses in boreal woods were relentlessly wet, could not race through canopies where the forest was deciduous. The particulars can be hard for later generations to confirm when fallen leaves were burned in the autumn or where surface burns could not generate sufficient charcoal to settle into ponds or where trees scarred by fires were cleared away during a more aggressive wave of settlement.

Still, there are iconic accounts from contact times that fit nicely into the above scenario. Those classic records have also prompted classic controversies, and since the suite of texts is not adequate to satisfy the demands made on them, they have been subjected to the scholasticism of endless parsing, glossing, and deconstructing, all interpreted through the prism of the viewer's values.

Begin with two that describe coastal Massachusetts. In his 1634 *New England's Prospect* William Wood observed,

> For it being the custome of the Indians to burn the wood in November, when the grasse is withered, and leaves are dryed, it consumes all the underwood, and rubbish, which otherwise would over grow the Country, making it unpassable, and spoil their much affected hunting: so that by this means in those places

where the Indians inhabit, there is scarce a brush or bramble, or any cumbersome underwood to bee seene in the more champion ground.[3]

("Champion" is a corruption of "champaign," a derivation from the Latin *campus*, which refers to an open plain.) Such descriptions were the norm along the coast. Thomas Morton elaborated in his 1637 *New English Canaan*.

The Savages are accustomed to set fire of the Country in all places where they come; and to burne it, twize a year, vixe at the Spring, and the fall of the leafe. The reason that mooves them to doe so, is because it would other wise be a coppice wood, and the people would not be able in any wise to passe through the Country out of a beaten path. . . . The burning of the grasse destroyes the underwoods, and so scorcheth the elder trees, that it shrinkes them, and hinders their growth very much: So that hee that will looke to finde large trees, and good tymber, must not depend upon the help, of a wooden prospect to finde them on the upland ground; but must seek for them, (as I and others have done) in the lower grounds where the grounds are wett when the Country is fired.[4]

("Savages" here derives from the Latin *silva*, tree or woods, by way of French, and refers to people who live in the woods rather than amid cultivated fields on the European model.)

There was enough burning going on that the newcomers had to burn around themselves "to prevent the Dammage that might happen by neglect thereof, if the fire should come neer those howses in our absence." Morton thought that the "Salvages by this Custome of theirs, have spoiled all the rest [of the countryside]: for this Custome hath been continued from the beginninge." Yet he also confessed that "this Custome of firing the Country is the means to make it passable, and by that meanes the trees growe here, and there as in our parkes: and makes the Country very beautiful, and commodious." Writing about New Netherland in 1656, Van der Donck reported that removing those fires had caused rapid reforestation, without which there would be "much more meadow ground."[5]

And further inland? Peter Kalm, one of Linnaeus's Apostles, noted how indigenes around Lake Champlain were "very careful" about escape fires, yet also observed that elsewhere "one of the chief reasons" for the decrease in conifer forest was "the numerous fires which happen every year in the woods, through the carelessness of the *Indians* [italics in original], who frequently make great fires when they are hunting, which spread over the fir woods when every thing is dry." Burning, that is, occurred where it could but was not indiscriminate, save where careless prevailed.[6]

Or consider the observations of Timothy Dwight IV, president of Yale College, describing expansive "barrens" in western New York, whose "peculiar appearance" he attributed to the fact that the "Indians annually, and sometimes oftener, burned such parts of the North American forests as they found sufficiently dry." Southern New England, "except the mountains and swamps," were covered with oak and pine, well adapted to such burning. The "object of these conflagrations was to produce fresh and sweet pasture for the purpose of alluring the deer to the spots on which they had been kindled." Dwight's QED came when he observed the consequences of removing those annual burns. "Wherever they have been for a considerable length of time free from fires, the young trees are now springing up in great numbers, and will soon change these open grounds into forest if left to the course of nature." He himself had witnessed many such examples.[7]

It would seem that the Received Standard Version of New England history, that the forest returned after clearing by Europeans, needs an earlier epicycle. The forest also returned after the vanquishing of the indigenes and the extinguishing of their fires.

• • •

Then European settlement created an undeniable wave of flame that, over several centuries, washed across the region and beyond. It helped that the immigrants came from climate, soils, and a biota similar to what which they encountered in America. They brought with them flora, fauna, diseases, fire practices, and an agricultural economy already preadapted to the conditions they encountered. In places—around the colony of New Sweden, for example—they fused to create frontier hybrids. Colonists learned "fire hunting" from the indigenes. They added landscape draining to slash-and-burn cultivation and herding to hunting. After the first pass of pioneering, fire flourished within an agricultural matrix that encompassed most of the Northeast.

European contact widened, deepened, and quickened the presence of fire. Fire technology and practices persisted, with some adaptations. Thanks to diseases and wars, the indigenes shrank. More land became available, and more was brought into production. But while intensifying burning and stressing the landscape, the fundamentals of fire ecology remained the same. The big reform was the introduction of livestock, which replaced fire hunting with fire herding, and so encouraged a different regimen of burning. Pioneers pushed fire where, by nature or indigenous economies, it would not have existed, and they pulled fire from where it had routinely flourished. They recoded the pulses and patches of burning to fit a broadly frontier society, and then an agricultural one. Those fires—much as settlement overall—passed through regional history like a flaming front.

Again, there are plenty of accounts, and even the origins of fire poetry (thanks to Robert Frost, who describes bonfires, burning fields, and blueberry patches). But the most apt might be those of Henry Thoreau, writing at the height of clearing, and just as the region was inflecting into an epoch of depopulation and reforestation, and equally important, as the locomotive was becoming the fire engine of choice.[8]

A hearth fire, he thought, was the "most tolerable third party." But he could see landscape fire with similar empathy. His summary view he wrote on June 21, 1850, two months after accidentally setting 100 acres of Concord woods on fire. Fire, he mused,

> is without doubt an advantage on the whole. It sweeps and ventilates the forest floor, and makes it clear and clean. It is nature's besom. By destroying the punier underwood it gives prominence to the larger and sturdier trees, and makes a wood in which you can go and come. I have often remarked with how much more comfort and pleasure I could walk in woods through which a fire had run the previous year. It will clean the forest floor like a broom perfectly smooth and clear,—no twigs left to crackle underfoot, the dead and rotten wood removed,—and thus in the course of two or three years new huckleberry fields are created for the town,—for birds, and men.[9]

He pondered the interaction of Indian fires and pitch pine (his favorite). He documented the endless sources of ignition in the cultivated countryside around him—the smoker, the sportsman, the debris burner, the campfires carelessly tended or abandoned, sassafras-collecting boys, farmers firing meadows, brush, fallow fields, and postharvest stubble, and increasingly the feral sparks of locomotives. He observed the prime time for fugitive flames to spread was spring, from mid-March to mid-April, when the leaves and grasses were dry, the unleafed trees let sun and wind pass freely, and dry cold fronts blustered through. He recorded how, for severe fires, "the men should run ahead of the fire before the wind, most of them, and stop it at some cross-road, by raking away the leaves and setting back fires." He noted the benign effects of fires, "how clean it [fire] has swept the ground, only the very lowest and dampest rotten leaves remaining," how "at first you do not observe the full effect of the fire," "how the trees do not bear many marks of fire commonly; they are but little blackened except where the fire has run a few feet up a birch, or paused at a dry stump, or a young evergreen has been killed and reddened by it and is now dropping a shower of red leaves." As for burning meadows, "I love the scent. It is my pipe. I smoke the earth."[10]

More and more, however, as former farmers decamped and the fields fell into the deep fallow of reforestation, the engines of third nature replaced the torches and campfires of the old order. The bad fires were those that followed railways, the Northeast's new lines of fire. In reflecting on the wildfire he had kindled, Thoreau rationalized that he "had done no wrong," and that "now it is as if the lightning had done it," for the fires were consuming their "natural food" as deer fed on browse. Besides, "the locomotive engine has since burned over nearly all the same ground and more, and in some measure blotted out the memory of the previous fire."[11]

And that has been the effect of industrial combustion overall. It has remade the countryside and erased the folk memories of the anthropogenic burning that had once helped power it.

• • •

Though condemned by European agronomists and foresters, fire flourished among folk living on the ground. In 1878 Franklin Hough gathered accounts of fire practices from resident-observers for his *Report upon Forestry*. Then C. S. Sargent plotted the pyrogeography for his 1880 census report on forestry, which showed the Northeast holding its own amid the other fire provinces of the country.

Meanwhile, a Great Depopulation was sweeping over the region and left half or more of the landscape fallow. The opening of the Erie Canal in 1825, then the railroads, both pushed and pulled demographic change. Transport made it easier to leave for better lands in the Ohio Valley and around the Great Lakes and cheaper to import farm products than grow them internally. From the mid-19th century to the mid-20th the process steadily drained population away from the countryside, then accelerated in the post-WWII era. In 1880 New England still had over 200,000 farms on 21 million acres. By 1940 those numbers had dropped to 135,000 and 13 million. By 1970 they had plummeted to some 20,000 and 5 million.

That left an ecological vacuum. Trees and shrubs once crowded into margins—white ("old-field") pine in particular—claimed the untended landscape. Fuels held in check by close cultivation and cropping by livestock overran the countryside. By the end of the century those woods had grown sufficiently to be harvested, which prompted another wave of fire, this time wildfire gorging on logging debris. Industrial slashing and burning in the mountains left denuded hillsides, which invited erosion and flooding.[12]

By now, inspired by such regional voices as Charles Sargent, George Perkins Marsh, and Gifford Pinchot, a doctrine of conservation argued for change. It was impossible to halt the logging, but it would be possible, given the region's fire environment, to stop the burning. Reflecting the progressive thoughts of the times, Sargent wrote that "fire threatens the forest at every stage of its existence, and a fire may often inflict as much damage upon a fully mature forest ready for the ax as upon one just emerging from the seed; and, as long as such fires are allowed to spread unchecked, there can be no security in forest property." Fire, he thundered, "is the greatest enemy to the American forest"—a conclusion repeated by Pinchot in his *Primer on Forestry*. The threat was no less apparent in Sargent's home state, Massachusetts. "Any attempt to improve the forest of the State is useless until they can be secured greater immunity from fire."[13]

Government at all levels made fire protection a priority. Some states like New York created nature reserves to ban fire and axe; some like Massachusetts gazetted state and town forests; some like Vermont and New Hampshire invited the U.S. Forest Service to acquire cutover lands in the mountains. Maine established a Forestry District to oversee its unsettled backcountry. The Weeks Act established federal grants to states to help. Its successor, the Clarke-McNary Act, expanded the range of watersheds available to the program. The fires seeped away.

But not before some of the worst wildfires in regional experience broke out. Rail opened up mountain forests and the recovered forest across abandoned lands encouraged another round of logging that left slash not seen since 18th-century landclearing; fires followed. The fallow land made for feral fires. Wildfire replaced the tamed fire of agricultural burning. By 1930, however, the era of breakout burns had largely passed. From 1908 to 1930 Massachusetts averaged roughly 40,000 acres burned a decade; by 1990, that fell to under 5,000. Vermont went from approximately 3,000 acres a decade to 300. New York burned 47,000 acres in the 1920s, and 2,500 by 1990. Codes regulated folk burning, agriculture found alternatives to field fires, locomotives ceased to spray sparks, fire protection systems toughened, land use spun away from fire-catalyzed occupations—all the usual prescriptions that work in temperate environments in which natural fires are rare came into play with reductions on the order of 10–20 in burned area.[14]

Increasingly, the Northeast moved from visible flames to remembered ash. Big burns (by regional standards) occurred after 50- or 100-year events like the 1938 hurricane (in which windfall replaced logging slash) or the early 1960s drought. The last great aftershock came, unexpectedly, in October 1947 along the coastal plain, mostly in Maine. That outbreak inspired a regional consortium among the states, the Northeast Forest Fire Protection Compact. What had been a national hotspot in 1880 became by 1980 a cold ember, quickly overgrown by new woods, its pyric history all but forgotten.[15]

Most observers felt about the banished fires as today's residents would the obliteration of Lyme disease. They were happy to remove fire from landscapes and stuff it into machines. Only later would the ecological effects of fire's mass removal slowly become apparent. Pine and oak suffered; wildlife suffered; shrubs like blueberry and rare forbs like poor Robin's plantain suffered. The fires that had animated those scenes did not have to come from nature to be significant: they just had to come. They were the fires of second nature, not first nature. When they no longer arrived, the lands that had long known them sickened. The flames they had depended on were now shackled in combustion chambers. A phase change had swept over the region's pyric history.

Its celebrated chronicle of fire had remained within the realm of second fire. Much as the natural landscape had been remade into a humanized second nature, so had fire. Second fire (as it were) encompassed all those fire practices humans distributed about the land, particularly those for which fire served as catalyst; people did not burn just to burn but as part of how they lived on the land. By the onset of the 20th century the fire regimes of the Northeast bore little resemblance to what might have existed under purely natural conditions. By the end of the 20th century, those fire regimes—for that matter, landscape fire—were mostly gone.

• • •

This time the shift did not represent changes and mutations within the realm of second fire but the wholesale replacement of second fire by another realm of combustion, the burning of fossil fuels. More and more, the Northeast burned lithic landscapes instead of living ones. The region had inflected into an age populated by machines that fed on fossil fallow, an era aptly symbolized by the internal combustion engine (ICE). A new ICE age spread over mountain, piedmont, and coast. Feeding on fossil biomass it freed up forest biomass. The ICE age of the Anthropocene promised to refashion the region as the Pleistocene's ice age had before.

Its anthropogenic fires had favored some species over others: it shuffled species and recycled biotas, but it had not destroyed them. People used a natural process to reshape natural materials. Second nature was first nature refashioned through the artifice of humanity, but through means and tools themselves remade from natural sources. Third nature reworked second nature, and it did so with means unlike anything in first nature, or with processes so reduced and isolated that they no longer did ecological work, even as surrogates. Third nature is a built landscape typically fashioned with asphalt, concrete, glass, steel, plastic. A stone fence was made from rocks removed from a field and relocated. A skyscraper was made from cement and metal burned from stone and ore that bore no interaction with its quasi-natural setting. The same has held for fire. Fire codes increasingly ban any form of open fire, even leaf burning. Open burning went from being plowed under to being paved over.

In 2009 David Kittredge wrote a commentary for the *Journal of Forestry* that spoke to the "fire in the East." He meant it metaphorically. The megafires of the West that drew public attention "do not *destroy* forests [italics in original]" but the sprawl that characterized contemporary land use in the east did. "Forests do not grow back after development." Through millennia of human use, through four centuries of European-style land conversion, the Northeast's forests had endured. Second nature had preserved forests, or where cleared it had retained the capacity for them to return. Third nature was different. A fire history might reify Kittredge's allusion. Third nature had its own combustion, not drawn from or dependent on the character of its surroundings. It was more ruthless, and potentially more ruinous, because it could deny the ability to rebound that had characterized the Northeastern forests since the ice had left.[16]

• • •

The fire story of the Northeast is the fire story of temperate Europe transplanted to the New World. But just as northern Europe had been the hearth from which European peoples and norms had birthed a second age of colonization, and from which modern science and the industrial revolution had been disseminated around the world, so was the Northeast for North America. What happened in the Northeast affected national fire norms and institutions.

That transfer of understanding led to many errors. Temperate Europe is not normative in fire-planet Earth; the Northeast is exceptional in fire-prone North America. The assumptions that what made sense in a region that lacked a climatic basis for routine fire would make sense in places like the Far West that had abundant natural fire, or like the Southeast had deep traditions of burning that intertwined natural and cultural fire was flawed. The 20th century showed how ruinous this presumption could be.

But there is a flip side to this story. However inappropriate its understanding of fire might be outside the Northeast, it suits the region itself. The fire practices of the Northern Rockies, or the Wichita Mountains, or the red hills of Florida can be equally maladapted in the Northeast. There is clearly a place for fire, but it will not be justified by appeal to wild nature or prescribed fire as an informing principle of land management. Fire will be anthropogenic; it will occur in second nature or be restored amid third nature; it will flare or subside within a philosophically awkward pluralism of peoples, purposes, and practices. The Northeast has its own fire rationale. It deserves a history that builds on its reality.

• • •

It's a place of mixes. It has long mixed people and nature. It has a mixed economy, a mixed forest, a mixed history, mixed land tenures, mixed aesthetics, mixed governance. Its history is kaleidoscopic; the pieces seem to endure or are metamorphosed in the way exurbanites reclaim colonial farmhouses or a software company repurposes a historic building; then they recombine. There are private lands with public purposes; there are public lands owned by private interests. Donations, land trusts, and conservation easements do what the federal lands do in the West. The grand public domain that has animated so much of American environmental philosophy and politics is absent, or one small voice in a choir. Instead of preserving wilderness, the Northeast is watching significant patches of its estate rewild.

To an outsider it can seem bewildering, a tangled bank of historical legacies and modern ambitions. What works in Montana or Florida doesn't in

Massachusetts or New York. It's a managerial mashup. That holds, too, for the region's fires, which after all must synthesize their surroundings. In parts of the public-land West, history can seem irrelevant. It's enough to manage fire through first principles in the here and now. In the Southeast history matters—the past is not even past, as William Faulkner put it—but through continuities in practice, particularly the tradition of anthropogenic fire, as with the oft-repeated stories about learning fire by helping a grandfather burn land. In the Northeast the scene can only be understood through a historical palimpsest of landscapes and ideas and institutions. The region has been remade by ice, then people, then newly (technologically) empowered people. The landscape story is one of change, resilience, and pluralism. So, too, is the way that history itself might be understood.

Today, outside southern New Jersey, the region has few significant fires and no fire crises that seem destined to influence national policy. The concern among the fire community is deciding what fire, if any, to reinstate amid the continual churn of land use. But just as the region's fire scene cannot be understood without understanding its history, so the history of the American fire scene nationally cannot be appreciated without the Northeast. You don't need big fires to have a fire history. You don't need conflagrations to have a fire problem. You only need to have people and nature interact in ways fire mediates. In the Far West fire history is in your face. In the Northeast it's more likely in your pocket.

Albany Pine Bush

MOST NORTHEASTERN firescapes exist within cultural landscapes. The Albany Pine Bush Preserve sits within an urban one.

Adjacent to its fragmented 3,300 acres are tract homes, condos, strip malls, an auto dealership selling Jaguars and Volvos, Walgreen's and CVS, a railway, Interstates 90 and 87, three nursing homes, medical offices, power lines and phone lines, a mattress store, a trailer park, a storage rental facility, Burger King and nail salons, gas stations, Subway and UPS, an Office Max and McDonalds, insurance brokers, engineering consultancies, gyms and massage parlors, a city dump (the highest point in Albany), a VFW hall, and an Italian American Community Center—an herbarium and birding list of contemporary American economic diversity. The Pine Bush Commission consists of the heads of two state agencies, representatives of the city of Albany, towns of Colonie and Guilderland, Albany County, the Nature Conservancy, and four members appointed by the governor. The Commission is charged with maintaining or restoring the pine bushlands, a fire-dependent ecosystem, which is the habitat for a constellation of rare species, including the federally listed Karner blue butterfly, which means the pine bush must burn.[1]

As Neil Gifford, conservation director, puts it, If you can burn here, you can burn anywhere.[2]

• • •

The Albany Pine Bush is a sandy plain, the residue of glaciers that spilled into Glacial Lake Albany. Historically it extended between Albany and Schenectady, stocked mostly with pitch pine and scrub oak amid a patchy mosaic of shrubs, grasses, and forbs. Ecologically, it's an inland equivalent of coastal barrier islands. Over the Holocene its fire regime has changed, but something like the contemporary scene seems to have persisted, as recorded in charcoal and pollen, for 4,000–6,000 years. With little dry lightning, the ignition source here, as throughout the region's barrens, must have been the indigenes. They had no reason not to burn.[3]

Early Europeans witnessed episodes. In the 1640s Adriaen van der Donck wrote a long account:

> The Indians have a yearly custom (which some Christians have also adopted) of burning the woods, plains and meadows in the fall of the year, when the leaves have fallen, and when the grass and vegetable substances are dry. Those places which are then passed over are fired in the spring in April. This practice is named by us and the Indians, "bush-burning" which is done for several reasons. First, to render hunting easier, as the bush and vegetable growth renders the walking difficult for the hunter, and the crackling of the dry substances betrays him and frightens away the game.
>
> Secondly, to thin out and clear the wood of all dead substances and grass, which grow better the ensuing spring. Thirdly, to circumscribe and enclose the game within the lines of fire, when it is more easily taken, and also because the game is more easily tracked over the burned parts of the woods.
>
> I have seen many instances of wood-burning in the Colony of Rensselaerwyck where there is much pine wood. Those fires appear grand at night from the passing vessels in the river, when the woods are burning on both sides of the same. Then we can see a great distance by the light of the blazing trees, the flames being driven by the wind, and fed by the tops of the trees. But the dead and dying trees remain burning in their standing positions, and appear sublime and beautiful when seen at a distance.

In 1796 Timothy Dwight observed that fires were particularly prone in pitch pine and oak scrub because the sandy soil meant they were typically dry. He speculated that New England's many barrens had been burned for a thousand years.[4]

But those fires didn't vanish with the passing of aboriginal Americans. The colonists often emulated their predecessors, or forged hybrid practices, in which they adapted burning to interact with livestock, introduced cultivars, and land tenure systems. To the old practices such as fire hunting, swidden, and berry production, they added new ones for large-scale landclearing and pastoral burning, and, prodigiously, with locomotives. Farms and fields disciplined burning within a cycle of cultivation; outside those fence lines, it burned without the interactions that had previously characterized it. There is some evidence that clear-cutting, plowing, and slash fires may have shifted the biotic composition. With industrialization the pine bush found new lines of fire along the rails, and ignition ran amok. The world's first passenger steam locomotive made its inaugural run between Albany and Schenectady on August 9, 1831. Its engine burned Lackawanna coal, which it then belched in palls of ash and embers across the countryside. The Albany Pine Bush (APB) found yet another way to burn.[5]

Then the pioneering fires of settlement and steam ended, and after old-field pine had been cut and its slash burned, the reaction known as conservation created formal programs to tame fire within the landscape, and ultimately to cull it out altogether. By 1900 it became policy to abolish fires from the pine bush. Towns established fire codes and enrolled wardens to fight wildfire. States mandated forestry bureaus to assist. Industrialization took people out of rural lands, or if they remained, moved combustion from human hands into machines. Eventually, New York State forbade burning in forests altogether.

For the APB the New York State Thruway and an infrastructure of water, sewers, and electrical power pressed home in the postwar era that suburbanization was imminent and added a sense of urgency to install urban-style fire protection. When Crossgates Mall opened in 1985, the prospect of the barrens being paved over moved from the hypothetical to the plausible. Still, the lands of the Albany Pine Bush remained relatively open. It continued to serve as a de facto landfill, an exurban and biotic dumping ground. Like vacant lots in cities, it attracted random fires; trash fires became part of its pyric mix. Some 65 large fires burned between 1935 and 1987.[6]

This was a pittance compared to the routine, landscape-scale burning that had shaped the APB over thousands of years. By the 1960s a new, more fire-free order was imposing itself. Flames flickered. Fire leached away. Fire protection succeeded. Fire-catalyzed biotas faltered. Without the pruning of open flame, pitch pine thronged on formerly plowed fields, oak thickened, aspen and invasive hardwoods like maple and cherry moved in. The pine barrens, long fabled as open and sunny, became impenetrable, a thicket of ecological gloom. A once-dappled scene of shrubs and grasses and flowers, a barrens savanna stocked with scattered pitch pine, was smothered under a tangle of woody plants and litter. Rare species became rarer. The Karner blue butterfly teetered on the edge of extinction. The pine bush was rapidly evolving into something else, which left the biota slowly strangling. It need ecological shock therapy. It needed fire.

• • •

But first it needed a controlling agency with a cause. A grassroots campaign emerged, and in 1973 some 472 acres, called the Pine Bush Unique Area, was set jointly aside by Albany and Guilderland and the Department of Environment and Conservation. Later, an additional 1,500 acres were added. The organizers recognized that they needed better information about the ecology of the pine bush and realized that fire was not simply a nuisance but a necessity. In 1986 they contacted the New York Field Office of the Nature Conservancy to assess the preserve's fire scene and possible operations to maintain it. The City of Albany, as part of a generic environmental impact statement (forced by the proposed expansion of the landfill), commissioned several scientific studies that addressed basic questions of fire history and habitat requirements. At the time it was not known what fire regime was suitable. For that matter, it was not known what minimum size and mosaic was required to sustain the biota, particularly the Karner blue butterfly.

In 1988 the New York State legislature created the Albany Pine Bush Commission as a public benefit corporation to coordinate action among the many owners within the preserve and the multiplying shareholders outside it. The Commission includes representatives from the three municipalities, the state Office of Parks, Recreation, and Historic Preservation and Department of Environmental Conservation, Albany County, the Nature Conservancy, and four members appointed by the governor. The Commission was state-funded through a real estate transaction tax but is not itself a state agency. Its charge was to preserve the pine bush habitat. The APB held 20 percent of all the species of greatest conservation need listed for New York.

Inevitably, scientists argued over particulars of data and interpretation. But a consensus developed (and thanks to a lawsuit became a legal mandate) that the preserve required at least 2,000 acres—more would be better—to retain ecological integrity; and those lands would have to burn. The TNC report, issued in 1991, made the case that the Commission should, as the classic Latin phrase put it, *festina lente*—hasten slowly. A solid program needed lots of test fires. It needed time to adapt to actual outcomes, not forecast ones. It would appreciate that fire could only do its work if the land was in a form that would allow fire to be controlled (a suitable fuel complex) while encouraging fire's full-spectrum catalytic effects (an adequate biotic matrix). All this was new. No one had tried to restore fire to the region's barrens and pine bush. No reference point existed that could serve as a beacon and index. They would have to invent the prescriptions and protocol. They took nearly 10 years before they launched serious operations. Hasten slowly.

By 2017 the preserve held 3,300 acres, gerrymandered through pine bush and suburb. They got some extra land as a trade-off for expanding the Albany dump. They got some more along the right-of-way of high-tension power lines belonging to National Grid (the power company did the first, heavy clearing). The preserve reckons it needs at least 5,300 to stabilize the scene. Of these, 200 (6 percent) are burned annually. But it's in the burning that the story gets interesting.

• • •

Fire is what its setting makes it. To get the right kind of fire APB has to fashion the right context. The combinations of treatments, and their sequencing, along with the timing and variations of firing, can resemble a Rubik's Cube in their complexity. Some of those "treatments" involve radical surgery. Restoring can mean rebuilding from the soil up (a demo patch from what was once a parking lot exists outside headquarters). It's ecological tough love.

The patches that make up the preserve vary in their deviation from historic norms. The ideal biomes are swathes of grassy savanna (rife with blue lupine) within the clusters of pitch pine-scrub oak savanna (rich with shrubs and forbs). Getting there can mean cutting and culling, drilling and filling, mowing and masticating, sowing and burning. Managers thin out pitch pine where it inhibits savanna and poses threats of crown fire. They clear-cut invasives like black locust groves, which not only shade out lupine but alter the soil biology and so must be extirpated by their roots, every fragment. They drill into birch and cherry and fill with herbicides. They brush-mow scrub oak. They plant lupine. They burn.

They've learned that not all fires are equal. Each has its place and time. Collectively, they have rules of assembly. Begin with dormant season burns after mowing oak. That will kill the oak but not burn up the stems. Burn next during the growing season so the lush growth will dampen the fire as it consumes the previously kill. To prevent too-rapid conversion—leaving refugia behind—no patch can be burned next to a recently burned patch. To keep down smoke each patch must be mopped up by the end of the day. The small-unit patches create lots of perimeter to defend relative to area burned, aggravated where that perimeter is a strip mall or office park.

The social preparations can be as demanding. Each burn requires 50–100 phone calls and involves mailing a ream of postcards. Managers answer to a commission of landowners and political appointees. Adequate staffing requires crews assembled from Department of Environment and Conservation Rangers, town fire departments, the Nature Conservancy, the U.S. Fish and Wildlife Service, and the National Park Service. Taking a landscape, however distorted it might seem to ecologists, "back to the Pleistocene," can arouse passions, especially among those who have grown up with the old pine bush and assume that the landscape of their childhood is the true and only state of nature.

The upshot is that a place that had no referent point has become, for the Northeast, the model for getting fire back into places that need it. It serves as the flagship for the New York State Fire Initiative, a metropole for repopulating colonies of Karner blues elsewhere, and the standard for restoring similar pitch pine-scrub oak sites like the Montague Plains in central Massachusetts. It can also stand as an object lesson for the rest of the country.

• • •

Really?

To fire officers accustomed to wildlands, not countryside, who calculate scaling in the tens of thousands of acres, who have few opportunities to prep burn plots beyond cleaning fuelbreaks, who backfire on an order of magnitude larger than the whole APB Preserve, the fire program at Albany Pine Bush can appear negligible. This doesn't feel like genuine fire ecology. It doesn't look like restoration burning. It looks like the wildland equivalent of colonial Williamsburg. Besides, APB sends its fire techs out West for experience; western fire staff don't cycle into APB.

They're wrong. A true fire ecology of Earth must include arable fields, woods pastures, and transhumant grazing regimes, as well as back-of-beyond wildlands. It must include the cultural heathlands of northern Europe and forest plantations in Brazil. It must include broken biotas in need of restoration. And it needs to include pine bush preserves. All landscapes are now, in some degree, cultural. You can't understand fire's ecology until you understand it in all its settings.

What APB contributes especially is to remind us that fire is deeply biological. Knowing fuel loads is useful for predicting fire behavior, but not, in such circumstances, how those fuel loads happen and what the effects of fire behavior are. They don't answer the questions about sustaining habitat; they don't say how to keep Karner blues prosperous. The biota doesn't exist within a matrix of fuels: the fuels exist within a biotic matrix. Fire alone won't drive out black locust or promote lupine, whatever the fuel loading. Set properly, it acts as an all-spectrum catalyst. The load of mycorrhizal fungi in the soil matters as much as the load of 10-hour fuels on the surface. Fire is an accelerant: the driptorch is a fulcrum to leverage other treatments: it interacts with its ecological matrix. Albany Pine Bush burns because fire does biological work that nothing else can.

What about getting burning to scale? Scaling fire practices doesn't only mean scaling up. It can mean scaling down. At some point, up and down, the results are not simply more or less of the same, but a change in kind. It means a toggle switch, not a rheostat. It means knowing how to reconfigure burn plans to satisfy the tiny. As America's estate continues to fragment, small-scale burning under tight constraints will become more of a norm. The Albany Pine Bush Preserve shows how it can be done.

The preserve expands our notions of what kind of institutions can govern landscape fire management. A solid program needs stable funding, political attention without political meddling, and hybrid agencies and collaborations. It needs a clear mission statement that can be made operational. It needs a fire culture, however eccentric it might seem to someone from the San Gabriels or the boreal realms of the Yukon Valley. The Albany Pine Bush shows what the Northeast requires for a successful fire program.

Paradoxically, it may be easier to transfer experience from the East to the West than from West to East. If it can work in the Northeast, it can probably work anywhere. It might even work in that confusing era that we call the Anthropocene, a future that won't be restricted to relict patches, however large, left from the Pleistocene.

The WUI Within

T HE NATIONAL FIRE PROTECTION ASSOCIATION was incorporated in 1896, the year before Congress passed an Organic Act for the national forests. In 1905 the Forest Service took over those reserved forests; two years later the NFPA began publishing its journal. In 1910 the Big Blowup in the Northern Rockies traumatized the USFS and catalyzed a national program of fire suppression. The next year the Triangle Waist Co. fire similarly bonded urban fire protection to Progressivism.

As those chronological pairings suggest, the world of fire was fissioning into two realms. One was public wildland, overseen by foresters. The other was the built landscape of city and industry overseen by engineers and architects, and this was the domain of the NFPA. Over the coming decades each grew separately, buffered by a still large rural setting, until postwar sprawl swallowed up that countryside, binding those two fire realms with asphalt velcro. Wildland fire overflowed into an urban fringe; structural fire protection scattered like embers into wildlands. The larger landscape displayed an increasingly fractal geometry of fire or to shift metaphors a non-Euclidean geometry in which seemingly parallel institutional lines might cross. The USFS had to cope with burning houses. The NFPA had to imagine codes and standards to govern fire protection amid wildlands.[1]

• • •

Still, those early tremors seemed peculiar to California, a fiery equivalent to its earthquakes. Even there no one outside wanted to "own" the issue. Instead, fire itself forced a fusion when flames in 1985 crashed through suburbs in Florida as well as California and consumed some 1,400 homes. Suddenly, the problem was big, national, and undeniable. Forest Service researchers, with western lands in mind, dubbed it the "wildland-urban interface (WUI) fire problem." The next year the USFS partnered with NFPA to establish a National Wildland/ Urban Interface Fire Program. Never had such a dramatic task begun with such a klutzy name.

The 1985 fires were the early signals of a long-wave drought that began to settle over the West. A whopping outbreak of fires struck California in 1987. The Yellowstone fires followed in 1988. When the ash settled, the wildland fire community was ready to turn from an obsession with wilderness fire to something seemingly less abstract and controversial. The WUI was the obvious candidate, and the National Fire Protection Association was the preferred partner. The NFPA could reach out to the urban fire services as the National Association of State Foresters could for wildland fire protection.

But the regrettably named WUI went beyond a fire of political convenience. More and more, the federal land agencies saw their mission distorted by the gravitational pull of houses burning along their borders. That telegenic spectacle deflected attention from fire's restoration, it redirected fireline strategies, it absorbed firefighting resources, it put fire costs on steroids. It threatened to turn fire as an integral part of land management into fire as another emergency service. And the problem got worse. The legacy of wildland fuels from long decades of fire exclusion, former rural lands now overgrown with houses and scrub, everything intensified by a stubborn shift to a drought-prone climate— all made the WUI into a black hole that threatened to suck everything else into its maelstrom.

The federal land agencies like the Forest Service tended to view the encroaching problem as an alien infestation, as though exurbs were an invasive exotic, though with an exoskeleton plated in wood. Such fires were neither in their mandate nor their experience nor their temperament. They were not equipped to fight them or to study them. The NFPA stepped into that gap. It brought its meticulous style of fire investigation to bear on the 1989 Black Tiger fire outside Boulder, Colorado, and in 1991 to the Spokane fires and the Oakland Hills fire. In 1993 the term *Firewise* was coined. Both fire communities, urban and wildland, began to interact. The NFPA participated in National Wildfire

Coordinating Group committees. The Forest Service sponsored research into how nominally "wildland" fires actually burned houses. The 1998 season with the International Crown Fire Modeling Experiment validated lab models, while the fires that year in Florida confirmed the urgency of the task. The national program, now known as Firewise Communities, scaled up.

The 2000 National Fire Plan brought serious money to the table. The next year saw a two-year pilot project underway, which then segued into a national program funneled through the state foresters. By now the program had many sponsors beyond the Forest Service, including the Federal Emergency Management Agency, the U.S. Fire Administration, the National Association of Fire Chiefs, and the National Association of State Fire Marshals, all with longstanding ties to urban fire services, while the Department of Interior added to the wildland side. Within a decade Firewise had over 600 communities enrolled and was targeting 1,000. Those formally enrolled measured only a fraction of its influence as hundreds of others incorporated reforms on their own from the Firewise example.

The I-zone fire, as Californians called it, was a borderland, a world of its own, yet one that forced each of its paired sides to internalize the other. In 2010, recognizing that fire protection had shifted from urban cores to the fringe, the NFPA created a Division of Wildland Fire Operations. The next year, when the federal agencies published their National Cohesive Strategy for Wildland Fire Management, they identified one of their three primary concerns to be fire-adapted communities.

•   •   •

The understanding of the I-zone, however, continues to be shaped primarily by the wildland fire community. The Forest Service in particular identified the problem, named it, and funded programs to address it. The service saw the WUI not only as a legitimate fire issue but as a threat to its larger mission in land management and fire restoration.

It did not want to absorb the WUI so much as to hand it over to others so it could attend to its core purposes. It saw the outbreaks as private-sector fires that the private sector should handle. It turned to NFPA to assist with structural fire as it did the Nature Conservancy to help with restoration; Firewise Communities was the political equivalent of the Fire Learning Networks. These were national programs, not federal ones. At the same time, the Forest Service handed over supervision of the National Incident Management System, which it had developed, to FEMA. It did not want to become an all-hazard emergency response operation: it was a land management agency. It regarded the WUI as an unwanted foundling abandoned on its doorstep.

Yet the NFPA offered more than a potential adoption agency. Its participation meant a chance to redefine the problem. The federal land agencies viewed sprawl as an encroachment, as a nasty problem that gnawed at its boundaries like a bark beetle infestation. They understood proper responses in terms of keeping that hazard at bay; the ideal solution was to simply zone exurbs out of the scene. It was equally possible, however, to view the issue from the other side of the border and define the WUI as a far-flung (if outrageous) extension of urban fire, as a species of city fires with peculiar landscaping. Until the early 20th century, American metropoli had routinely burned; then that melancholy chronicle of conflagrations stopped. How, exactly, had that happened? Might those same lessons extend into exurbs?

Viewing the WUI as urban fire could redirect the quest for remedies. It shifts focus away from wholesale landscaping and onto the structure itself, what became the "home ignition zone." It emphasizes exposure, the relationship between structures rather than between individual structures and their interstitial woods. It targets combustible roofing, open eaves, faulty glass windows, attic window mesh. It looks to questions of egress, or how firefighters might get in and civilians get out. It suggests that insurers would not be vital players. It emphasizes the voluntary adoption of model codes. It proffers a political as much as an engineering exemplar for protecting assets from unwanted fire.

Perhaps surprisingly, granted the urban fire services' passion to eliminate fire of any and all sorts, the NFPA understood that fire would happen. It did not seek to promote fire as a means of urban renewal as the USFS longed to reintroduce fire for ecological regeneration; but it accepted that fire was inevitable. In its own way it sought means to live with fire. For the Forest Service a fire-adapted community was one that could coexist with the wildland fires needed by its surroundings, that did not interfere with the agency's larger goals for land management; ideally it was one that was never built at all. For the NFPA a fire-adapted community was one that could survive wildfire. Its codes sought to eliminate as much of the hazard as possible. It scrutinized the house and its immediate, direct-contact landscaping. It regarded efforts to redirect or shut off the flow of sprawl as quixotic. Houses would happen. Houses would burn.

• • •

In 1981 NFPA moved out of congested Boston to a campus-style complex in Quincy, adjacent to the Blue Hills, a prime patch of public open land acquired by the Metropolitan Parks Commission in 1893. As one might predict, it attended to landscaping as carefully as it did the interior workspace. The main building wrapped around a constructed pond framed by planted shrubs and trees. This was a place built to code, from its choice of window panes and door handles to its selection of flower-bed flora. It was also a case study in why the wildland-urban interface, or as some preferred, the American intermix, was so potent. No less august an institution than the National Fire Protection Association had planted its headquarters squarely in the I-zone.

There was not much threat from fire, not only because Boston is not an intrinsically fire-prone place but also because the setting and structure had been meticulously sculpted to deny fire of any kind a presence. But the fact that the scene existed at all testified to the social drivers behind the WUI. This is where many Americans wished to live. They wanted to see nature, they wanted a buffer of privacy, they wanted a secure setting. Given a choice they would spread out rather than build up. The NFPA was no more or less than the society that sustained it. Its behavior encoded the culture's values; and they were norms and expectations other than those of the engineer's workbench. Until those values changed, perhaps with an overturn of generations, the WUI would match wilderness as the defining landscape of cultural interest.

The drivers were powerful because they came from within. The NFPA gave physical expression to that fact, too, when it constructed within its hilly headquarters a waterfall that splashes from the foyer down to the lower level where, outside massive windows, the pond resides. Flowering plants line the staggered falls, golden carp swim at the bottom. If, at 1 Batterymarch Park, a particle of the city had been brought to the country, it was equally true that a patch of nature had been brought into the built world. The NFPA had internalized the WUI. That's why the problem will not disappear soon, why the NFPA had to create a division of wildland fire, and why the wildland fire community needs its awkward liaison with the urban fire services.

The View from Bill Patterson's Study

START A fire story in Florida, and it will eventually, sooner rather than later, lead to prescribed fire. Start one in California, and it will end, inevitably, it seems, with suppression. Start one in New England, and it will likely lead to Bill Patterson.

There is no one in the regional fire community he doesn't know, and nothing to advance the cause of fire management he has not assisted with. Hundreds of fire practitioners have trained under him. The bulk of the region's fire scientists have studied under him or have referenced his work. He collected the background data and wrote early fire plans for the major national parks. He carried driptorches to the region's archipelago of prime burning sites. He worked with the Nature Conservancy to establish a regional fire program. He carried the torch of New England fire to a national audience.

New England's spidery roads are a nightmare for someone raised out West, full of overlays, legacy routes, name changes, and destinations that often no longer exist. They are a historical artifact as much as a system for transportation, and in this they resemble the region's fire network. But no matter how many turns and missed turnoffs you take, you seem to end at Bill Patterson's study.

. . .

William A. Patterson III was born in St Paul, Minnesota, on July 2, 1945, grew up on family stories from his grandfather, William A. Patterson Sr., about logging the Great North Woods, in this case second-growth aspen, and knew early that he wanted to be a forest ranger. He didn't, but he went one better.[1]

The family moved often when he was young. His father, William A. Patterson Jr., had graduated after three years at the Naval Academy in 1943, became a naval officer, then, after the war, a carrier pilot until a medical discharge ended his career. Despite the moves, the family returned in summers to a hunting cabin built by William A. Sr. in northern Minnesota. The woods and what had happened to them impressed the young boy. In 1956 his father was transferred to Hingham, Massachusetts. The family arrived a year before the Great Plymouth fire burned in a rush, with 150-foot flame lengths, until it finally expired against the Atlantic shoreline. "I have this vivid memory," Bill recalls, "of being driven to scout camp through this absolutely blackened landscape." In high school he was an average student and considered the "quietest boy" in his senior class, the kind of personality that might find life in a lookout tower attractive. In fact, he built a fire lookout tower out of toothpicks.[2]

Inevitably, he went to college with the idea of being a forester. He attended Maine, and eventually came to appreciate how the frontier of logging had moved from New England to the Lake States. He notes his grandfather acquired his lands by paying the back taxes owed. That land had originally been granted to "a Dakota-Sioux halfbreed in late 1864 and eventually sold to William D. Washburn (US Rep and Senator for Minnesota, founder of Pillsbury Co., first President of Soo Line RR)." The pines had been clear-cut in 1883; the slash burned, then returned mostly to aspen. His grandfather started cutting the aspen. It's now a tree farm. In 2005, Bill III's son (Bill IV) returned to Maine to oversee the management of more than 300,000 acres of forest land for the Nature Conservancy.[3]

The University of Maine-Orono required that Bill join an army ROTC program; but Bill followed family precedent and enlisted in the Naval Reserve in 1964 (eventually being commissioned an ensign upon graduating from Maine in 1967). There wasn't much fire in his program of study. The closest he came was a study of the effects of burning slash on soil properties for an honors thesis. When he returned to Minnesota for a master's, he found an active group of fire scientists. *Fire ecology* was a term finding its legs, and he heard its background hum, even as he pursued new disciplines—limnology and paleoecology. By now forest ecologists were discovering that fire was not so much destructive as transformative, that its removal could be as disruptive as its unwanted presence; and palynologists were discovering that all that charcoal in their cores was not simply noise, the junk DNA and lint of paleoecology, but a rich source of information about fire history. His thesis work under Henry Hansen focused on Itasca State Park, a place with a robust chronicle of fire. In 1969 he earned an MS in forestry.

The navy finally called, and from 1969 to 1972 he spent his time first aboard ship and then in Saigon working with an intelligence unit as part of the Vietnamization program. When he returned to the States, he picked up where he had left off, pursuing dissertation research at Itasca, combining paleoecology with modern disturbance ecology. In eerie echo of his grandfather, part of his work involved the killing of aspen (Agent Orange had been used since 1967 as part of herbicidal slashing, followed by burning). By now the first text on fire ecology, edited by C. E. Ahlgren and T. T. Kozlowski, was out (1974), and the hum had become a buzz. Still, what Bill knew about fire he had learned indirectly by reading *A Sand County Almanac* and Loren Eiseley and listening to lab banter and seminar speculations, even as an unhappiness with national wildland fire policy that had simmered for years was beginning to boil over. In 1968 the National Park Service recanted the 10 a.m. policy in favor of a policy of fire restoration. The U.S. Forest Service, too, stutter-stepped toward a new policy.

By the spring of 1976, with the prospect of an end to his dissertation and to work at Itasca, and with jobs scarce, he worked for the Minnesota State Planning Agency on potential impacts of a proposed copper-nickel mine; the project lasted two years. When a position opened at the University of Massachusetts-Amherst in urban forest ecology, he applied, then scrambled to complete his PhD in time to meet the hiring deadlines. Modestly—it's an instinctive trait with him—he claims he didn't know the subject, but the hirers saw enough in him to waive the details. By the time he arrived in the fall of 1978 the U.S. Forest Service had also rechartered its fire program.

None of that mattered much to New England. For the region, large fires were a memory, not an inspiration. The fire revolution didn't seem to matter much to Bill Patterson either. He spent parts of his first four summers from 1979 to 1982 on the Seward Peninsula in what was becoming the Land Bridge National Monument and later in the Noatak National Reserve; later, he returned for Noatak Biosphere Reserve. Fires were something that happened out West, or deep in the Alaskan bush, or back in the heyday of Lake States slashing. Fire restoration didn't seem relevant to Amherst, Massachusetts, or to someone hired to teach urban forestry.

Then he came to a fork in the road, which quickly turned into a roundabout. In the spring of 1981 Bill decided to offer a grad seminar on fire, for which he could use fields in the Quabbin Reservoir for prescribed-burn training. In May 1981 he conducted his first prescribed burn. (He notes wryly that he did not evolve into prescribed fire from suppression because he has never fought a wildfire.) With fire ecologists thin on the ground, that was enough to interest the National Park Service, then mandated to write fire plans for all their holdings, even those in the Northeast.

The program started quietly with a study of Acadia National Park, the most famous site of the 1947 fires. But its scope expanded to include Cape Cod National Seashore, Fire Island National Seashore, Gateway National Recreation Area, and Saratoga National Historical Park, which wanted to burn to keep the scene roughly as it had been when the Continentals defeated the British in October 1777. In 1985, with the help of a graduate student native to Cape Cod, David Crary, he established at Cape Cod a complex array of plots to measure the effects of burning and mowing, both during the growing and dormant seasons, and at intervals of one to four years, all with controls. Where his ancestors had fueled wildfire with logging slash, he proposed to stoke prescribed fire with field data. Probably only the plots at Konza Prairie reap as much information. Camp Edwards on the cape needed to quell its artillery-sparked wildfires and sought his assistance in substituting prescribed burning. The Massachusetts Audubon Society asked him to help burn some grassy plots on Nantucket; the Nature Conservancy wanted help on Martha's Vineyard; and the program migrated to the Elizabeth Islands. He worked with Ron Myers to nurture a burn program at the Albany Pine Bush Preserve. From there he helped birth burn programs at several sites in New Hampshire and Maine (including Waterboro Barrens) and finally at the Montague Plains Wildlife Management Area in central Massachusetts. Each site fed on the others. When the National Advanced Resource and Technology Center (later, NAFRI) wanted to include an eastern regional representative in its long-running course on fire and ecosystem management, it turned to Bill Patterson. He taught from 1991 to 2010.

Suddenly, he had tentacles into most of the combustible parts of New England, what might be termed the Patterson plexus; he had a topic, fire restoration, that galvanized him; he had a cause. Better, he had a teachable cause. He began to offer a one-unit undergraduate course adequate for red carding

(S130, S190) to complement the grad course in forest management. Over the years the numbers added up. With few studies predating their work, Patterson's students laid down the basic matrix for fire science in New England. He spent a sabbatical year as a Bullard Fellow, spending most of his time at Harvard Forest, with a pilgrimage to Tall Timbers Research Station, the hearth of the fire revolution. It was as though a mental magnet had strengthened, which quickly aligned all the fragments and filings pertaining to fire he had acquired over the years. He was bringing to the Northeast the field trials and basic fire science that Tall Timbers had done for the Southeast.

When asked, he points to his students. But everyone knows the fount. In 1991 the Nature Conservancy presented him with its President's Stewardship Award. The next year was the turn of the New England Wild Flower Society, with its Conservation Award. In 2010 the NPS Northeastern Region honored him with its Natural Resources Award for Research, followed by the NPS national Fuels and Ecology Award. In 2012 the Association for Fire Ecology (AFE) bestowed its highest honor, the Herbert Stoddard Lifetime Achievement Award. That was particularly gratifying because he wasn't a member of the AFE (it had been founded toward the end of his career) and because it harked back to the charismatic prophet of prescribed fire, a man whose career had led to Tall Timbers.

In October 2010 he learned that the odd twitchings and numbness he increasingly felt were the product of a brain tumor. He retired into emeritus status. The surgery was successful, though the tumor turned out to be neither benign nor malignant but anomalous, which forced him into radiation therapy. He recovered, accepted a continuing appointment, without pay, in which he has taught basic National Wildfire Coordinating Group fire courses and saw his final grad students to their degree. Whatever is happening he seems, somewhere, to be in the mix. His bustle remains undiminished, his enthusiasm still infectious.

In April and May 2017 he attended the Northeastern Natural History Conference in Connecticut, traveled to a workshop on pine barrens in Long Island, joined a field tour to Montague Plains for U.S. Fish and Wildlife Service regional fire management officers, then hurried to a Silviculture Institute to teach regional silviculturalists about fire use, and finally drove to the Albany Pine Bush Preserve for a North Atlantic Fire Science Exchange Monitoring workshop. In spring 2017, with a former Mormon Lake hotshot as a TA, he qualified 26 undergraduates as NWCG certified firefighters, more than in any year since he came to UMass. "Seems people know I am willing to 'work' for free," he dryly notes, "and I am as busy as I can be (but not as busy as I once was)."[4]

• • •

The panorama from his study on the slopes of Brush Mountain in Northfield is a cipher for how he sees fire. He notes that the indigenous peoples called it *Mish-om-assek* (Rattlesnake) Mountain, and the early colonists Brush Mountain. Those labels only make sense if the land was regularly burned. The closed forest canopy that blankets the slopes today support neither brush nor rattlesnakes. In the distance, by a snatch of the Connecticut River, where it bends, you can see the Montague Plain Wildlife Management Area, where colleagues and former students are trying to restore the pine bush biota.

The near view is of course actually a very deep view through history. It testifies not only to fire's longevity on the land but to active burning by aboriginal Americans. The far view is actually a vision into the future. It takes special sight to imagine in the tangle of invasives, dog-hair pitch pine, lost lupine, all amid power lines and a land often hard used, a healthy, reinstated pine bush community.

For all his commitment, Bill appreciates the niche character of New England fire. It matters, though not in the way free-ranging fires in the Bitterroot Mountains or horizon-reaching pastoral burns in the Flint Hills or chaparral fires crowding South Coast suburbs do. The last nationally significant conflagration was Maine's in 1947. The last serious wildfires in Massachusetts broke out near Plymouth in 1957, with a smaller echo in 1964. The last regional complex aligned with the drought years of the early 1960s. He makes no prophetic claims that the Dark Days are about to return or that Maine is primed for a pyric eruption. But he knows that fire has been a persistent feature for millennia, that the last ice sheets left geomorphic patches favored by pitch pine, scrub oak, and fire-loving shrubs, and that they need fire to thrive. And he knows, as any serious naturalist in the region does, that humans have shaped those regimes. People burned, and by land-use practices they have promoted fire-favored species. Strikingly, the distribution of pitch pine and scrub oak are more likely to conform to once-plowed fields than to relic patches of the Pleistocene.[5]

It can be a hard sell. New England may have once been known as the Burned-Over District, but that adjective is a past participle, not a label for the present. There are critics on all sides. Fire scientists trained in the West, which is where most fire schools reside, still want wild lands untainted by human contact. Ecologists in the Southeast accept anthropogenic burning but anchor their philosophies with such fire-obligatory species as palmetto and longleaf pine. New England naturalists can accept landscapes as a legacy of human habitation, and they recognize the value of fire for pitch pine and blueberry, but preserved landscapes are few, fire can be unpredictable, and they imagine landscapes as akin to heritage buildings. Better to replace their inner workings than burn them to the ground and rebuild. Fire seems a specialty tool like a pruning hook, not a broad-spectrum ecological catalyst. Nobody likes smoke. Regional ecologists are more likely to unite in suspicion over fire than in its promotion. Fire might be okay in the kitchen but not at the table amid the conversation.

• • •

He remains, at core, a teacher. Research grants are opportunities to instruct students. Prescribed fires are opportunities to train burners. Demonstration plots are occasions to teach the public. His students are his true legacy, which makes talking to him about himself difficult because he invariably shunts the conversation to his students, who universally refer to him as "Dr. Patterson." It's an odd title—an honorific, really—for someone so congenitally soft-spoken and unconcerned with the usual emblems of status. But it may be apt for someone who favors the whole over the pieces. "Personally," he says, "I'm more interested in communities than in individuals. I'm more concerned that all of human society survives than that individuals do. Because we can't any of us live forever."[6]

The community over the individual. It's a sentiment Bill Patterson believes. I'm not sure, however, that the New England fire community shares his belief with equal conviction, because there is at least one individual without whom that community might barely be said to exist.

Westward, the Course of Empire

Westward the course of Empire takes its way.
The four first acts already past,
A fifth shall close the drama with the day:
Time's noblest offspring is the last.

—Bishop George Berkeley (1685–1753), on
Britain's North American colonies

IN SEPTEMBER 1919, still recovering from his war wounds, Ernest Hemingway went to the upper peninsula of Michigan to fish. He later wrote a two-part short story, "The Big Two-Hearted River," based on that experience, published in 1925. It opens with the panorama of a burned-over landscape.

> The train went up the track out of sight, around one of the hills of burnt timber. Nick sat down on the bundle of canvas and bedding the baggage man had pitched out of the door of the baggage car. There was no town, nothing but the rails and the burned-over country. The thirteen saloons that had lined the one street of Seney had not left a trace. The foundations of the Mansion House hotel stuck up above the ground. The stone was chipped and split by the fire. It was all that was left of the town of Seney. Even the surface had been burned off the ground.[1]

The fire, he realizes later, must have come "the year before." He scans the countryside, "burned over and changed," such that even the grasshoppers are sooty, but decides "it did not matter. It could not all be burned." He seeks out unburned patches, a refugia for the soul. He finds a clean stream with good trout.[2]

The burned-over woods are a metaphor, an external manifestation, of the internal state of the protagonist, Nick Adams, burned out by the war. His fishing expedition is an attempt at recovery. Revealingly, he leaves the river when it splits and enters the dark, quiet waters of a swamp. He might try the dark place later, after he had regained competency and sanity. "There were plenty of days coming when he could fish the swamp."[3]

But the burned-over landscape was not just what literary critics call an objective correlative. It was an objective reality. For decades the region had been ruthlessly cut and fecklessly fired. The reckoned fire year, 1918, had seen massive burns, one of which had incinerated Cloquet, Minnesota. And those fires were the final tremors of an earthquake of axe and torch that had transformed the landscapes throughout the Lake States, even seemingly remote settings such as Seney. They had slicked off cutovers and scoured out cold swamps. Now the landscape, like Hemingway's traumatized hero, was struggling to recover. In fact, a few patches had survived. Rehabilitation would take a lifetime. The scars were deep. Even by the time Hemingway killed himself in 1961, recovery was slow. The vanished woods had banished fire. The wreckage of axe and torch remain today.

• • •

The Lake States were where the Northeast went to slash and burn after it had felled and fired its own woods. Westward, as Bishop Berkeley wrote, had the course of empire moved from England to New England, and now the course of an American empire continued westward from New England to the Lake States. The southern lands of Wisconsin, Michigan, and Minnesota were prairie or oak savanna. Their northern lands were boreal forest, the western expression of the Northern Forest that clothed much of New England.[4]

More closely, Michigan, Wisconsin, and Minnesota were where New York, the Empire State, and Maine, the premier logging state, extended their patterns of settlement. They were places and times of extravagant, feckless, promiscuous, and abusive logging, which made them places and times of extravagant, feckless, promiscuous, abusive, and lethal burning. What the axe scalped, the torch incinerated. What the ice had wrought on bedrock and watershed, fire and axe did to forest and wetland. They violently remade the landscape in ways that would echo through generations.

Once the Great Lakes opened to transportation, the process began, relying on waterways and spring freshets to carry logs cut over the winter. It was the pattern that had prevailed in Maine and much of eastern Canada; it was the same pattern that, at the same time, was felling forests around the Baltic Sea. The world woodland was being cut much as whales were hunted and bison slaughtered to the edge of extinction. This was not folk clearing, but industrial felling. There was little to stay the woodsman's hand. The perception—the belief—that the woods rolled endlessly onward made qualms about exhausting the stock and notions of conservation quaint. There was always more beyond the next height of land. There was little to stop them.

After the Civil War, the pace picked up. Steam opened up the interior of the Northern Woods. The volume of logs rose; the slash left in their wake was staggering. As soon as one plot was cut out, companies moved on to another. Cut out and get out. They passed over the conifer forests like locusts over fields of wheat. Bad as the era of axe and fire was in New England, it was worse around the Great Lakes. The forest was larger, the climate more continental, lightning more potent, the onslaught of axe and flame more vicious and sudden, the counterbalances fewer and less sticky. The Great Lakes forest had a natural fire regime more vigorous than New England's. It had a broad transition biome from savannas. Minnesota was almost as large as New York and Maine combined. It had a river and lakes network for moving logs that made Maine's seem quaint. There were no checks or balances, either social or governmental. There was no federal presence willing to intervene in the name of conservation. The roll call of American conflagrations bulges obscenely during the era of Lake State burns, with spikes at 1871, 1881, 1894, 1908, 1911, 1910, and 1918. It was a Gilded Age, an era of robber barons and untrammeled capital, a time that V. L. Parrington later characterized with a macabrely apt phrase as "the Great Barbecue."

The companies insisted cynically they were performing a public service by opening the dense woods to farmers, that they were bringing the capitalism of civilization to the benighted frontier, that they had to cut out and get out to stay ahead of the fires that gorged on moraines of slash. "If it [the North Woods] is to be saved," the *Lumberman's Gazette* argued, "it must be cut as fast as possible." Once begun, there was no pause possible; fires would take what the axe didn't. After big burns was, the *Detroit Post* insisted, the ideal time to push settlement. "These lands offer the best inducement for new settlers. These lands are now in such a condition that they are all ready for sowing wheat, merely requiring the harrow to be used upon them, in case there is not time to plow." The fire had completed the onerous work of clearing; it had fumigated and fertilized. No pests and vermin since they were extinct, no threat of future fires since the land was incinerated, high wages since everything would need to be rebuilt. There was even a small trade in memorabilia and trinkets. There was no reason to delay.[5]

• • •

Ironically, the primary check was the blowback caused by fires.

Here lies a paradox created when fire ecology met fire history. Free-burning fire was elemental in the dynamic of the boreal forest. Most white pine stands grew on old burns, nearly all birch and aspen followed flames. Fire had coevolved with the Great Lakes forest. What changed was not a sudden immersion of flame into a landscape that had rarely known it, but an orders-of-magnitude increase in its breadth and intensity. The coevolving, check and balance of forest and fire broke down. There was more to burn and less to halt it.

The reason is that the logging companies did not pick up after themselves. The massive debris fields they left behind fed massive fires. The slash dried and burned on a scale and at times unprecedented in the region's natural history. The track of westering Americans was a trail of burning, not just figuratively but literally, a long swell that had its tempests and tsunamis, and these overwhelmingly crowded around the Great Lakes. From the 1860s to the 1930s, overlapping the wildland fires that became notorious in the Northern Rockies, these were the Great Fires that gave state-sponsored conservation its lurid pulpit.

Institutions to cope with them were feeble, where they existed at all. The Lake States came halfway in the national narrative, as they stood halfway across the country. They came between private exploitation and public protection. They could look east to New England for one kind of state response, and west to another, more federal model. The political economy of the time favored development, which meant privatization, unchecked capital, and the rapid conversion of wildland to farms. But the newcomers had no knowledge of local landscapes or long-acquired folklore to know where to cut and burn and where to leave well enough alone. They left that knowledge in Europe or the backwoods of New England. Only luck decided whether fires burned hugely or died out in plowed fields and cold swamps.

The wreckage of woods and towns and lives did not go unrecorded. In 1871, the Black Year, as the *Chicago Tribune* called it, the immense fires in Wisconsin were paired with the great Chicago fire that burned simultaneously; *Harper's Weekly* printed engravings still used today. The 1881 Michigan fires pushed the

Red Cross into civilian disaster relief for the first time, and the Weather Bureau sent Sgt. William Bailey to inquire into meteorological conditions, leading to the first published use of the term *fire storm*. Newspapers reported the horrible, thrilling, lurid accounts of folk fleeing, of settlers rescued moments ahead of the flames by trains, of desperate flight into rivers, lakes, and marshes. Most farms failed and left trashed landscapes, but the sagas end with the trees felled and the soil plowed. The ecological and economic wastage was slower, less visible, less amenable to heroic narratives of heroes and villains. But fires were, and so fires gave light and power to the story.[6]

That outburst of flame inspired one of the two literary traditions of settlement fire (the other being the prairies). That literature spoke not only to fires that were more savage than those to the south, but to people, educated, literate, outraged and willing to document where the saga of American pioneering had turned to the dark side. We don't know all the big fires of the era, but we know those that became great because the literate class—priests, schoolteachers, ministers, journalists—wrote down what they had witnessed and often photographed them, like ghastly battlefields in a misguided war on nature. No fire has inspired so many books as the 1871 Peshtigo burn. New books continue to be written for Hinckley, for Baudette, for Cloquet. No other region, save perhaps Southern California in the post-WWII era, has so rich a documentary record. The first enduring images of forest fires in America are those that illustrated the grotesque bestiary of flame loosed and gorged on the overdosed combustibles around the Great Lakes. The memory still lingers in the 21st century in Jim Harrison's *True North* (2005), where the young scion of an old timber baron who had "laid waste" to the Upper Peninsula marvels that "the grandeur of the destruction had been mythologized in story and song" before he seeks to purge that past from his future.[7]

•   •   •

These were frontier fires, a part of the inevitable violence (in this case environmental) by which the land and its occupants changed. All parties recognized this fact, and they all appreciated that the fires would fade away as the frontier passed. Once converted from wildland to farm and town, the big burns would vanish from the land. The wild would be domesticated. Slash would become fallow, and wildfires, field fires. The issue was whether the ride would be a waterfall or a rapid, whether society, specifically the state, should intervene to quiet the axe and quench the torch and extend the process in a less ferocious form or whether it should let the frontier pass as rapidly as possible. A few conflagrations, like the occasional stock market crash, were the price of progress. The Great Barbecue had flames before it had coals.[8]

This was the Northeast saga on steroids. What happened over two centuries in New England occurred in a few decades in the Lake States, occurred in an environment more prone to explosive burning, occurred without the folk mores and local traditions that, here and there, had braked the process even in Maine or the state interests that had in New York led to the creation of forest preserves. So, too, the Lake States looked to the Northeast for inspiration in quelling the havoc, again transitional in its juggling of state and federal actions. And so, also, its lands, like those of New England, underwent a cycle of abandonment and recovery. With the big burns as backdrop, forest conservation became an issue as the Lake States began to organize state forest bureaus and acquire forest lands for the state to protect or reforest. Forest commissions were impaneled, forestry schools founded, and forest bureaus authorized.[9]

The project became a major political movement during the Roosevelt administration. In 1905 he transferred the existing forest reserves from the Interior Department to the Forest Service in the Department of Agriculture. In 1907 he doubled the size of the national forests. In 1908 he hosted the famous Governors Conference on conservation; the next year, the ambitions swelled into the North American Conservation Conference. Continued by his successor, President Taft, the first national forests for Minnesota and Michigan were created in 1908 and 1909. The next year, the fabled Year of the Fires, the Lake States Forest Fire Conference was held in St. Paul—the first such gathering in American history. By now serious timber companies recognized that they could not continue amid the threat of wildfires; in Wisconsin and Michigan they agreed to pool efforts into a Northern Forest Protective Organization. Industry, the states, and the federal government would have to collaborate or flame would take the last stands and prevent any hope of regeneration. In 1911 the Weeks Act established a mechanism for federal-state cooperation, prompting a surge of state forestry bureaus. The more expansive Clarke-McNary Act in 1924 broadened the range of support. The final surge of federal interest came during the New Deal, when tax-delinquent lands became targets for acquisition and the CCC provided a workforce to fight fire and replant woods.

Meanwhile, the amount of cutover, burned-over, and abandoned lands spread like cancer. Such lands were not merely unproductive: they were a menace, a point of pyric infection. They would not mature into rich farmlands—that would happen on the former savannas and prairie to the south; and without formal protection, particularly from fire, they would never regenerate. Instead the states inherited vast swathes of cutover that loggers left rather than pay taxes on the stumps. These became the nucleus for state forests, as much on the model of the Adirondacks and Catskills as on the newly gazetted Superior, Huron, and Marquette National Forests. Wisconsin lagged. In 1910 voters had approved purchase (at fire-sale prices, as it were) for state forests, but then developers pushed back, and in 1915 the Wisconsin Supreme Court voided the act. Not until 1924 did citizens effectively reinstate it.

Still, there were new lands yet to cut, and the ax kept up its work. When the prime white pine was gone, loggers started in on jack pine and aspen, and switched from saw timber to pulp, like miners sluicing through tailings or hunters gathering the bones of bison for fertilizer. But finally too many lands were stripped, and even the organic soil burned away. The communities, too, burned off, or rotted away. The only strategy left was abandonment. The axe went to the still-extant big trees of the Northwest and the Southeast.

The two years after Clarke-McNary were bad for fires but worse, economically, for farming. When drought and Depression struck in the 1930s, opposition collapsed. In 1931 nearly a million acres burned in Minnesota alone. The old mirage of converting forest to farm evaporated. Farms would stay south, forests would return north, and this time the state would keep fires out of both. However the future might evolve, it would include fire protection.

When the Laurentide ice had left, as its vast burden melted away, the land had slowly rebounded upward. So when the heavy hand of settlement lifted, the biotic landscape began to rise from the ruins. The land went into rehab. As in New England the forest, much of it a different forest, returned.

• • •

The oracle of the New England countryside, Henry Thoreau, found a Lake States echo in Aldo Leopold, and *Walden* with *A Sand County Almanac*. Leopold's shack, like Thoreau's cabin, has become a tourist destination, if not a pilgrimage site. The land ethic has become as vital for modern environmentalism as civil disobedience for the civil rights movement. When Leopold acquired the worn-out land that become the farm, with the chicken coop that morphed into the shack, the Lake States landscape was probably at low ebb. Not content to argue only for preservation in the guise of wilderness, Leopold worked at restoration. The land ethic applied to both. The tools that had unraveled landscapes could, with intelligent care, begin to wind them back together.

His personal project coincided with wholesale government efforts to the same end. State and federal agencies acquired more abandoned land. The Soil Conservation Service stabilized soils. The Resettlement Administration helped the relocation of farms from the woods to the prairies. The CCC planted trees, reflooded some wetlands, erected lookout towers, and fought fires. The infrastructure for settlement segued into one for reconstruction. As land passed from laissez felling and firing to organized agencies, fire seeped away. There were fewer people committed to burning and more empowered to put them out. After the war years, the Forest Service became a conduit for surplus military equipment. The region established an equipment development center at Roscommon to assist the conversion from front lines to fire lines. The economy shifted from raw commodities to services, particularly recreation. Big burns ceased to be routine. The great fires remained seared into collective memory, and celebrated in museums, but nothing like the community-consuming megafires returned. When the Forest Service sought to promote a laboratory for forest fire research, as it was doing in the Southeast, Northern Rockies, and Southern California, there was little enthusiasm, and the idea withered away. The problem fires were a bad nightmare from the past. The national narrative for fire, like the old loggers, had moved on.

Revealingly, Aldo Leopold, then at the shack, died on April 21, 1948, when he spotted a fire on a neighbor's land. It had started in trash, the domesticated slash of a tamer time. A flank of the fire moved toward marshlands and flashed toward a patch of pines he had planted. Neighbors gathered to save the neighbor's house, while Leopold took a backpack pump and worked the errant front. He collapsed with a heart attack. The flames burned lightly over him and inscribed the final entry on his notebooks with scorch marks. The next year *A Sand County Almanac* was published posthumously.[10]

His death by fire came midway between the holocausts of the past and the problematic burns of the future. Thirty years had passed since the last monster fire had threatened Cloquet. Twenty-eight more years would pass before bad fires again returned to Seney.

• • •

The region's fires did not suddenly cease. They remained common, but as with the Northeast, they had few opportunities to blow up and run freely through trashed landscapes. Wisconsin, Michigan, and Minnesota created prominent bureaus to manage the states' public forests and formidable fire control apparatus to throttle fire out of the rural landscape. The biotic partition of logging from farming quelled the most provocative causes. But the old boosterism that had invited new settlers to seize the burned lands because the fire threat was over had, in macabre ways, proved true. The new regimes kept bad fires small. The region banished its fires into memorial plaques and county museums, while the country boxed up that awful heritage and put it into archives.

The land healed. The forests came back. Large swathes were now public land, either state or federal, and these felt the impress of an environmental movement that bubbled out of the 1960s and was codified in the 1970s. Public lands began to specialize, spinning off wildlife refuges, national parks, and wilderness—the most celebrated, at Boundary Waters, not far from disaster fires in 1908 and 1910. Fire returned. They were modest by historic standards, like fires smoldering in peat that pop up when the winter snow melts off. But institutions, however reformed, were in place this time, not the wistful hope of reformers. Scientists probed the past for fire scars and charcoal in lake varves, extending the region's fire history beyond the eruptive era. Fire officers were less obsessed with suppressing big burns, which seemed unlikely, than with restoring good fires.

The best-known wildfires were, in fact, escapes from prescribed burns—the core emblem of the national fire revolution—gone bad on public lands. Two in particular frame the 1970s. In mid-May 1971 a holdover from a prescribed burn blew up on the Superior National Forest. The Little Sioux fire rushed across 14,628 acres and stunned fire managers not accustomed to fires of this size, even if it was two orders of magnitude smaller than the historic burns. The decade closed with an escaped prescribed fire on the Huron National Forest to promote habitat for the endangered Kirtland's warbler. Some 30 hours and 24,000 acres later, the fire had burned 44 homes and structures in the hamlet of Mack Lake and killed one firefighter. Both fires led to model case studies by Forest Service fire researchers. Both caught national attention. But neither was likely to stake claims that the region was again in the vanguard of national fire reforms.[11]

Rather, the episodes appeared as cautionary tales, of what can happen when a region that had (to national interests) lost its fire capacity, had managed to dampen rural burning and to quickly kill wildfires, tried to adopt cutting-edge practices for a new era. Rightly or wrongly, the Lake States remained on the periphery. Their most active fire management lay in restored prairies, not a recovered North Woods.

There seem no unique fire problems that the rest of the country look to the Lake States to resolve. The fires that preoccupy the region are local expressions of national themes—wilderness, WUI, prescribed burning, fatalities. The Lake States is the critical habitat for none of them. The core region for prescribed fire is Florida. The core for backcountry burning is the Northern Rockies. The core for wildland-urban fringe fires is California. What the Lake States bring to the fireline is history.

• • •

After the wave of slashing and burning that swept over Seney, developers moved in to encourage farming. A land company cut ditches to drain wetlands and swamps. Farmers plowed and burned, and then, as their forbearers had in New England, left. Better to abandon than pay taxes on ruined land. Those lands fell to the state, which in the case of Seney wanted to off-load the burden onto the federal government by proposing the feds convert it into a wildlife refuge.[12]

Amid the New Deal, conservation programs flourished. The upshot was Seney National Wildlife Refuge. The Works Progress Administration and CCC moved in and began systematically rehabilitating what earlier visitors had systematically trashed. The wetlands reflooded. The woods grew back. Wildlife, particularly migratory birds, returned. And so, too, slowly, probingly, feeling its way back like black bears and badgers, did fire.

Fire had been away a long time. The lavish slash that had fueled its riotous explosions was gone. Loggers, farmers, promoters, transients—all those who had set those old fires had left. Fire protection had spread over the peninsula and, in the postwar years, mechanized. Some 25,150 refuge acres went into legal wilderness. It would take time for spark, woods, and weather to recombine. No one lamented the lost conflagrations, but wildlife specialists did worry about the complete remission of burning. Some fires would happen, and some fires were useful, some were the emergent properties of new opportunities on old landscapes. Eventually, fire would return in avatars, wild and prescribed, that suited its new circumstances.

In 1976 they all came back, all at the same time, all in the same place. Seney had not suffered a major drought in 40 years or a significant fire for 70. Outside the refuge, a wildfire started on the Lake Superior State Forest. Within, on July 7, the refuge staff kindled the Pine Creek prescribed fire, an experiment in habitat maintenance, originally planned for 40 acres but, after Department of Natural Resources protests, shrank to one acre, which smoldered stubbornly in dry muck. (Intermittently since the 1940s, with uncertain results, the refuge had tried some controlled burns.) On July 30 lightning started a fire, later named Walsh Ditch and treated as a prescribed natural fire, in a recently designated wilderness area. Three categories of land, three management goals, three fires.[13]

The Michigan Department of Natural Resources suppressed, with mechanized equipment, the wildfire. The Pine Creek prescribed fire slowly burned on, gnawing through dense regeneration and peat until, with the assistance of the DNR, the refuge stalled the fire at 200 acres. The Walsh Ditch fire, monitored intermittently by air, swelled to 1,200 acres and seemed to gather strength for a rush outside its sanctuary. The refuge sought help from the DNR to contain it while also requesting national assistance in what escalated into the largest mobilization of the year and became the most expensive fire in Department of the Interior history to that time. The imported fire team elected to burn out the entire wilderness area. Eventually, the Pine Creek fire was drowned after diverting a stream with bulldozers. On September 7 the Walsh Ditch fire was declared controlled. But the drought deepened, the smoldering burn found fresh fuels, the fire blew up and out of the refuge. Arson fires appeared. The demobbed crews were sent back. When snow finally quenched the flames, the Walsh Ditch fire had blackened 72,500 acres. Fish and Wildlife Service reviews suggested that the ecological effects of the fires had, overall, been positive.

The experience of 1976 was deeply inconclusive. Fire managers knew they could not simply suppress every fire, and they knew, too, that many landscapes needed the right kind of fire. Yet history was an unreliable guide—what kind of restored fire suited this kind of recovered land? Ecology understood the basic principles, but fire is a particular event at a particular place, and it was unclear what the outcomes would be. (They were mostly surprising.) The cost of the outbreak was obscene, not only in money but in political attention. With several other stumblings to follow, the fires put the FWS on the path to a national reformation of its program. Revealingly, the catalytic event occurred in Florida, not the Lake States.[14]

Still, the summer's fires were synecdoche for the region. With the recovered biota had come a recovery of fire. In New England the public lands that serve as a prime habitat for free-burning fire were small and scattered, and outside sandy soils and swathes of true boreal woods, not inherently fire prone. In the Lake States they were large enough to support big burns, and there were opportunities for fires in the spring and fall of most years, and increasingly summers. The historic burns had gorged on the carcass of scalped lands. The new fires had to incorporate ecological goods and services; the birder's binoculars, the ecologist's transect, the rod and gun of hunter and fisherman, the open space of the recreationist.

No one today looks to the Lake States for insights into the critical fire issues of our time. But the region holds important lessons about how to manage fire in rehabilitated lands, about fire management collectives where the states have equal or greater powers than federal agencies, about the character of the workforce contemporary fire management requires, about what kinds of fire are appropriate. To date, efforts to import intact the systems of the Southeast, Northern Rockies, or California have stumbled. The Lake States require their own solutions.

Yet in an era that promises a future of fusion fires, of hybrids between simple suppression and prescribed burning, the region might enjoy another renaissance because it is, historically and geographically, a transition zone. The Lake States stand midpoint in America's pyrogeography—geographically, between Northeast's private land tenure and the Northwest's heavily public lands; historically, between the era of laissez-faire cut-out-and-get-out and state-sponsored preservation. The Lake States have pieces that exist (and have purer expression) elsewhere. What it offers uniquely is a chronicle of how those pieces have come together.

In times of uncertainty and rapid change, the generalists survive. The variety that prevents the Lake States from imposing a model on the rest of the country may prove their greatest asset in a future that promises changes as radical in outcomes, if more subtle in means, than occurred in the settlement past. The nation's fire strategists could do worse than take a few days and follow Nick Adams into a resurgent Big Two-Hearted River.

SLOPOVERS

Missouri Compromise

WHEN HIS WESTERING finally took Daniel Boone to the Femme Osage district beyond of St. Louis, the Missouri territory remained a marcher land, an unsettled locale between frontiers, of which there were several, each seemingly incommensurate, yet crossing one another like a braided stream. One frontier was political, the division between slave state and free. One was environmental; here the eastern woodlands thinned and the western grasslands thickened.[1]

And one was historical, the place where the old trans-Appalachian frontier ended and the trans-Mississippi frontier began, ready to sweep across the wide Missouri to the Pacific. A few long hunters followed the northward trek of Lewis and Clark along the Missouri River to found the Rocky Mountain fur trade; more trended south into the Ozarks, where they recapitulated the world they had known all their lives. In the winter of 1818–19 Henry Schoolcraft traveled through the hills and recorded an account that could have applied to this whole peculiar generation.

> The settlement at Sugar-Loaf Prairie consists at present of four families. . . . These people subsist partly by agriculture, and partly by hunting. They raise corn for bread and for feeding their horses previous to the commencement of long journeys in the woods, but none for exportation. No cabbages, beets, onions, potatoes, turnips, or other garden vegetables are raised. Gardens are unknown.

Corn and wild mats, chiefly bear's meat are the staple articles of food. In man-ners, morals, customs, dress, contempt of labour and hospitality, the state of society is not essentially different from that which exists among the savages. Schools, religion, and learning are alike unknown. Hunting is the principal, the most honourable and the most profitable employment. . . . Their system of life is, in fact one continued scene of camp-service.[2]

A single generation, Boone's, liberated by the American Revolution, had made a long hunt from the hinterlands of the Atlantic to the western tributaries of the Mississippi.

It was their sheer westering that had sparked these proliferating frontiers. What had been separated, joined. What had found common cause among the British colonies in expelling Britain now split over who would control the West. The progeny of that Great Migration came to rest in Missouri. Those new lands destabilized the old political equilibriums, particularly between slave and free states. The entry of Missouri into statehood nearly stressed the system to the breaking point and forced an accommodation, the first of several. The Missouri Compromise of 1820 was enacted the year Boone died.

Something similar may follow America's fire revolution that raged from the mid-1960s to the mid-1970s. Insurgent groups had united against a common foe—in this case a hegemonic commitment to suppression. But as new lands became available for absorption into the new order, one committed to fire's restoration, they could quarrel about means and ends. America's fire polity split into two dominant creeds. One looked to wilderness as a guide and tolerated human activities insofar as they led ultimately to the removal of human presence in favor of fires that could free-range as fully as wolves. The other looked to working landscapes for which fire remained an implement for hunting, herding, logging, and other forms of sustenance that serve human economies. There was little common ground between them: any land, it seems, must ultimately sub-scribe to one or the other. The lines between those two visions, often with legal and political sanction, are rigidly drawn. This time the national polarities do not align north and south but east and west. The wilderness ideal remains firmly anchored in the public domain of the West; the working landscape, in private ownership for the most part, or on the public lands providing recreational ser-vices, in the East, especially the Southeast.

Missouri sits between them, a middle ground—middle geographically, mid-dle thematically, middle politically. It remains fundamentally a landscape of the border, settled when the public domain was being sold off or handed out as quickly as possible ("doing a land office business" was a phrase with literal punch). Over the past few decades this landscape has become again unsettled, a frontier in the environmental contest between the wild and the working. Out of it perhaps is emerging a new Missouri Compromise.

• • •

The Ozarks form a modest uplands, spanning southern Missouri and northern Arkansas, grading into foothills eastward along the Ohio River and westward into eastern Oklahoma. Its core is a granitic dome, long ago leveled, and then raised again into a shallow plateau. That uplift entrenched the major rivers, complete with meanders, and it kindled a new era of erosion that dissected the plateau into an intricate lacework of hills and hollows.[3]

They constitute a distinct landscape for fire. Compared to the Great Plains, they broke down the capacity of fire to free-range. The bluffs, the spring-fed streams, the ravines—all fragment the ability of wind-fetched flame to soar untrammeled. Compared to the eastern plains, etched primarily by streams, the stony rims add texture to the terrain, thus doubling the resistance offered to fire's spread. It was possible in the East to amass burned patch by burned patch into extensive prairie peninsulas and barrens, particularly on karstified limestone like the Pennyroyal that seasonally removed streams. But the topographic texture of the Ozarks fractured even those features into smaller parcels, many of which were less readily fired or given to grasses. Early observers thought the biota similar to the prairies and the terrain similar to Appalachian hills. In his journal of travels Henry Schoolcraft described the routine burning of uplands and slopes.[4]

Both biotic realms, western prairie and eastern woodlands, thrived in the Ozarks but in different settings. The rolling uplands were savanna woodlands; the ravines held the thick forest, tucked away from wind-driven flame. Perhaps a third of such woods was shortleaf pine; the rest, a mixed oak-hickory hardwoods. Dry lightning is rare. Fires are set by people, and like people they have to struggle to overcome the tendency to split and diminish any movement through the hills into ever-tinier tributaries, a kind of reverse stream, splintering into rills and springs of fire as the process proceeds deeper into the plateau.

As the entrenched rivers deepened, and then meandered, mesas were sometimes left within oxbows, which further eroded into a still deeper isolation, what became known locally as "lost hills." Geographically and historically, the Ozarks became themselves a lost hills. Geologically, they stand as an outlier and muted echo of the southern Appalachians, much as the Black Hills do for the Northern Rockies. Ecologically, they are a triple junction, where the oak woodlands, the southern pines, the tallgrass prairies converge.

• • •

The Ozarks are not prime farmland, although there are bottomlands that qualify, but the interior was largely shunned by colonizing agriculturalists. It knew the usual sequence of prehistoric inhabitants, from Archaic to Woodland peoples, before feeling the outer touch of the Mississippian civilizations. It lay on the margins of those cultivating civilizations that claimed the humid bottomlands of eastern North America, raising maize and building mounds. While relics remain to testify of these various occupations, those peoples themselves had gone, perhaps through that mysterious collapse that swept away so many societies across 14th- and 15th-century North America, from the Anasazi to the Hohokam to the Mississippian Oneota. Throughout, the Ozarks were likely occupied seasonally, part of an annual cycle of hunting and foraging—a Barrens in the hills. They abounded with game from turkey to bison, deer, and elk. By the time exploring naturalists arrived, and trees in the mid-17th century began recording fire scars, permanent occupants had vanished. Their fires left with them. The Ozarks became a fire sink.[5]

That changed in the early 19th century when the Cherokees, dislocated by the border wars in the southern Appalachians, began to arrive. They found a kindred landscape, well suited to their economies of hunting, forest farming, and foraging, but one they set about fashioning into still more usable forms, for which fire served as a universal catalyst. They were joined by long hunters and their families, who were, by the reckoning of most observers, indistinguishable in their land use from the displaced natives.

The record of burning ticked upward; and when drought overlay the hills, it became widespread. The burning dappled the Ozarks with prairie pockets and barrens, balds and glades, and where the prevailing westerlies could blow freely, as on the uplands, oak savannas emerged of varying purity. Early observers reported that "both the bottoms and the high ground" were "alternately divided into woodlands and prairies," that it was overall "a region of open woods, large areas being almost treeless," and that the prevailing cause of this action was fire, for "it was common practice among Indians and other hunters to set the woods and prairies on fire." Later naturalists like Curtis Marbut concluded that the open character of the scene was "without doubt, wholly or principally due to the annual burning of the grass." Carl Sauer later summarized the record by noting that some fires were set in the spring or fall to improve grazing, and thus draw game to preferred sites, that some were set "to drive game toward the hunters," and that such fires were mentioned by almost every early writer "as the cause of the prairies." As domestic livestock replaced wild ungulates the burning persisted.[6]

The record of burning waxed with each surge of immigrants and waned when they decamped. Yet even when thriving, their flames could not propagate everywhere. They constantly ran into ecological baffles and geologic barriers. On that roughened terrain the swells of flame that rolled with the westerlies from the plains broke, like a storm surge against a rocky isle, splashing forward but with spent momentum. Something more would be needed to overcome those internal checks—more people, greater biotic leverage, more firepower. An 1828 treaty sent the Cherokee to Oklahoma. Their forced removal meant a forced eviction of fire. But already a new wave of colonizers was probing into a land partly broken to an agricultural halter before lapsing into fallow. The newcomers preferred to hunt rather than herd, and to herd rather than plant; they had coped with the oak woodlands and barrens since they breached the Appalachians. They were a loose-jointed, restless society that worked best when moving and became troubled when stuck. What spared their settlements from full ruin was the intrinsic dynamic of the frontier. It struck, broke, and moved on, leaving to others the tedious task of gathering up the ecological shards and remaking landscapes into viable habitats.

In Missouri the earliest settlements clung to major riverways, which served as routes of transport and trade. But the broad Missouri River that bisected the state also defined its two biotic realms, the prairie loess to the north and the forested highlands to the south. Vast prairies were not landscapes a backwoods society favored: they were a place for the plow, not the long rifle. The French clung to the rivers; Germans sought out bottomlands and modest hillsides; the Scots-Irish pushed into the interior, where they could hunt, trap, put down maize plots, and loose their herds to fatten on the abundant grassy glades that served as ready-made pastures. In brief, the newcomers favored places akin to those they knew. The floodplains were fever-ridden and prone to cholera; the highlands allowed the newcomers to scatter, as though the frontier were tem- porarily suspended. When asked why they settled the Ozarks rather than the farm-lusher plains, the pioneers said simply they liked the hills. They resembled the frontier they had tracked across.

• • •

As did those before them, they began claiming the land by remaking it in their own image. They hunted, they gardened, they turned out hogs, goats, horses, and cattle onto the hills as an open range, and they burned. The numbers of fires increased, rising with populations of people and their herds. Livestock granted biotic leverage, amplifying the effects of fire and more than replacing the fast-hunted indigenous fauna. Free-range grazing, in particular, invited free-range burning. Soon every hollow and hillside found its match.

Not only fire's numbers but its sites and seasons changed. The Cherokees had preferred setting autumn fires associated with fall hunts on the uplands. This had the added benefit of forcing game to find winter forage in the bottoms and canebrakes, closer to encampments. The newcomers, with livestock to sustain them, mostly burned in the spring, not wishing to strip the uplands of winter forage and pushing for a quick flush of fodder to plump up the stock after a lean winter's fare. This altered fire regime modified the composition and dynamics of the Ozarks landscape.

Still, such ecological nuances were secondary to the sheer increase in numbers of fires and their propagation throughout the countryside. People and their stock overwhelmed the internal checks that had held fire to grassy patches between bluffs, creeks, and southerly exposures. The rains were good enough to keep growing something, and a fire-catalyzed economy kept the land constantly kindled. The fires filled out every nook and cranny. This repeated firing and quenching tempered the Ozarks into a hardscrabble landscape. The fires worsened as a logging rush after 1890 replaced shortleaf pine with slash, and as oak thickets replaced savanna woodlands, and as more and more of the flora broke down into biotic rubble and rock. Visiting the hills, Aldo Leopold concluded that many people burned simply to shield themselves from all the burns others were setting. The Ozark candle was burning at both its ends.[7]

By the 1920s the Ozarks were a shambles. To Sauer's mind they were less retarded than the Appalachians of eastern Kentucky and Tennessee, and he distinguished within them "between the farmer of the larger valley and the farmer of the hillside or small cove," but the people remained relatively isolated and backward, and their lands were a mirror of their sinking circumstances. He thought that parks, forest reserves, and recreation were among the best options for the future.[8]

Its chronicle of fire again records a decline, this time not because people had left but because they had stayed, and in fact multiplied along with their livestock, for the land could no longer grow enough to support fodder for both slow-combusting herds and fast-combusting flames. That old economy of frontier burning had no new lands to move into. Resprouting oaks took over sites once under pine, feeding hogs in ways that pine roots could not. Pasture degraded. Erosion worsened. The felled forests left a scalped and furrowed dome. Then drought and Depression forced another emigration, and state and federal governments intervened to acquire significant tracts of land—a new, reversed round of treaties, as it were—and they imposed doctrines intended to evict fire from the land. Even as the biota rebounded, fires diminished in number and shriveled in size. The fire history of the Ozarks once more tracked its human history.

• • •

Over the past century the Ozarks have experienced another cycle of migration, another reformation of landscapes, and another long-wave cycle of fire. Neither emigration nor immigration is as complete as those of the 19th century, and the emergent landscape is fragmented, with large patches still mired in the old order. But the long rhythm of burning is unmistakable. From 1581 to 1700, the mean fire interval in the southeast Missouri Ozarks was 15.8 years; from 1700 to 1820, 8.9 years; for 1820–1940, 3.7 years; but since 1940 it extends to 715 years. Across some 500 years the landscape for burning had blossomed and then withered.[9]

By the 1920s the Ozarks were breaking down, and they crashed during the drought and Depression of the 1930s. Once the orgy of cutting passed, and its slash had burned, fires thinned, and those that survived weakened, due to the sheer accumulation of the human presence. There wasn't enough to burn in the old way. Then people began decamping, lands fell into tax delinquency, and the flinty stubbornness of Ozark political culture cracked. The removal of the human hand created a new frontier, as land changed ownership or acquired a new cover, or both. Missouri came late to conservation, but it came with hard-wrought compromises that bequeathed an institutional steadiness.

Between 1929 and 1933 the General Assembly authorized the federal purchase of forest land under the Clarke-McNary Act. Soon the U.S. Forest

Service acquired 1.3 million acres to make the Mark Twain National Forest. The election of 1936 established a Missouri Conservation Commission, later renamed the Missouri Department of Conservation (MDC), which oversaw forestry, fish, and wildlife, and began acquiring lands of its own, apart from state parks. Decade by decade, slower than activists wished, but with a steady if oft-spasmodic tread, the institutional apparatus for state-sponsored conservation matured. In the 1960s the National Park Service entered seriously into the consortium with the Ozarks National Scenic Riverways. In 1976 and again in 1984 voters approved a sales tax devoted to conservation programs. By 2000, over 13 percent of Missouri, capturing all the critical conservation elements, was lodged in protected public landscapes.[10]

Thoughtful observers had long agreed, however, that genuine conservation could only follow from grassroots popular support, not elite control over bureaus. The private sector controlled most of the land and would determine the grand mosaic of Missouri habitats. A spectacular fusion of private ownership and public service commenced when Leo Drey, beginning in 1951, began developing the immense Pioneer Forest. Eventually his holdings grew to 160,000 acres, all dedicated to sustainable forestry through selective cutting and intimate knowledge of its intricate mosaic of sites. That experiment helped establish a pattern, if sometimes grudging, of cooperation between private and public sectors. Subsequently, donated land, private reserves (such as those belonging to the Nature Conservancy), and conservation easements have expanded the realm of rehabilitated hills.[11]

Together they formed a mixed economy of ownership, some private, some state, some federal, but all working landscapes. The national forests housed CCC camps, which set about stabilizing soils, replanting hillsides, and stopping fires. Peck Ranch, in particular, became a showcase; and when oak decline threatened to spread from Arkansas, the MDC was willing to log off 17 million board feet to halt it in its tracks. MDC brought back the turkey, and even exported its thriving flocks to a dozen other states. But the most potent measure was fire control. The MDC made fire protection a foundational program on the assumption that ending the biota-stressing flames would allow the land to recover. Pioneer Forest banned burning of all sorts.

In the early years locals often resented the new order: they regarded fire lookout towers darkly as prison watchtowers, and told of soaking a rope in kerosene, setting it afire, and dragging it behind a galloping horse through the woods.

Federal foresters were so exasperated that they arranged for anthropologists to study the local residents as one might Inuit or Trobriand Islanders. But the tide had turned, and that creaky pioneer culture could no more hold its own ebb in place than Knut could stand against the sea's advance. The last blast of fire came in the spring of 1953 when an insurgent outbreak of fire affected an estimated 80 percent of the Missouri Ozarks. Thereafter fires stayed on private lands, or if they strayed onto the public estate were quickly rounded up. In 1967 Missouri at last banned open-range grazing. That reduced the primary motive for continued folk burning to vandalism, a kind of flaming graffiti, unmoored from its economic piers. The old regime collapsed.

Yet after the sighs of relief passed, after the land had recovered sufficiently to regrow pine, oak, and a Midwestern scrub, another, if predictable, fire problem emerged. There was not enough fire to make the landscape habitable according to the definitions of the society that was reclaiming the Ozarks. Left to themselves the Ozarks would ecologically transform into something people couldn't use and didn't like. The region had been settled on a roughly Midlands border pattern, though a footloose piedmont and mountaineer model replicated neither New England village nor coastal-plains slavocracy. The imprint of those origins endured, as did many progeny of the pioneering generation and a peculiar political culture.

The rebuilt Ozarks remain a working landscape, not a wild landscape. But "working" has acquired a new definition. It means recreational, not subsistence, hunting; biodiverse habitats, not open-woods pastures and gardens; sustainable logging, not landclearing and long fallowing; exurban visitors, not backwoods pioneers. It means fire practices suitable to such ambitions—not a restored flaming front, rolling over the hills in a wave of settlement, but a patchy rehabilitation in which varied fires catalyzed diverse habitats. It means a hard slog of fire reintroduction, feeling what flame might do in the new order. As everywhere, foresters resisted, still locked in ancient blood feuds with open-range graziers, land-scalping loggers, and fire-promiscuous ruralites. But over the past 20 years they, too, have converted or retired from the scene.

The emerging Missouri consensus features a mixed economy of land ownership and purposes and a fire regime for the working landscapes of a service economy. These are not the practices of the Wild; nor does that vast corpus of fire science devoted to free-burning flame in the Wild hold much pertinence. These are landscapes with people at their core: people set fires, people determine how fire behaves, people decide what species fire will promote or contain, people carry fire across the political roughness of land ownership and the historical roughness of a new era. A new generation wants fall colors from hickory, sycamore, maple, and the 22 species of oak in the state. They want turkey, and otter, and bear. They want prairie patches high with native grasses and thick with forbs. They want glades not overwhelmed by brush and cedar. They want clean streams for floating—the Ozarks National Scenic Riverways was the country's first such protected complex. They want habitat for the endangered Bachman's sparrow, and as a grail quest, enough restored shortleaf pine to reinstate the red-cockaded woodpecker. There are pressures for wilderness, too, but they are tiny tiles (23,000 acres) in a vaster mosaic.[12]

• • •

To those who consider expansive wilderness as the paragon of nature protection, the Missouri model will seem slow, flawed, and exhausting. They will want immense public estates, and will long for administration by agency fiat, presidential proclamation, and court rulings. Conservation in Missouri has proceeded differently, never moving much beyond a close-argued public opinion, which has required that advocates convince the body politic, not a court or an agency chief. This means a more tedious pace, often lagging national headlines, but a more thorough political legitimacy. Elsewhere, what one administration can declare a roadless area, another can delist. But in Missouri, public opinion, not simply an appointed official, has approved the measures, and what the public has granted after long deliberation it is not likely to cede casually. Conservation in Missouri must work through multiple owners, varied ambitions, and a deeply plowed field of politics. But when it comes, as it has, it speaks with democratic authority as tenacious as matted oak roots.

The wilderness ideal conveys a purity not only of nature but of politics. It works best on empty public lands. It seeks to deal only with the administering agency, not with the whole messy muddle of civil-society politics, which may not be trusted in the end to make the right choice. And it demands a science as seemingly pristine as the nature it aspires to, one stripped of human agency. That type of politics won't work in Missouri, nor will its science. Both must begin with the anthropogenic landscape, subsequently modified to suit local tastes, not with eco-utopian visions in which humans have vanished and the torch left with the last of the Oneotas.

Across most of America our fire policies, and environmental controversies generally, continue to polarize between the wild and the working. Abolitionists remain intent on banning people from preserves, while traditionalists are keen to defend a way of life whose time has passed and that can seem antiquarian, or even ethically repugnant, to much of the national citizenry. As the founding conflict spreads into new landscapes, the prospects ripen for a low-grade civil war, as each side pulls the middle ground apart, forcing it to choose one polarity or the other, all of one or all of its rival. This time, the contested frontier lies not to the west but between the two fires of either coast. In the 19th century relentless expansion wedged a social fissure into a political chasm. In the 20th the growth of environmentalist agendas amid a rapidly unsettled landscape threatened to do the same.

•　•　•

The new Missouri Compromise is an idea, an organization, a practice, and a moment of history. As an idea it puts people into fire behavior as propagators, it puts anthropogenic fire into ecosystems as perpetuators, and it puts cultural landscapes into the pantheon of protected places worthy of restored fire. As an organization it demonstrates how a consortium of researchers, landowners, agencies, and nongovernmental organizations might pool resources to create the heft and momentum to put good fire back into the land and how to carve a place at the table for a region not typically invited to the banquet. As a practice it shows how complex is the task, and how little a putative natural order can serve as a guide.

But it may be as a moment of history that the Missouri Compromise may contribute most enduringly. It shows how to create, if not a microrevolution, then at least an insurgency, against the Establishment. It's easy to forget how startling the assertions of oak, fire, and history were when they were announced. In the early 1990s the Ozarks, like the oaks, were an outlier, an interpretive anomaly, a freak of fire history. Its fire record contributed nothing to fire science. It seemed to have no genuine fire history—fire was an annoying feature like ticks and gnats, not a vital one—because no one had looked for it as anything but ecological vandalism. Now the issue is how to do fire, not whether it belongs. Once again, people are carrying fire across the pixels, this time the rough terrain of fire history.

Two centuries ago the inclusion of Missouri into the union forced a political compromise. Today, for the American fire community, the Missouri experience suggests the contours of a compromise about how to expand the frontiers of fire management without splitting the larger premises behind the project. It shows how to reconcile working landscapes with environmental ideals. This time Missouri is not a centrifugal frontier that threatens to pull the competing factions apart but a centripetal middle that shows how the center might hold.

Fire and Axe

The First and Second Timber Wars

Axe cuts forest.
Fire burns axe.
Forest covers fire.

THE NATIONAL CHRONICLE of historic conflagrations tracks the spoor of settlement. For most of a century, from the 1825 Miramichi fire to the 1918 fires that blasted through Moose Lake and Cloquet, Minnesota, the big burns feasted on the slash left by landclearing. The North Woods migrated from Maine and New York to the Lake States and then to the Northwest. The outbreaks were most lethal where logging and settlement colluded and rail intensified the slashing and burning. When the Lake States began to falter as its great pineries disappeared into mills and flames, the timber industry looked to the southeast, where the longleaf would soon be cut away, and to the northwest, where immense woods had survived the pulse of blowups throughout the 19th century or had grown to maturity in their aftermath.

The Pacific Northwest thus claims a transitional phase in the national narrative between the era of laissez-faire landclearing and that of state-sponsored conservation, and later that phase change within state programs from conservation to preservation. Its 20th-century history can be framed by two timber wars. The first timber war hinged on concerns over a timber famine and wildfire. Conservation promised to reduce waste and havoc and ensure a young forest for the future. Fire control would protect existing and growing woods for the axe. The second timber war pivoted over old growth and a fire deficit. Preservation argued to reduce cutting, lessen environmental damages, and ensure the continuance of old forests. Fire management would work to promote good fires as well as prevent bad ones.

• • •

The first timber war was a classic political brawl of the Progressive Era. Thanks to cut-out-and-get-out practices, the nation's forests were being felled far faster than new woods could be found and old ones regrow. The rallying cry for reformers was "timber famine," which served at the time as peak oil has for recent generations. Peak timber arrived in 1910 and 1911—Gifford Pinchot declared in 1910 that "the United States has already crossed the verge of a timber famine so severe that its blighting effects will be felt by every household in the land." The industry found it hard to consolidate and insisted that more cutting was the only way to stay ahead of the inevitable flames. Conservationists favored reserving more land in national forests, where the cut could be regulated, or outright nationalization; and they observed that the reckless cutting was the cause of those terrible fires. Neither side could overcome or silence the other. But both feared wildfire, and on that one point of agreement they eventually worked toward a compromise program of cooperative forestry.[1]

Fire protection was the dynamic weld that bonded state government with federal, and that soldered them both to industry, either through special taxes or through private fire protective associations. The first flowering appeared in Idaho in 1906, and quickly jelled after Idaho's Forest Law a year later that mandated either private protection or a tax for the state to do it, and so became known as the "Idaho idea." But it was elsewhere in the Northwest that the idea became firmly institutionalized into a tripartite condominium of fire partners. In 1909 representatives from private protection associations in four states (with Weyerhaeuser as linchpin) met with the Forest Service to discuss fire issues. Out of that gathering came a consortium, the Pacific Northwest Forest Protection and Conservation Association, later renamed the Western Forestry and Conservation Association, that committed to joint programs in fire prevention, public education, legislative lobbying, and mutual assistance. At its first meeting in 1910 it elected E. T. Allen, a district forester with the Forest Service, as permanent secretary. What had become a toxic quarrel among themselves over the axe evolved into a collective fight against their common foe, wildfire. Importantly, fire protection was intended to regulate the axe, not remove it.[2]

On westside forests agencies worked to replace bad slash fires with good ones. With little lightning and scant interest in folk burning outside pastures and fields, it was the axe that created the conditions for fire outbreaks. Slovenly slash invited feral fires. The solution was to organize that slash and to mandate burning it; those burns were themselves subject to legal restraints and approved conditions (as measured by relative humidity); and an apparatus was created to attack those fires that escaped. After fires the axe returned to harvest standing boles and cut fuelbreaks. Instead of regulating the axe, the interested parties poured their efforts into controlling the fires that too often followed and threatened the future of the industry. It helped that timber lands were roughly divided among public and private ownership.

On eastside forests the problem was different, but the final solution was identical, the elimination of free-burning fire. Here folk burning by settlers, shepherds, and indigenes mingled with abundant lightning to saturate the pineries with mostly surface fires. After logging moved in, these cutover lands, too, became subject to obligatory slash burning. In the working understanding of the agencies, fires that crept through the understory were no different from those that soared through the crowns. Both threatened future forests and invited what officials considered a social tendency toward "lawlessness." With adequate fire protection, however, forests were considered insurable. They could be regrown in the expectation of a future harvest without the risk, as William Greeley once put it, of being swept away on a windy afternoon. A footloose agency could stay and stabilize.

Westside and east, controlling fires could not be segregated from controlling people. Since human engagement with the land pivoted around the timber industry, so, too, was most people's engagement with fire. Fuel management meant treating logging slash; fire management meant burning slash and stopping fires that threatened plantations; smoke management meant handling smoke from pile fires. To a remarkable extent, the project succeeded in its goals. Out of the Siskiyou National Forest came the 40-man crew, progenitor to the interagency hotshot crew. At Winthrop, Washington, appeared the first smoke-jumper base. From the Western Fire and Conservation Association came a model for collaborative fire protection, even among blood enemies. The Pacific Northwest became one of the powerhouses of America's fire establishment.

• • •

The second timber war boiled over during the 1980s and 1990s. A new environmentalism more concerned with preservation than conservation furnished a social movement and legislative levers. A New Forestry, best articulated by Jerry Franklin, gave it scientific credibility. Snags, windfall, and dead-and-down trunks were no longer just biotic debris, fuel hazards, and lost timber, but legacy structures that helped inform whole forests; fires were not existential threats but an essential process of regeneration. Forestry should seek to maximize all of a forest's assets, not simply its board feet.

The second timber war had several proxies. The first was smoke. Smoke in the valleys was nothing new: residents complained long and hard about it in the 19th century, and it was a charge often directed at the transhumant shepherds (mostly Basque) who moved flocks up and down the Cascades. The reward for tolerating a very long, very wet winter was a dry, clear summer, unless smoke saturated the skies. Fortunately most slash burning occurred in winter, when rains could help cleanse the scene.

As cutting accelerated, particularly on public lands, there was more slash to burn, which meant more smoke to linger in the valleys, increasingly converted to a service economy and filled with urbanites, many of whom hated the ugly and damaging clear-cuts, and sought leverage to reduce or eliminate them. If it was bad to breathe secondhand smoke from cigarettes, it must be bad to breathe the wood smoke that seasonally poured into cities and suburbs. Besides, if you controlled smoke, you might control the industry that produced it because without slash burning it would not be possible to replant on logged sites nor to contain the fires that would gorge on the combustibles and that threatened the uncut standing stocks and replanted, highly vulnerable plantations.

The search for alternative ways to manage the mountains of "fuel" left by the axe so obsessed fire agencies throughout the region that it became a national phobia. For most of the 1960s, 1970s, and into the 1980s the problem of wildland fuel meant slash, and most prescribed burns were slash fires. The smoke campaign spread into agricultural fires as well—burning was widespread for grass seed production. Gradually, smoke was strangled out of the scene. The Willamette Valley began shutting down field burning in the 1980s, culminating in legislation that capped the allowable acreage at 40,000 acres (from 180,000), and in 2010 shrank that further to 25,000 with a goal of complete elimination. Spokane followed suit, slashing acreage by 30 percent in 1998, with the goal of abolishing it altogether. Until the digital revolution arrived, the issue of smoke from forestry and agriculture went to the heart of the region's political economy because without slash disposal industrial logging could not succeed. An attempt to haul off ("yard") "merchantable material" failed because no market existed and the practice damaged soils. So desperate were foresters to rid themselves of slash that one fire officer proposed in *Fire Control Notes* that it simply be buried.

The more powerful proxy was the Endangered Species Act (ESA). The marbled murrelet and especially the northern spotted owl seemed to require old-growth forests to survive, and as those forests were felled, the owl faced extinction. With the ESA as a fulcrum, environmentalists threatened the economic model of the entire industry, which sought to liquidate the stocks of old-growth ("decadent," in the language of foresters) woods before reseeding to plantations. Every value—ecological, aesthetic, even other economic assets—the public might have in those woods was funneled into one output, timber. The ESA was, in fact, nested within many pieces of environmental legislation that challenged the character and consequences of postwar logging on air, water, land, and life. The reliance on large-area clear-cuts as a harvesting strategy threatened them all.

The controversy over the northern spotted owl marked the onset of hostilities that announced the Northwest's Second Timber War. The fighting paused when the 1994 Northwest Forest Plan brokered a cease-fire. The plan caused the timber harvest on federal land to plummet; the era of postwar logging reached a second peak timber. The plan also forced fire agencies to reconsider how to protect those landscapes without appeal to the solutions that had more or less worked over most of the 20th century. Once more, green covered black.

But apart from the Northwest, change had come to fire protection. Simple fire control had metamorphosed into a more pluralistic fire management amid a revolution that sought to restore good fire, and that tended to lump slash burns with bad fires. While industry and the U.S. Forest Service might rebrand slash fires as "prescribed burns," that linguistic sleight-of-hand fooled no one seriously invested in the controversy. Instead of sanitizing slash burning, it only tainted broadcast burning for more ecological purposes. And to close the triangle, about the time the Northwest Forest Plan was promulgated, fires began to mutate into more virulent expressions. They burned hotter, they burned bigger, they burned more severely. By the turn of the century the term *megafire* was replacing *blowup* as the stock expression for a big burn. An old plague, thought banished, had reemerged with vehemence.

This time the Pacific Northwest's contribution to the national fire narrative was indirect. It demonstrated the power of the environmental movement, and it helped to disable the Forest Service, still the keystone agency in the national infrastructure for fire. If the Forest Service faltered, so did the country's fire system as a system. The Pacific Northwest had helped propel Forest Service leadership in fire protection. Now it helped unwind it.

• • •

There was a sense, among many observers, that this emergent fire plague might bring the sulking, snarling rivals together as the Western Forestry and Conservation Association had a century earlier. Outside the industry the conviction grew that fire had to be managed on a landscape scale, that the only way to contain bad fire was to substitute good fire, and that no one agency could do the job alone. Collaborative forest restoration replaced cooperative forestry. The number of shareholders multiplied to include a civil society of nongovernmental organizations, citizen groups, and nonprofits, along with many federal agencies and interested tribes. But it did not, 20 years after the Northwest Forest Plan, include industry. That left it sated with ideas and starved for funding.

The sticking point remained the trees and how, or if, they should be cut, either before fires and after fires. Bitter controversies remained over how forests should be managed and to what end; considerable public unease persisted with clear-cuts and their ecological (and aesthetic) aftershocks. Instead of fire protection bringing the parties together, the controversies over the axe worsened the prospects for fire management. In many landscapes it was believed that pretreatments were necessary before good fire could be reintroduced, but those "mechanical" treatments could look a lot like logging, or its kissing cousin, thinning, and might require roads, and maybe some kind of forest products industry to help pay the costs. An appeal to "fuels" as a defining metric for fire management looked like silviculture under an assumed name and labeling as "prescribed fire" what to ordinary folk looked like axe-enabling slash burning only deepened suspicions that foresters were playing a shell-and-pea game. The 1995 Timber Salvage Rider, tacked on to an emergency supplemental appropriations act, which allowed for logging after fires, thus evading the restriction of the Northwest Forest Plan, only further poisoned the association. Instead of fire management calming the quarrel over the axe, the quarrel over the axe threatened to spill over and contaminate fire management.

In this new dispensation the tenets of Progressive Era reform would be challenged as well as the practices and policies it had promulgated. In the first timber war it was assumed that science would inform management, and that experts grounded in that science would apply solutions in—ideally—a disinterested way. In principle, granted enough resources, it would be possible to get ahead of the problem. In 1911 William Greeley asserted that firefighting was as amenable to scientific study as silviculture. As chief forester during the 1920s he made cooperative fire protection a centerpiece of his tenure before decamping to become executive secretary of the West Coast Lumberman's Association (a revealing commentary on the gravitational pull of industry).

In the second timber war scientific research had been fundamental in establishing the requirements of the northern spotted owl, but the protest had followed from a change in cultural values that pitted owls and old growth against board feet, and it was less vital in fire's management. The sense grew that, outside of communities and municipal watersheds and some select biotic sites of high values, we would not get ahead of the problem. Much as a phase change in social values had led to the owl research, so it was likely that fires would lead and the science follow. Fire officers look to managed wildfire to do the heavy lifting and put good fire on the ground. This is what a rational compromise looks like today. One wonders if public skepticism will stir the ashes as hillsides bristling with burned boles replace those with burned stumps.

The first timber war wanted wildfire extinguished and would tolerate slash burns to dampen the fuels that powered them. It ended with a condominium among the contestants. The second timber war wanted good fire promoted, decided that slash burns were not among them, and argued, if reluctantly, that managed wildfire was the best means to boost burning. New cooperative arrangements, now called collaborations, abound but they lack the binding common cause of their predecessors, not to mention the political and financial sinews. Industry enters the program as a paid service, not as a paying constituent. The second war rests with a truce rather than a victory.

• • •

To track this historic evolution, consider three fires, each roughly 70 years apart, each typical of what a big fire meant in its day.

The 1868 fires raged from the Olympic Peninsula to the California redwoods. Probably over a million acres burned from August through September. Hundreds of small fires—landclearing, camp, and incendiary—swelled into conflagrations as meteorological bellows drove the east wind. Fires burned around Victoria, Seattle, Olympia, Yaquina Bay, and Coos Bay, where a single fire blasted over 125,000–300,000 acres and lent its name to the whole complex in Oregon. Near the mouth of the Columbia a clearing fire escaped, a backfire was lit by a panicky neighbor, the two fires merged under the breath of the east wind, and despite efforts by considerable numbers of volunteers "an advancing line of fire extended from the very edge of the bay to the mountain tops." Smoke smothered valleys, even the long fetch of the Willamette. Navigation slowed.[3]

Yet despite their immensity, they occurred in country not yet fully settled or logged, and the loss of life and property was less than their size suggests. The outbreak occurred 40 years after Jedediah Smith first probed into the country, 25 years after the first great migration to the Willamette Valley, and nine years after statehood. There was no organized effort to fight it, no argument over what to do in its aftermath. It came and went with the east wind. It appeared as another trial to be faced by sturdy pioneers along with floods and grasshoppers. It blew up half a century before the industrial axe bit deeply into the westside forests and redefined Oregon's economy. The only timber war that existed lay within the contest over land generally, between American newcomers and Northwest indigenes. It was part of a cycle of fires that spanned the settlement era.

The 1933 Tillamook Burn came 23 years after the founding of the Western Forestry and Conservation Association established the terms of fire protection as an alliance of necessity and 22 years after the Weeks Act established the terms of federal-state cooperation, later upgraded by the 1924 Clarke-McNary Act, part of an alliance of necessity between the industry, the states, and the federal government. Slash burning was mostly domesticated, disciplined by law and scientific prescriptions, and those overseeing the process even extended bans to the very act of logging when conditions veered into the extreme. On August 14, amid a general woods closure, a company decided to yard one last log, or a careless spark lodged in the piles, or an aggrieved logger set a spite fire—the exact origin remains ambiguous—flame got into logging slash west of Portland, with the east wind shooting through the Dalles like a bullet down a gun barrel. The slash burn blew up. Over the next 24 hours some 10,000 acres an hour burned through the crowns. Smoke blotted out the midday sky. Debris landed on ships 500 miles at sea. The Oregon Department of Forestry mobilized its staff and those of its industrial cooperators. The U.S. Forest Service called up its usual forces and this time added the massed labor of the newly created Civilian

Conservation Corps; this was, in fact, the CCC's first trial by fire. Untouched, the Burn blew to the Pacific, some 330,000 acres in all.

This, however, was only the first iteration of the Tillamook Burn saga. What followed was an expression of the region's dialectic between axe and fire. Industry mustered to begin an immense exercise in salvage logging, while the CCC felled miles of snags for fuelbreaks, and together they cut roads throughout the hills. The Burn reburned in 1939. Another wave of salvage logging followed. It burned again in 1945. More salvage, culling ever fewer trees, but the scene still seemed inexhaustible. Then came the last of the six-year-jinx fires, in 1951. This time mechanized equipment could be brought to bear, and there was less and less for the flames to feed on to burn. Between them fire and axe had gutted a third of a million acres.

Still, this was timber country and a logging economy. All parties viewed the savage Burn as an environmental and economic disaster. Citizens amended the state constitution in order to replant the hills. School children were bussed to help—such was the social consensus on what should be done. With modern fire control growing more muscular, and with better discipline over slash, the Tillamook Forest regrew into one of the epic stories of American forestry. What had begun as a competition between axe and fire to see which would consume the great woods had become an alliance, as both were regulated into the rhythms of a postwar economy. The Tillamook Burn cycle proper lasted 18 years, but the era it embraced spanned another 50.

Even as that new forest was emerging, so was a new service-based economy, an environmental movement, and a fire revolution. Together they wrote a modern analogue to the Tillamook Burn cycle. The first outbreak, the Silver Complex, a cluster of fires in southwest Oregon, racked up some 200,000 acres. An overview of the regional fire scene noted that it was "the most severe fire in the last 50 years, and one of the two worst in the last 120 years, yet the acreage burned was only 30% of the average acreage historically burned by wildfire in Oregon." Over the course of 140 years landscape burning had shrunk and big fires had become not just rare but anomalous. The simple dichotomies between green and black and between public and private were less useful as explanatory schemes.[4]

In 2002 the Biscuit fire reburned much the same landscape, then spilled out for more. It began from a lightning bust that started five fires on July 13 within the Kalmiopsis Wilderness. Two monster fires—the Rodeo-Chediski fire in Arizona and the Hayman in Colorado, both the largest on record for their respective states—had already sucked in most of the nation's suppression forces, and fires were popping up in northern California. There were scant resources to spare for something not threatening communities or municipal watersheds; and possible private contractors had not been adequately vetted by the Oregon Department of Forestry and were not usable. A spate of lightning fires in a legal wilderness gazetted out of a fire-dependent landscape had a weak claim on suppression resources when the nation was at Preparedness Level 5 (the maximum allowed). The fires grew together, and then grew larger, and kept on growing. By August the complex had outgrown the Kalmiopsis Wilderness, the Siskiyou National Forest, and the state of Oregon, as flames crossed into California. Now it commanded national attention and even international, as firefighters were drafted from Mexico and trained fire officers from Canada, New Zealand, and Australia. Some 7,000 firefighters were on lines by mid-August. Burnout operations aroused concern over their severity. Costs went to a ballistic $150 million. So egregious was the expense that the Government Accounting Office (GAO) was asked to investigate. (The GAO found no malfeasance, only confusion and clumsiness amid perhaps a dose of bad timing.) Environmental controversies worsened when the Forest Service proposed selective salvage logging. In the final reckoning the fire burned a whisker under half a million acres.[5]

But money was a proxy fight for the old contest between fire and axe. Those who wanted the land logged argued that the axe would prevent big fires, or at least not cede such economic losses to flame; the movement to reserve federal forest lands for wilderness only led to uncontrollable wildfires. Those who favored fire management as a vehicle to promote ecological goods and services argued that the interruption of natural, or at least historic fire regimes, much in the name of protecting commodities like timber, had caused the buildup of fuels that powered the burn.

Rather than resolving the controversy, the Biscuit fire could be diverted to argue for each side's fundamental philosophy. It became notorious for its cost and divisiveness, not for its role in innovation or as the symbol of a new consensus. It epitomized what was wrong, not how it might be corrected. It was not a fire that clarified a strategy, it was a fire that condemned the existing system. In 2017 the Chetco Bar fire reburned much the same landscape once again. The Silver Complex had mutated into the Chetco Bar megafire. The Tillamook's 6-year jinx had become the Siskiyou's 15-year curse.[6]

The 1868 fires were barely fought at all. The Tillamook Burn was fought as hard as agencies could from start to finish, though their counterforces were puny compared to that of the fire and its stratospheric pyrocumulus. The Silver, Biscuit, and Chetco Bar fires could call up firepower unimaginable to earlier generations, but it was unclear that full-bore suppression was either possible or desirable. The first ignitions that led to the Biscuit fire were left to themselves in the Kalmiopsis; then, when they bolted out of those legal bounds, crews conducted vast firing operations primarily along the Illinois Valley, sending towering columns of smoke that terrified residents. Helping power those burns was a buildup of fuels, a quasi-natural slashing caused not by the axe but by the absence of fire. Probably half the final area was burned within historic ranges of intensity; many of the most savagely burned sites were the outcome of burnout operations amid deep drought compounded by unnatural levels of fuels amid groves of fire-sensitive tan oak that had thickened in the years of fire protection. If not so clearly anthropogenic in its causes as the Tillamook Burn, the Biscuit fire showed the continuing interaction of nature and culture. This was hardly a natural fire in the sense of burning in historic patterns with historic outcomes. Fire operations notably expanded the fire's final perimeter.

The early arguments were over the fire and its management. Once the flames were controlled, the argument turned to the axe. In 1933 salvage logging began on an epic scale, remaking the slopes and, with a broad social license, foresters replanted a new forest. Some 69 years later, that project had been inverted. After the Biscuit burn it was proposed to salvage-log burned trees to the order of 67 million board feet, all outside the Kalmiopsis Wilderness and perhaps 1 percent of the total timber affected. By now, however, the axe had become anathema to significant camps of partisans. The Forest Service was taken to court, and protesters tried to block roads. The courts sided with the agencies and the logging proceeded. There was no effort to replant outside the logged sites. There was, in truth, no consensus about what had been done or what to do next. There was no agreement over what the proper theme of the fire should be. A default narrative laid the issue at the feet of climate change.

The cycle of great fires in the 19th-century Northwest had aligned with heroic narratives of settlement and rebuilding. They were great fires not just because they were big but because they provoked their societies to rise to their challenge. The Tillamook Burn cycle repositioned that storyline but not its narrative structure or its moral subtext. Such fires were the dark villain that tested the temper of the hero. The newer cycle more resembles the profile of an antihero. The Biscuit fire had left a sour taste. Of the three monster fires of the 2002 season it was the one that sparked no honored legacy. It was a negative exemplar: the fire that showed how fires that had once brought society together were now wedging the pieces apart. It was the fire everyone seemed to wish would go away, and it helps explain how the region could claim national interest without national leadership. Then the Chetco Bar burn arrived, in eerie sync with a 15-year cycle. It seemed to confirm that these were the fires and these the issues that would define the new millennium for the region unless its fire community could find an alternative fire or some other narrative through which to refract it.

An Ecological and Silvicultural Tool

Harold Weaver

T'S NOT UNUSUAL for someone to be imprinted with the landscape of his childhood and to calibrate fire from that remembered baseline. This was certainly true for the conservative revolutionaries at Tall Timbers, and for many who sought to restore prairie, longleaf, and even pitch pine. But it is odd to the point of quirkiness that that childhood should begin at a camp devoted to hydraulic mining in a side valley of the Blue Mountains. Yet that is what happened to Harold Weaver, who spent his adulthood trying to protect the ponderosa pine forests he had known so intimately in his youth at Sumpter, Oregon.

Harold Weaver is the oft-overlooked member of that great triumvirate that argued in the 1960s for controlled burning. Ed Komarek had Tall Timbers Research Station, and could host fire ecology conferences that ranged the world, and published proceedings to libraries everywhere. Harold Biswell had the University of California-Berkeley, demo sites at Hoberg's Resort and Redwood Mountain, and student acolytes. Harold Weaver worked for the U.S. Indian Service (later, Bureau of Indian Affairs, or BIA) on tribal lands in Washington, Oregon, and Arizona. He was quiet—a forester and an administrator, not a scientist—who published his observations in professional journals, preferring to take photographs, mostly while wandering the woods barefoot. Being in backwater sites, he was able to experiment and observe in ways that would have been unthinkable in the Forest Service; if he did not receive much support, neither did he meet much obstruction. When he left a reservation, its prescribed fire program generally withered. Yet in one critical respect he had leverage that the other prophets of controlled fire didn't.

He was a credentialed forester. He could publish in the *Journal of Forestry*. He could speak, colleague to colleague, on fire and forests in ways that Herbert Stoddard and Ed Komarek, wildlife biologists, or Harold Biswell and S. W. Greene, rangeland scientists, could not. Harold Weaver had professional standing. That didn't mean foresters would agree with him or even listen closely. It did mean that, in a professional sense, they were willing to let him talk.

• • •

He grew up in the town of Sumpter, tucked deep in the Blue Mountains west of Baker City. His father, Amos Weaver, was a partner in a cluster of placer claims whose special contribution was to begin hydraulic mining operations "at the earliest possible date in the spring, after winter snow could be cleared from about five miles of water ditch and wooden flumes along steep mountain sides." When the water flowed, it was directed against gold-laden gravel "near the Blue Mountain summit, washing the sludge through sluices." The project continued "day and night" until around July 4 when the snowmelt was exhausted and the flow became inadequate.[1]

Those childhood memories stayed with Weaver. He recalled the "new, bright green needles" of the western larch, which remained one of his favorite trees. He recalled family picnics and camping trips, once the "furious storm of mining" ceased. He remembered gathering mushrooms "in a recent burn" and, later, huckleberries. From about the age of 12, he was "permitted considerable latitude in exploring and roaming with my dog and .22 rifle." Most of the lower elevation forest was mature ponderosa pine, and with help from a narrow-gauge railway, it had been cut over. Still, pockets remained of "mature pine with open, park-like, pinegrass covered forest floors."[2]

He decided to make the woods his career. By the time he graduated from high school, with "interludes" in Indiana and Southern California, those magical forests of the Blues were drawing him back. He enrolled in the forestry program at Oregon State College, spending summers cruising timber in eastern Oregon and California. He graduated in 1928, soon afterward joining the Forestry Department of the U.S. Indian Service, which sent him to the Klamath Indian Reservation in southern Oregon.

He stayed with the agency for all of his professional life. In 1933, a year after he married Jessie ("Billie") Gray, he was transferred to the regional office in Spokane to oversee the CCC projects on reservations. From 1940 to 1948 he worked at the Colville Agency, then moved to Phoenix, Arizona, as Area Forester. Three years later he accepted a position at the Washington, D.C., office as assistant chief, Branch of Forest and Range Management. He returned to the Northwest in 1954 as area forester, stationed at Portland. He retired in 1967, remaining in Portland until Billie passed away in 1978, before moving to Jasper, Arkansas. He died in 1983.

This is not the career of an agency pariah. It's the story of a man who was given serious responsibilities by the BIA, who moved up the bureaucratic food chain, who retired where and when he wanted, with honor. The one anomaly was his passion for prescribed fire. Whatever misgivings the profession and agency had about it, those concerns did not affect his career, nor did they stop him from implementing programs wherever he went. The folklore of the fire revolution sings a saga that tells how its prophets were denounced and scorned by the fire Establishment before they finally succeeded in overcoming their critics. That's not the biography of Harold Weaver.

• • •

As a child in the Blues, he had seen fires, which seemed benign compared to the havoc of hydraulic mining and clear-cutting, and had foraged among their scars for berries and mushrooms. At Oregon State, however, he was taught ("thoroughly imbued" with) the utter "incompatibility of pine forestry and fire."[3]

The clash between field and classroom worsened on the Klamath reservation, when he met older woodsmen who regarded fire exclusion as foolish and mistaken, though none could answer academic forestry's charge that even light fires prevented regeneration, that routine surface burning traded the forest's future for today's convenience. Later, a forester and entomologist, F. Paul Keen, "shocked" him by echoing the same concerns as the "nontechnical" woodsmen, and then took him into the woods to show the antiquity of fire scars amid early-growth ponderosa. That demonstration Weaver then continued on his own "throughout most of the western states," particularly in ponderosa pine regions, which were also where his assignments with the BIA took him. Then he witnessed a wildfire in 1938 on the Warm Springs Reservation that swept through "many thousands of snags and windfalls" in beetle-killed ponderosa.

Yet some saplings survived that harsh pruning, flourished, and grew amid a setting of very low fire hazard. That trial by fire convinced Weaver that fire was an ecological process fundamental to ponderosa pine forests and that "fire, under proper control," could be a useful tool.[4]

He published his first report in 1943. It mattered that he framed the issue in terms amenable to forestry. Controlled fire was a silvicultural tool. It reduced fire hazards, promoted growth, pruned thickets—did what foresters tried to do with other methods. It could, in the hands of foresters, assist fire protection and timber production. That didn't mean the profession accepted his notions: most didn't, and the BIA insisted on disclaimers. ("This article represents the author's views only and is not to be regarded in any way as an expression of the attitude of the Indian Service on the subject discussed.") But that year the U.S. Forest Service allowed the Florida National Forest to use controlled fire, along the lines of the "prescribed" burning proposed by Paul Conarro a year before. Still, Weaver's article was a controversial enough notion that the journal *Journal of Forestry* solicited a formal commentary from Arthur A. Brown, member of the Society of American Foresters and head of fire research for the Forest Service.[5]

Brown conceded that "Mr. Weaver offers some challenges to fire-control policies of public agencies that deserve careful consideration." He worried that Weaver left himself open to criticism by taking in too broad a territory; by speaking to one species when others that, in the absence of fire, could overtake it might prove more economically valuable; by urging a tool, fire, so rife with "ifs" that its effects were hard to predict; by proposing a long-term burning program in place of a temporary boost in fire control that might be enough to produce the desired results. Mostly, Brown concluded, with bureaucratic caution, that "the answer requires considerable prior research."[6]

Mostly, Mr. Brown sought to reframe the title of Mr. Weaver's article from "ecological" to "silvicultural" factors. "In conclusion," he wrote, "a word on the general philosophy that I believe has controlled to date."

> To serve society, the forester must substitute harvesting by logging for nature's method of harvesting by bark beetles and fire. To do that he *must* intervene in the old natural cycle. The first urgent step was to control fire and insects. With nature's harvest reduced, there is an opportunity for system in the second step, which is management of the forest by methods of cutting. With both under full control, which has not yet been attained, there will much room for refinement of method.[7]

Fire control now, control by axe later—a national theme but one that resonates especially well in the Northwest. The argument over fire use was not allowed to spill over out of forestry into realms of ecology, wildlife, rangelands, and so on that the most ardent critics of fire suppression had promoted. Weaver's challenge was treated as an internal quarrel among a band of brothers. Yet without that fraternal badge it is doubtful Harold Weaver would ever have been heard at all.

• • •

The first article, "Fire as an Ecological and Silvicultural Factor in the Ponderosa-Pine Region of the Pacific Slope," contained the gist of all that followed. Weaver believed "that periodic fires, in combination frequently with pine beetle attacks, and occasionally with other agencies, formerly operated to control the density, age classes, and composition of the ponderosa-pine stands." Yet 30–40 years of fire exclusion had "brought about changes in ecological conditions which were not fully anticipated, and some of which seem to threaten sound management and protection of ponderosa-pine forests." The proliferation of litter, windfall, and understory thickets had created conditions similar to postlogging slash; Weaver's broadcast burns were promoted to reduce hazard over that dispersed debris. He concluded that "progress in converting the virgin forest to a managed one depends on either replacing fire as a natural silvicultural agent or using it as a silvicultural tool." Weaver found "little evidence of success" in attempts at the first option, and "far too little thought and research" in applying the second. He followed his thesis with examples and evidence.[8]

The examples continued through the next 16 years. In 1947 he argued for "Fire—nature's thinning agent in ponderosa pine stands." In 1955 he considered "Fire as an enemy, friend, and tool in forest management." In 1956, taking his case to a wider public in *American Forests*, he noted that "Wild fires threaten ponderosa pine forests," and so needed tame surrogates. In 1957, back to the *Journal of Forestry*, he framed the case in more contemporary language, "Effects of prescribed burning in ponderosa pine." And in 1959 he summarized his career's conclusion with a case study of the Warm Springs Indian Reservation. The ecological damage evident in the pine stands was the result of too much grazing and too little fire. He repeated his fundamental argument: if foresters could not replicate fire in its ecological fullness, they would have to use fire.

His years in Arizona drew particular attention. The program at Fort Apache got front-page, two-photo treatment in the *Arizona Republic* under the headline,

"Fire Tested as Forest Friend." The journalist, Ben Avery, had visited an experimental burn in the Bog Creek area and compared it with a wildfire near McNary, "fed by old rotting logs and limbs in a cut-over area." The contrast was convincing. "We always have opposed forest fires. We also have been against sin. But when fire is used as a tool, the evidence of its usefulness was there to see." When the newspaper reported on Weaver's transfer to the Washington Office, it noted that "his experiments in utilizing controlled fire as an aid in silviculture . . . have won wide attention."[9]

When the fire revolution arrived, its partisans, eager to find examples of successful burning, honored Weaver as a forgotten prophet and identified Fort Apache as the stellar example of prescribed broadcast burning in the West. But he was not unknown: it was rather that his ideas had not gone beyond his experiments and most did not survive his administrative transfers elsewhere. In 1967, nearly thirty 30 years after the fire on Warm Springs that triggered his curiosity, he was asked to recapitulate his thoughts for the 7th Tall Timbers Fire Ecology Conference. He drew primarily on his experiences in the Colville Indian Reservation in Washington and the Fort Apache Reservation in Arizona. Later, by invitation, he visited Yosemite and Sequoia-Kings Canyon National Parks to comment on their fledgling prescribed fire programs.

In 1972 the Tall Timbers Research Station organized a task force to inquire into the state of burning in the Southwest and invited Harold Weaver to join them. The objective of the tour was twofold. One, it wanted to highlight what could be accomplished in the Southwest's ponderosa pine. ("More controlled or prescribed burning has been done on these three Reservations (mainly the Fort Apache) and over a longer period of time than in any other forested area of the western United States." The primary purpose was hazardous fuel reduction.) Two, it sought to show what could happen if that burning ceased. Weaver had left in 1951; his successor, Harry Kallander, who built out the experiments into a steady program, retired in the late 1960s. The program's momentum kept it alive, until, with a change in tribal leadership, it began to slide into decay; and wildfires returned with a severity not seen before. The 1971 Black River fire and savage Carrizo fire were the precipitating events for the task force.[10]

After those tours, Harold Weaver, well retired in Portland, receded as a personal presence in the fire revolution. He never broadcast those experiments as part of a wider program of fire restoration. He let others add his experiences to the campaign. What endured were his string of publications from 1943 to 1959.

. . .

The strongest voices of the fire revolution came from the Southeast. Harold Biswell transferred that chorus to the West Coast, and prairie folk argued for fire in their tallgrass swales. Most of suppression's critics were wildlife biologists, ranchers, wilderness enthusiasts, and park managers. Foresters joined, reluctantly, then, in the Southeast, avidly, to support longleaf and loblolly pine plantations. The movement stalled crossing into the arid West, where it stirred unwanted memories of light-burning and "Paiute forestry." Harold Weaver mattered because he was both a westerner and a forester

Since the late 19th century, foresters had controlled state-sponsored fire protection. While forestry's European founders did not regard fire as part of their script—fire control was a precondition for forestry, not an ongoing charge—fire protection was what made public forestry powerful in the United States. It gave forestry agencies something visible to do. It gave them a story that could unfold within days, not over generations. It simplified the message of forestry-led conservation in ways the public could understand. Early on, American foresters embraced systematic fire protection as their contribution to global forestry.

They resisted messages from outsiders. In the 1940s and 1950s forestry was agog with industrial logging as the national forests opened up. Foresters doubled down on fire protection as vital to spare that resource for the axe. When they, as a group, converted to environmental sensibilities, they read Aldo Leopold, a forester (Yale, 1909). When, collectively, they contemplated a change in fire policy, they searched for an equivalent figure and lit on Harold Weaver and his string of at-the-time-eccentric articles in the *Journal of Forestry*. They had a prophet from within, and they embraced him.

There is a Pacific Northwest story here, too. In Weaver's work controlled burning didn't challenge the axe: it enabled it. The deeper protest, the sense that fire had an ecological purpose beyond silviculture, had to wait until the second timber wars and the emergence of New Forestry. Fire and axe, green over black—even amid the fire revolution, the Pacific Northwest held to its originating traits.

Kenai

THE KENAI PENINSULA is a promontory, a beacon, and a portal. It's a geographic portal to the interior, a historical portal to Alaska's past and perhaps its future, an ecological portal to Alaskan fire. That's another way to say it's a transition point. Its eastern half is mountainous, its western a lowland plain. Its forest is a hybrid, both maritime and boreal. It's half wild and half semideveloped. It's as good a port of entry as any for a fire survey of Alaska.[1]

What may most characterize the Kenai, however, is its absence of a reliable baseline. There is no fixed point by which to measure the changes that are recorded. The climate has mutated continually, and recently seems to be passing through a major inflection point. The habitats have changed and may be in the process of conversion. The species have changed; even the moose that were the reason for creating the refuge may be relative newcomers. The human history has changed and is currently passing through a phase change not merely in demography but in the mode of settlement. The fires have changed; there are even reports of lightning fires, which were unknown previously.

All in all, the Kenai is famously active, but it's unclear whether that's because it has more happening or whether, being close to Anchorage and historic points of colonial contact, it has more records. What is clear, however, is that it has few obvious reference points. Apart from the grossest, geological markers there is no restore point for management, no anchor point for narrative. The Kenai offers a landscape expression of mindfulness. It's a continual now.

• • •

The Kenai compresses Alaskan pyrogeography and pyrohistory into a penin-sula. Compared with most of Alaska, the Kenai fire story is known with some thoroughness. It has coastal forests, mixed interior forests, tundra, muskeg, and Alaska's three spruce, Sitka, white, and black, and in keeping with its transitional character, a hybrid, the Lutz. Sitka spruce burns rarely. White and black spruce burn as their understories and ambient conditions permit. What it hasn't had historically is the pyrotechnic busts that lightning kindles in the interior.

Charcoal cores extend back 13,000 years. They record three long waves of biomes and associated fire regimes. An early tundra era had the longest return interval; a woodland era shortened it; a black spruce era, surprisingly, lengthened it again. Over the past 300 years, which is the age of the oldest spruce, research suggests an average fire return interval of 89 years, plus or minus 43 years. For black spruce the interval is 79 years, for white spruce 200-plus years, and for mixed spruce forest 170 years. Very few sites in Alaska can boast of such detail. The record spans the entire contact era, predating even Bering's first expedition.[2]

What makes the chronicle especially intriguing is that there are few known lightning fires, and those in recent years. The resident indigenes, the Kenai-tze Athabascans, have no oral tradition of lightning, much less of lightning fires. Rain is common, thunderstorms almost unknown. Instead, the Kenai's fire history over the past few centuries is one of anthropogenic fire meeting what increasingly has been a directly or indirectly anthropogenically influenced landscape. That makes the Kenai different from the interior, but perhaps not so different from the interior's future.

On his third voyage, in October 1778, Captain James Cook sailed past Kenai into what became Cook Inlet. He spotted smoke, which (as was typical) he interpreted as a sign of human inhabitants. There were landscape fires, likely from Native sources, 70 years earlier in 1708, and later in 1862. Cook's reports on sea otters—one of the few items that China would trade for—set off a fur rush. Russian *promysleniki* (trappers) from the eastern movement of the fur trade across Siberia followed. In 1787 the Russian American Company estab-lished a post on Kenai, Fort Nikolaevskaia (near the mouth of the Kenai River). Prolonged contact with the Natives usually led to a demographic decline pre-cipitated by disease, coercion, and social breakdown. Certainly this is what hap-pened in the Aleutians and along the Pacific Northwest; the particulars on the Kenai are unknown. How this might have translated into fire is murky. Peoples dependent on caribou tend not to burn deliberately, except in small patches or along traplines, because lichen is both the primary surface fuel and the prime winter forage, and it takes decades to recover from a fire.

The chronicle of known fires beat on—1801, 1828, 1833, 1834, 1849, 1867, 1874, 1888, and 1898, and from unknown sources in 1871, 1883, 1891, and 1910. The arrival of Europeans and Americans coincided with the departure of the Little Ice Age. The discovery of gold sparked a rush in the 1880s, which continued into the 1890s, probably leading to as many sparks as nuggets. That was followed by coal discoveries. Roads and railways sought to run from Seward to Anchorage—more slash, more sparks, more fires, though most were confined to the broad corridor defined by those rights-of-way. The railroad became a particular source of irritation, indifferent as it was to cleaning up slash or preventing fires (it was hemorrhaging money).

The landscape adjusted. Wolves vanished. Eagles had a bounty placed on them. Salmon nearly went extinct locally. Caribou disappeared between 1906 and 1917. (Visiting in 1952, Starker Leopold and Fraser Darling attributed the extinction to widespread fires that had wiped away the arboreal lichens that supplied the herds' winter range, the same lichens that powered most spruce burns.) The fires that likely drove the caribou out also enticed moose in; in general, more burns, if the patches are not too widespread, means more moose. Moose apparently expanded into the big-mammal vacuum, finding fresh fodder in the new postburn landscapes, before nearly collapsing under the onslaught of hunting, much of it for trophies. Requests for game protection led to the Kenai National Moose Refuge, established in 1941, nine days after Pearl Harbor. Following the Alaska National Interest Lands Conservation Act (1980) the mission of the refuge was broadened into the maintenance of general habitat for many species, though moose have remained a charismatic core.[3]

• • •

The modern fire history of the refuge began when landscape fire met industrial fire.

In 1947 road crews constructing the Sterling Highway started the Skilak fire that swept over 300,000 acres. As an early season burn it was not notably severe, but it remade a large swathe of the refuge. The year was, for the young Alaska Fire Control Service, a trial by fire, well beyond the Kenai, but the aftershocks stayed in Alaska. A decade later the Kenai boasted the first oil and natural gas discovery in Alaska, galvanizing a new mineral rush with its attendant roads and sparks, this time fueled by and for internal combustion. Some 1,500 miles of seismic survey lines were cut through the refuge; roads were hacked everywhere; a rectangular block east of Soldotna (six townships in all) was officially removed by Congress from the refuge and privatized. In 1969, amid a deep drought, two abandoned campfires from oil exploration crews led to serious burns in August. The Russian River fire blackened 2,570 acres and the Swanson River fire 79,000 acres. This time the fires scoured deeply into the soil and led to type conversion, and this time they rumbled through the American fire establishment.

The Swanson River fire became a national story. It was the most expensive fire to suppress in American history to that time. It coincided with the Bureau of Land Management's fast-emerging fire program and with the operational opening of the Boise Interagency Fire Center. It prompted alarms over the character of industrial fire suppression as some 50 bulldozers worked the lines (so many dozers were available because of the oil field development, which brought roads and a build out of settlements). The eroding tracks left behind scarified the soil with effects that outlasted any biotic effects of the burn; a new combustion order was literally impressing itself on the old. Suddenly, Alaska was not simply a place to which wildland fire protection could be sent. It was a place with the power to unmoor the national system. An important symposium, Fire in the Northern Environment, followed. Alaska began to look more like the rest of the country in its fire problems.[4]

However the various pieces had come together in previous times they began to disaggregate. Research hints at an inflection point around 1968 when the creeping consequences of global warming seemingly crossed a threshold on the Kenai. Between 1985 and 2000 bark beetles killed nearly a million acres of spruce, mostly white and Lutz, the longest outbreak in North American history, its epicenter at the Caribou Hills. The Kenai biota began reshuffling: white spruce and hemlock spread outward, shrubs moved up slopes, peatlands began drying, yielding to shrubs and spruce, and, thanks to beetle kill and salvage logging that opened the soil to sunlight, bluejoint grass expanded, its dense rhizomic roots crowding out young trees. The old burns, especially the 1947 fire, had reestablished black spruce that 60 years later were primed to burn as crown fire again. People who had lived in small villages, who had once harvested fish and marine mammals, and the occasional big game, were replaced by the exurbs of Anchorage and an industrial society fueled by oil. The eastern lowlands went into the Kenai Wilderness (68 percent of the entire refuge). The western lowlands rubbed against Alaska's version of sprawl, acquiring a 175-mile border of towns, strip malls, and feral cabins strung along the Sterling Highway. (During the 2009 Caribou Hills fire, crews discovered over 200 cabins not on official registers, just tucked away in the woods.)[5]

Three trends began to harden, then to align like tumblers in a lock. One, the landscape dried, as it had since the end of the Wisconsin glacial, quickening after the Little Ice Age, and inflecting again around 1968, reconfiguring the biota into new patterns. The massive bark beetle outbreak was one likely outcome; the appearance of black spruce and shrubs on formerly sphagnum-moss peat is another (covering some 60 percent of the peat since 1950), and so is the propagation of grasses. Two, ignitions have increased. More people, more starts, deliberate or accidental—this is to be expected. The surprise, however, is the appearance of lightning fires in areas for which there is no historic record or traditional recollection. Three, the ability to respond is shrinking. Wilderness on the east prevents active measures, so does road-inspired sprawl on the west. Human settlements have thickened and splashed outward, trailing sparks like beer cans and shell casings. Roads cut for oil exploration became points of entry; unregistered cabins sprouted like morel mushrooms. Within the remainder of the refuge, treatments must not harm the habitat and species for which the refuge was established. Experiments during the 1960s in mechanically improving moose habitat proved expensive and inconclusive and are not likely to be repeated. Still, the threats were too great and close at hand to overlook. A small fuels program began working the town edges. The erosion of budgets—the Fish and Wildlife Service, the Alaska Department of Forestry, which handles fire protection—leaves little surplus to experiment and small margin for error.[6]

The fire scene picked up in the 1990s. In 1991 the Pothole Lake fire burned 7,900 aces; in 1994, the Windy Point fire, 2,800 acres; the 2004 Glacier Creek fire, 6,900 acres. The 2005 Irish Channel fire, kindled by lightning, burned 1,100 acres of mountain hemlock—a doubly unprecedented event. The fires began to move south. They became scary in 2007 when the Caribou Hills fire, started by sparks from a cabin resident sharpening his shovel on a grindstone, burned 55,000 acres and 197 structures. In 2014 the Funny River fire burned both biotic borders, one along the outskirts of Soldotna, the other into alpine tundra. Had a shaded fuelbreak not been put in years earlier and the originating wind not blown from the north, the consensus is that the fire would have crashed into town. The next year the Card Street fire struck near the same area and burned three houses before swarming into the receptive refuge.[7]

Both the Funny River and the Card Street fires began in that anomalous rectangle that oil and gas discovery caused to be excised from the refuge. The two realms of combustion were interacting in ways that no one would have predicted. So we can add fire to the roster of ecological processes that are affecting the Kenai not so much by themselves but in unexpected synergies with the others.

• • •

The issues that plague the Kenai are those typical throughout fire-prone Alaska; the pieces just have somewhat different dimensions and combine in peculiar ways. Maybe the surest assessment of what is happening is that the Kenai is reasserting, in tongues of flame, the sense in which it is transitional, that it functions as a portal. It's the warmest and wettest subregion of the state, save the coastal southeast. Here the coming order of fire may be first entering Alaska. It's a future that appears to promise more fire with less control.

Almost uniquely over the past 40 years, Alaska has succeeded in keeping fire, good fire, on the land. But the constraints are growing to shrink the area available for ecological burning, and the fires that are coming are not, by traditional standards, unequivocally good. They are burning differently. They are catalyzing the effects of change in ways that may not be restoring, or even maintaining, existing fire regimes, so much as kindling novel ones. This may be exactly what the land needs as the wave train of changes continues—fires as a constant series of ecological jolts and tweaks that lessen the grand shock of cumulative change. Or they may serve as an accelerant to those changes, shifts that people will be sorry to see.

The refuge knows it must keep its fires (and their smokes) out of the towns. They know the public expects a moose refuge to have moose. They know the prime movers behind change are beyond their control. But with so much in flux, it is not simple to identify future desired conditions or to specify prescriptions to achieve them. Without a clear, usable past, they are left with a series of ad hoc adjustments, hoping that their actions and fires will ease the refuge into a usable future.

But then it's not clear what that future *should* look like. On the Kenai the past seems less like a prologue than an endless present. The future may be unsettled in ways the past doesn't foreshadow. This is not a stable place that is now changing, but a changing place that is poised to change faster. In that respect, the Kenai fire scene might well stand for most of Alaskan fire. Or for that matter, for the Alaskan Anthropocene.

North to the Future

Pleistocene to Pyrocene

I N 1898, as Klondike fever raged, the U.S. Geological Survey published a map of Alaska to show the known sites of mineral wealth. The most prominent locales, of course, were the gold fields. But not far behind were coalfields. Fossil fuels have been the black gold of Alaska since the beginning of American rule.

Twice those lithic landscapes have entered the national narrative of environmentalism. In 1909, after Teddy Roosevelt had left office, Gifford Pinchot, then chief forester, picked a fight with Secretary of the Interior Richard Ballinger over the leasing of Alaskan coal lands. The particulars of the leasing were murky at best, but the coal was only a means (to Pinchot's mind) to address larger questions of how to manage the nation's estate after Roosevelt's departure. The resulting controversy forced President William Howard Taft to fire Pinchot, precipitating a major split in the Progressive movement and the Republican Party, and leaving a bitter legacy for conservation. At the time Alaska was two years away from status as a bona fide territory.

Sixty years later it was Alaskan oil from the North Slope that ignited a national uproar. The oil could only come to market by constructing a pipeline from Prudhoe Bay to Valdez. Alaska had been a state for 10 years and suffered from a feeble economy in need of defibrillation. The controversy proceeded along classic Alaskan lines between those who wanted the Wild and those who wanted a Wild West. Those who didn't want the Arctic drilled for oil, those who worried about oil spills from a metal tube traversing cold, seismically active terrane, those who feared ecological disruption, especially for migratory wildlife and from roads branching into every nook and cranny—all wanted the pipeline stopped. Those who wished for Alaska to develop along the lines of the Lower 48, who desired a chance to advance a new frontier, who sought a state budget that could provide for basic services—all wanted the pipeline built. The process stalled until 1971 when the Alaska Native Claims Settlement Act resolved the issue of land tenure, and it found a greased legislative rail with the first OPEC oil embargo.

The energy crisis ensured the pipeline would be built. The long controversy, however, meant it would be designed to withstand the known hazards of the Alaskan landscape and the threats a pipeline might pose. It was constructed to withstand earthquakes, permafrost, and intense cold. It allowed the movement of fauna and the migration of caribou. And it was hardened against the heat posed by a boreal forest fire. Along its wending way the two grand realms of combustion meet.

• • •

The pipeline divides the state geographically and historically, and during its construction, it divided it politically. It provoked major controversies that have defined the economic geography and political economy of modern Alaska. It laid down the basis for Alaska as a modern petro-state. The oil industry accounts for a third of the state's economy; and between 80–90 percent of state revenue comes from oil taxes, rents, and royalties. This much is widely understood. Less appreciated is that the pipeline symbolized, and makes possible, a combustion divide. It segregates two eras of Alaskan fire history.[1]

On most popular maps, and in the public imagination, two great paths cross Alaska. One reenacts the past, one leads to the future. The Itidarod trail runs from Willow (near Anchorage) to Nome, and it's the site for an annual dogsled race that harkens back to Alaska's first mineral rush. The other wends from Prudhoe Bay to Valdez, and through it flows the oil that lubricates the modern Alaskan economy.

Those two routes can also stand for two paths of Alaskan fire. The Klondike Gold Rush, which eventually spilled over most of the state, was rife with fires. Fires cleared the woods to expose outcrops, fires melted permafrost to reach placers; fire was life, for without it there was no resistance to the killing cold. The pipeline speaks to another kind of fire, without which modern Alaska could not exist. Fossil fuels heat homes, run power stations, fuel vehicles, and fill (or not) the coffers of the state legislature. Thanks to oil, Alaska has no income tax, no sales tax, and in most years sends a rebate to citizens. If the flow of those fuels falters, the state suffers. If something extinguishes the fires it feeds, the state would collapse.

Both fires—those that burn living landscapes and those that burn lithic ones—continue; and both are projected to increase in coming years. The linkage between them is worth exploring. There are instances of direct competition, where flames burn along the pipeline, and where smoke from wildland fires has forced the turbines running pumps to shut down. Mostly, however, the interaction is indirect.

The society that oversees wildland fire management is a profoundly fossil fuel–based civilization. Industrial combustion supplies the vast bulk of its energy needs. That society runs on machines, literal fire engines, that burn fossil fuels. It no longer exists on a subsistence level because its industrial fire economy can supplement what it can produce locally by importing goods and services from around the globe. Industrial combustion makes possible its wildland fire program. Fire management operations run on trucks, engines, pumps, helicopters, aircraft, even driptorches—all burning gas—that get firefighters or water to the fireline. Fire management policy is made in offices lit by electricity, heated and cooled by electricity or gas, and over desks with telephones and computers powered by off-site dynamos burning coal or oil. State firefighters are paid with the revenue derived from oil and coal leases. Emergency firefighter crews choose between a traditional economy based on open flame and one that houses fire in machines.

But there is a still deeper interaction, which lies in fire's capacity to mobilize carbon. Greenhouse gases liberated by burning taiga and tundra are joining those far more pervasive gases spewing out tailpipes and smokestacks, and together they are unmooring the climate that human society has adapted to over the past 6,000 or so years. Projections suggest that the atmospheric warming subsequently created will alter—must alter—the existing arrangement of fire regimes. Most models posit an increase in factors that will dry fuel; many also suggest an uptick in lightning. The nightmare scenario holds that the resulting big burns will set up a positive-feedback loop such that the large carbon-storing woods will convert to little grasses and shrubs, freeing thus more carbon. The two grand realms of combustion will no longer compete so much as collude. The Anthropocene will become a Pyrocene.

• • •

Yet once again, Alaska is different.

Much as its Natives upset the traditional dialectic between economy and ecology, so its immense tundra is destabilizing the traditional discourse about the two prevailing forms of combustion. The third party here is organic soil. Those soils constitute huge carbon sinks, whether as boreal peat or embedded in permafrost. They unbalance the usual calculations of competition and collusion.

There are, in fact, three fuels in play. One lies above the surface—the woods, shrubs, grasses, mosses, lichens. One lies far below the surface in the form of ancient biomass, sequestered in the sediments of deep time. And one is the shallow subsurface biomass still lingering from the Pleistocene, whose planetary frost and thaw cycles left big reservoirs in the not-too-distant ground. If liberated by fires, or climatic warming, or an intricate choreography of warming and burning, they could push the planet quickly into a tipping point from which a return might be impossible. Those frozen reserves are so immense that they transform a combustion dialectic into a braided narrative. How these three fires amplify, dampen, and leverage one another is the evolving story of fire in Alaska.

It is unclear whether fires in tundra are increasing. Soil charcoal hints that fires have occurred in the past, though nothing, over the past 5,000–7,000 years, on the scale of the 2007 Anaktuvuk burn on the North Slope. It's too early to know whether such outbreaks are the harbingers of a new regime or simply reflect the short memory and lean data sets of recording Alaskans. The calculations of what quantities of carbon might be released if the burning infects the land into a combustion contagion are terrifying. What is clear is that the Alaskan fire scene may change in ways that escape the grasp of our present anecdotes and algorithms.

It may be that the permafrost is another Pleistocene relic that has lingered for millennia, surviving when wooly mammoths and giant ground sloths and other emblems have vanished. One after another, the relics of the Ice Age are going. The glaciers are receding, the arctic pack ice is shrinking, the annual snowpack is smaller. Permafrost survived because it was insulated. But if fires and warming peel back that layer of organic insulation, it will go the way of Pleistocene ice generally. A Fire Age will have driven off the last vestiges of the Ice Age.

Alaska may be America's great refuge for ice, as it has been for wolves and brown bears. What should not be forgotten is that the ice has been a check on fire, and fire on ice. This is not so much a planetary dialectic as a dialogue. In the 1970s climate scientists warned of a coming ice age. The Milankovitch cycles still spun, the Earth's oceans and continents were still aligned favorably to leverage snow into ice sheets, and 80 percent of the past 2.6 million years had been glacial—there was no reason to think that our brief interglacial would persist for much longer. It had already lasted longer than models predicted. The ice was coming. The Little Ice Age was a warning shot off the bow that a big ice age was inevitable.[2]

It didn't happen, and now it seems it can't. We've halted the ice. But in stopping ice we've unleashed fire. We're a fire creature: we can exist without ice but not without fire. But we may be knocking away all the constraints that have traditionally kept fire within the bounds of usefulness. The Alaskan fire scene may be where we drive the last of the Pleistocene ice into extinction, or whether, alternatively, we recognize that we are also a creature of the Pleistocene and may be unwisely turning our firepower against ourselves.[3]

Like the rest of America, Alaska seems caught between two fires. Unlike the rest of the country, it still has room, within limits, to trade space for time. Those choices will undoubtedly involve national as well as Alaskan politics, as they should, since those choices will affect us all.

HERE AND THERE

Our Coming Fire Age: A Prolegomenon

AT NIGHT, viewed from space, the cluster of lights looks like a supernova erupting in North Dakota. The lights are as distinctive a feature of nighttime North America as the glaring swathe of the northeast megalopolis. Less dense than those of Chicago, as expansive as those of Greater Atlanta, more coherent than the scattershot of illuminations that characterizes the Midwest and the South, the exploding array of lights define both a geographic patch and a distinctive era of Earth's history.

Nearly all the evening lights across the United States are electrical. But the constellation above North Dakota is made up of gas flares. Viewed up close, they resemble monstrous Bunsen burners, combusting excess natural gas released from fracking what's known as the Bakken shale, named after the farmer Henry Bakken, on whose land the rock formation was first discovered while drilling for oil in the 1950s. In 2014 the flares burned nearly a third of the fracked gas free. They constitute one of the most distinctive features of the U.S. nightscape. We might call them the constellation Bakken.

While the flares rise upward, the fire front is actually burning downward into the outgassing drill holes as surely as a candle flame burns down its tallow stalk. The flames are descending as rapidly as their fuels are rising. They are burning through deep time, combusting lithic landscapes from the geologic past and releasing their effluent into a geologic future. Eerily, the Bakken shale dates to the Devonian, the era that records the first fossil charcoal, our earliest geologic record of burned material. Its gases will linger through the Anthropocene.

In 1860 the English scientist Michael Faraday published a series of public lectures in which he used a candle to illustrate the principles of natural philosophy. Fire was an apt exemplar because it integrates its surroundings, and it was apt, too, because in Faraday's world, fire was ever present. Every nook and cranny of the human world flickered with flames for lighting, heating, cooking, working, and even entertaining. But that was starting to change. By then, Britain had 10,000 miles of railways and the United States had 29,000. Those locomotives demanded more fuels than the living landscape could supply. Engineers turned to ancient landscapes—to fossil biomass, notably coal—and they simplified fire into combustion.

Today, a modern Faraday would not use a candle—probably couldn't because the lecture hall would be outfitted with smoke detectors and automated sprinklers, and his audience wouldn't relate to what they saw because they no longer have the lore of daily burning around them. For a contemporary equivalent he might well turn to a fracking flare, and to illustrate the principles behind Earthly dynamics he might track those flames as they burn down through the deep past of fire and humanity.

•   •   •

Earth, water, air—all are substances. Fire is a reaction. It synthesizes its surroundings, takes its character from its context. It burns one way in peat, another in tallgrass prairie, and yet another through lodgepole pine; it behaves differently in mountains than on plains; it burns hot and fast when the air is dry and breezy, and it might not burn at all in fog. It's a shapeshifter.

The intellectual idea of fire is a shapeshifter, too. The other elements have academic disciplines behind them. The only fire department on a university campus is the one that sends emergency vehicles when an alarm sounds. In ancient times, fire had standing with the other elements as a foundational axiom of nature. In 1720, the Dutch botanist Herman Boerhaave could still declare that "if you make a mistake in your exposition of the Nature of Fire, your error will spread to all the branches of physics, and this is because, in all natural productions, Fire . . . is always the chief agent." By the end of the 18th century, fire tumbled from its pedestal to begin a declining career as a subset of chemistry and thermodynamics, and a concern only of applied fields such as forestry. Fire no longer had intellectual integrity: it was considered a derivation from other, more fundamental principles. Just at the time open fire began retiring from quotidian life, so it began a long recession from the life of the mind.

Fire's fundamentals reside in the living world. Life created the oxygen fire needs; life created the fuels. The chemistry of fire is a biochemistry: fire takes apart what photosynthesis puts together. When it happens in cells, we call it respiration. When it occurs in the wide world, we call it fire. As soon as plants colonized land in the Silurian period about 440 million years ago, they burned. They have burned ever since. Fires are older than pines, prairies, and insects. But nature's fires are patchy in space and time. Some places, some eras, burn routinely; others, episodically; and a few, only rarely. The basic rhythm is one of wetting and drying. A landscape has to be wet enough to grow combustibles, and dry enough, at least occasionally, to allow them to burn. Sand deserts don't burn because nothing grows; rainforests don't burn unless a dry spell leaches away moisture. Biomes rich in fine particles such as ferns, shrubs and conifer needles can burn easily and briskly. Landscapes laden with peat or encumbered with large trunks burn poorly, and only when leveraged with drought.

As life and the atmosphere evolved, so did fire. When oxygen enriched during the Carboniferous period around 300 million years ago, dragonflies grew as large as seagulls, and fires swelled in like proportion such that 2 to 13 percent of the era's abundant coal beds consist of fossil charcoal. When grasses emerged in the Miocene, they lavished kindling that quickened fire's spread. When animals evolved to feast on those grasses, fire and herbivores had to compete because that same biomass was fodder for each. It could go into gullets or up in flames but not both.

Today, ecologists refer to landscape fire as a disturbance akin to hurricanes or ice storms. It makes more sense to imagine fire as an ecological catalyst. Floods and windstorms can flourish without a particle of life present: fire cannot; it literally feeds off hydrocarbons. So as atmosphere and biosphere have changed, as oxygen has ebbed and flowed, as flora and fauna have sculpted biomass into new forms, so fire has evolved, morphing into new species.

Yet there was one requirement for fire that escaped life's grasp, the spark of ignition that connected flame with fuel. Ignition relied on lightning, and lightning's lottery had its own logic. Then a creature emerged to rig the odds in favor of fire. Just when hominins acquired the capacity to manipulate fire is unknown. But we know that *Homo erectus* could tend fires and, by the advent of *Homo sapiens*, hominins could make fire at will.

A revolutionary phase-change all around. Until that Promethean moment, fire history had remained a subset of natural history, particularly of climate history. Now, notch by notch, fire gradually ratcheted into a new era in which natural history, including climate, would become subsets of fire history. In a sense, the rhythms of anthropogenic fire began to replace the Milankovitch climate cycles which had governed the coming and going of ice ages. A fire age was in the making.

Earth had a new source of ecological energy. Places that were prone to burn but had lacked regular ignition (think Mediterranean biomes) now got it, and places that burned more or less routinely had their fire rhythms tweaked to suit their human fire tenders. Species and biomes began a vast reshuffling that defined new winners and losers.

The species that won biggest was ourselves. Fire changed us, even to our genome. We got small guts and big heads because we could cook food. We went to the top of the food chain because we could cook landscapes. And we have become a geologic force because our fire technology has so evolved that we have begun to cook the planet. Our pact with fire made us what we are.

We hold fire as a species monopoly. We will not share it willingly with any other species. Other creatures knock over trees, dig holes in the ground, hunt—we do fire. It's our ecological signature. Our capture of fire is our first experiment with domestication, and it might may well be our first Faustian bargain.

Still, ignition came with limits. Not every spark will spread; not every fire will behave as we wish. We could repurpose fire to our own ends, but we could not conjure fire where nature would not allow it. Our firepower was limited by the receptivity of the land, an appreciation lodged in many fire-origin myths in which fire, once liberated, escapes into plants and stones and has to be coaxed out with effort.

Those limits began to fall away as people reworked the land to alter its combustibility. We could slash woods, drain peat, loose livestock—in a score of ways we could reconfigure the existing biota to increase its flammability. For fire history this is the essential meaning of agriculture, most of which, outside of floodplains, depends on the biotic jolt of burning to fumigate and fertilize. For a brief spell, the old vegetation is driven off, and a site is lush with ashy nutrients, and—temporarily—imported cultivars can flourish.

In 1954, the U.S. anthropologist Loren Eiseley likened humanity itself to a flame—spreading widely and transmuting whatever we touch. This process began with hunting and foraging practices but sharpens with agriculture. Most of our domesticated crops and our domesticated livestock originate in fire-prone habitats, places prone to wet-dry cycles and so easily manipulated by fire-wielding humans. The way to colonize new lands was to burn them so that, for a while, they resembled the cultivars' landscapes of origin.

Yet again, there were limits. There was only so much that could be coaxed or coerced out of a place before it would degrade, and there were only so many new worlds to discover and colonize. If people wanted more firepower—and it seems that most of us always do—we would have to find another source of fuel. We found it by reaching into the deep past and exhuming lithic landscapes, the fossil fallow of an industrial society.

Instead of redirecting or expanding fire, the conversion to industrial burning removed open flame, simplified it into chemical combustion, and stuffed it into special chambers. Instead of being constrained by the abundance of fuels, anthropogenic fire was constrained by sinks, the capacity of land, air, and ocean to absorb its byproducts. The new combustion was no longer subject to the old ecological checks and balances. It could burn day and night, winter, and summer, through drought and deluge. Its guiding rhythms were no longer wind, sun and the seasons of growth and dormancy, but the cycles of human economies.

The transformation—call it the pyric transition—was as disruptive as the coming of the aboriginal firestick and fire-catalyzed farming, but it was more massive, much faster, and far more damaging. Amid the upheaval catalyzed by industrial combustion, some landscapes burned to their roots. Seasonally, skies were smoke palls. Frontier settlements vanished in flames. The pyric transition runs through fire history and Earth's pyrogeography like a terminator.

Eventually, as the new order prevailed, as it wiped flame away by technological substitution and outright suppression, the population of fires plummeted, leading to ecological fire famines. The transformation might have left Earth with too much generic combustion, and too much of its effluent lodged in the atmosphere, but the industrialized world also left too little of the right kind of fire where it's needed.

Promoting the steam engine developed with his business partner James Watt, in the late 1770s, Matthew Boulton boasted to James Boswell that they sold what all the world wanted—power. In 1820, a year after Watt died, Percy Shelley published *Prometheus Unbound*, in which he celebrated the unshackling of the unrepentant Titan who had brought fire to humanity. By then, the use of coal, and later oil, was liberating a generation of New Prometheans.

This newly bestowed firepower came without traditional bounds. For a million years the problem before hominins had been to find more stuff to burn and to keep the flames bright. Now the problem became what to do with all the effluent of that burning and how to put flame back where it had been unwisely taken away.

The new energy revolution leveraged every activity, like fire itself creating the conditions for its spread, each reinforcing the other. But the collateral damage in the form of wrecked landscapes could not be ignored. Engineers sought to keep fire within the machines, not loosed on the countryside. Countries, particularly those with extensive frontiers, public lands, or colonial holdings, sought to shield their national estate from fire. They set lands aside to shelter them from promiscuous and abusive burning and strove to control fires when they occurred.

State-sponsored conservation had considerable currency among progressive thinkers. When Rudyard Kipling wrote "In the Rukh" (1893), a story that explained what became of *The Jungle Book*'s Mowgli after he grew up, he had him join the Indian Forest Department and fight against poaching and "jungle fires." Only later would the paradoxes become palpable. Only later would overseers realize how hard they would have to struggle to reinstate fire for its ecological benefits.

But the flames were only the visible edge of a planetary phase-change. The slopover that followed once Earth's keystone species for fire changed its combustion habits is best known for destabilizing climate. But humanity's new firepower has a greater reach, and the knock-on effects are rippling through the planet's biosphere independently of global warming. The new energy is rewiring the ecological circuitry of the Earth. It has scrambled ecosystems and is replacing biodiversity with a pyrodiversity—a bestiary of machines run directly or indirectly from industrial combustion. The velocity and volume of change is so great that observers have begun to speak of a new geologic epoch, a successor to the Pleistocene, that they call the Anthropocene. It might equally be called the Pyrocene. The Earth is shedding its cycle of ice ages for a fire age.

The traditional view of North Dakota, as of the Great Plains generally, divides it into humid east and arid west with the border between them running roughly along the 100th meridian. It's a division by water but it works for fire as well. It also marks a potential boundary between Pleistocene and Anthropocene.

• • •

For Pleistocene Dakota, look east to the prairie pothole region. It's a vestigial landscape of the ice sheets. The retreating ice left a surface dappled and rumpled with kettles, drumlins, eskers, potholes, kames, and ridges that slowly smoothed into a terrain of swales and uplands. The swales filled with water. The uplands sprouted tallgrass prairie. Those ponds make the region a vital flyway for North American waterfowl. But keeping the wetlands wet is only half the management issue. The birds nest and feed in the uplands and, being clothed in tallgrass prairie, the uplands flourish best when routinely burned. Few of these fires start from lightning; the only viable source is people, who followed the retreating ice and set fire to the grasses. Those fires are themselves relics of a bygone epoch. They annually renew the living landscape that succeeded the dead ice.

For Anthropocene Dakota, look west to the Bakken constellation. Not only is it a symbol of industrial combustion, but a major source of greenhouse gases and a catalyst for land-use change and all the rest of the upsets and unhingings and scramblings that add up to make the Anthropocene. The flares speak to the extravagance of industrial fire—burning just to burn in order to get more stuff to burn. It's both a positive feedback and an eerily closed loop that accelerates the process and worsens its consequences. Instead of seasonal waterfowl, vehicles powered by internal combustion engines traverse the landscape ceaselessly.

East and west represent two kinds of fires and two kinds of future for humanity as keeper of the planetary flame. One is a Promethean narrative that speaks of fire as technological power, as something abstracted from its setting, perhaps by violence, certainly as something held in defiance of an existing order. The other is a more primeval narrative in which fire is a companion on our journey and part of a shared stewardship of the living world.

Sometime over the past century, we crossed the 100th meridian of Earth history and shed an ice age for a fire age. Landscape flames are yielding to combustion in chambers, and controlled burns, to feral fires. The more we burn, the more the Earth evolves to accept still further burning. It's a geologic inflection as powerful as the alignment of mountains, seas and planetary wobbles that tilted the Pliocene into the cycle of ice ages that defines the Pleistocene.

The era of the ice is also our era. We are creatures of the Pleistocene as fully as mastodons and polar bears. Early hominins suffered extinctions along with so many other creatures as the tidal ice rose and fell. But humans found in the firestick an Archimedean fulcrum by which to leverage their will. For tens of millennia we used it within the framework bequeathed by the retreating ice, and for more than a century we have been told that we thrived only in a halcyon age, an interglacial, before the ice must inevitably return.

Gradually, however, that lever lengthened until, with industrial fire, we could unhinge even the climate and replace ice (with which we can do little) with fire (with which we can seemingly do everything). We can melt ice sheets. We can define geologic eras. We can, on plumes of flame, leave Earth for other planets. It seems Eiseley was right. We are a flame.

Coming to a Forest Near You

WHEN CALIFORNIA LAST had a drought this severe, the United States was on the cusp of a reformation in how it managed fire on public wildlands. Fifteen years earlier an insurrection had boiled over into a full-fledged revolution that moved the federal agencies from a policy committed to fire's control to one dedicated to fire's management. They would restore a natural process to natural sites. They would reclaim humanity's oldest technology to its rightful place in working landscapes. Hot fire would reemerge as the newest of cool tools.

But the summer of 1977 was not quite the time nor California the right place. The U.S. Forest Service was still months away from officially publishing its reformed policy, and the National Park Service, building on a decade of experimentation, continued to fuss over a manual that would guide planning across its diverse holdings. Besides, the West was in drought; California's was brutal; even the two Sierra parks that had pioneered the national campaign to reinstate fire, Sequoia-Kings Canyon and Yosemite, were battening down for an anticipated siege. Now was not the time to play with fire.

The fires, when they came, were fought with every engine, hotshot crew, and air tanker the country could muster. In August I joined a National Park Service crew drafted from the western parks to fight the Marble-Cone fire in the Ventana Wilderness outside Big Sur. Wilderness be damned, bulldozers scraped lines across ridgetops in a vain attempt to halt a mammoth blaze that threatened the watershed of Carmel. That fire continued, eventually becoming at the time the second largest of record for California. Other fires followed. They didn't cease until the Honda fire overran Vandenberg Air Force Base during the winter solstice.

The season was a classic call-out: the firefight on a subcontinental scale. It confirmed what the 1970 blowout had demonstrated, that no single agency could by itself cope with the big burns, that the future of fire protection depended on interagency sharing, or what was known as total mobility. It also pointed to what officials hoped would be the future of wildland fire, that fewer fires would need to be fought on this scale because we could let remote ones burn themselves out (and celebrate the biological work they did) while substituting our own quasi-tamed fires for wild ones. Before the next fire season arrived, the policies and institutional reforms to make that happen were in place. America's great cultural revolution on fire had arrived.

Then it stumbled. The fires that were expected to strike California in the coming seasons were the inevitable outcome.

* * *

In retrospect it's easy to see how America's western firescapes evolved into an alloy of the big, the costly, and the feral. How prescribed fire became too complex, too expensive, and too laden with liabilities. How concerns with firefighter safety pushed fire agencies to pull back under extreme conditions and hostile settings. How legacy fuels powered more savage outbreaks. How climate change removed the temporal buffers and exurban sprawl the spatial ones that had granted previous generations room for maneuver. How political gridlock over public lands paralyzed agencies like the Forest Service. How costs argued to go big with escape fires and multiple ignitions. How the only fires and smokes that were allowed were, paradoxically, wildfires for which there was no culpable agent. How, in sum, fire agencies have (if silently) ceded an illusion of control, have surrendered beliefs that they can get ahead of the problem, and have sought, with big-box burnouts and point protection, to turn necessity into opportunity.

Widen the historical aperture, and it is possible to see the present scene as a logical, though not inevitable, span in a narrative arc that began a century ago. For 50 years following the Big Blowup of 1910 the country, under the aegis of the U.S. Forest Service, had sought to remove fire. For the past 50 years, under a consortia of institutions, it has sought to restore fire as fully as possible, for the most part under a doctrine of fire by prescription. The first strategy only destabilized biota after biota. The second has failed to reinstate fire regimes on anything like the scale required. Both ambitions, however, have shared the belief voiced by William Greeley in 1911 that fire control was as surely a matter for scientific management as is silviculture.

Suppression had never been—could never be—as successful as advocates wished. Neither did restoration fail as utterly as critics might claim. In the Southeast prescribed fire has become a foundational practice, in principle on par with suppression. While no one, not even Florida, burns as much as they want or should, the practice is so accepted there that controversies turn on what season or mix of burns should be applied. In the West prescribed fire has succeeded in select grasslands and in prairies overrun with shrubs, but elsewhere it exists as a boutique practice. The belief so prominent during the early years of the 1960s fire revolution that tame fire might replace or complement natural fire has faltered. We aren't going to restore landscapes to a prior golden age or fashion new ones to our liking for a golden age to come. We can't classify fires by some rational schema that nature will recognize. We can't separate the various practices of fire management into bureaucratically correct categories. Science can't solve what are, fundamentally, political issues.

It's also easy to see where current trends are heading and what we might do to improve the outcomes. The emerging issue is how to cope with whatever wildland fires appear on the land. Their source of ignition doesn't matter—fire is fire, as the mantra goes. Nor does it matter much where they originate. Unless they threaten high-value assets like exurbs or sequoia groves, it makes little sense to dictate responses based on legal locations. There will never be enough money, enough political capital, or enough space for maneuvering to get ahead of the threats or to subject wildland fire to the categories of managerial theories. We will have to deal with what is coming at us.

• • •

Instead, fire management has become a mashup. A given fire might have parts fought, parts boosted by burnouts, parts loose-herded, parts monitored, parts burned one day and extinguished the next. The trend is clear. Suppression will concentrate on the point protection of critical assets whether the enclaves be ecological or exurban. Wildland fires will be confined within big boxes of land-scape whose perimeters are then burned out. Area burned will explode. Costs per acre will decline. The West's wildlands will continue to cascade from one record year to another with very mixed outcomes. Some fraction of the burns will be more severe than we will be comfortable with. Some fraction within fire perimeters will remain unburned. The rest will exhibit burning more or less within what we might consider the slippery range of historic variability.

Such burns, and the burnouts that accompany them, will constitute the bulk of prescribed fire in the West, provided one ignores the inconvenient fact that they are not really prescribed. They just happen. They are a fusion of wild and set flames. They are not directed toward the prescriptions of a golden past or those of an imagined golden future. They offer resilience, not restoration. They implicitly acknowledge that humans, while fire's keystone species, are not in control. We can, within limits, start, stop, and shape fire and we can certainly disrupt fire regimes, but we will only truly be in charge within built landscapes; the more meticulously those places are shaped, the greater our grip. We have pretty rigorous control over what happens within a diesel piston. We have very little over free-burning flame in wildlands where our reach exceeds our grasp.

• • •

Such developments do not mean we can do nothing or that research is irrelevant. But what might we do? and what topics merit systematic investigation?

Begin with burning out. On too many fires, the most severely damaged sites are those from backfires. We can do better. Particularly when firing occurs a ridge away from the main burn or from secure lines under less than emergency conditions, there are good reasons to consider these counterfires as a variant of prescribed burning. Surely our fire behavior knowledge is sufficient to prevent ruinous aftershocks. We could spot-burn and chevron-burn rather than strip-fire. We could adjust burning according to what would be de facto prescriptions. The know-how and technology exist.

Those defensible perimeters can look a lot like the preattack zones of an earlier age. We could construct them in advance so that they meet not only considerations of fire behavior and firefighter safety but make sense in terms of ecological benefits: we could design them to be burned with consequences beyond their simple value as counterfires. We are already doing some of this with strategic treatments for fuels. Rationalize the process a bit further, and you have a matrix of burning blocks. When the fire revolution began, those blocks would have been the basis for prescribed fire. Now they must accept mixes of fires, a fusion of wild flames and burnouts. That mashup is what prescribed fire in the West is starting to look like.

Among large-perimeter fires, the interior will present a dappled texture of sites burned to their roots, of sites unburned, and of sites variously touched by flame. This is an ideal matrix for postburn or recovery firing. Some of those new burns and reburns will happen immediately; some, over a sequence of years. It sounds counterintuitive to argue for more fire as soon as the ash cools, but that postburn firescape is often a perfect setting in which to prescribe burn because it offers safe perimeters. You can burn against the black. You can burn in different seasons. You can burn, in patches, over large areas. Call the immediate process salvage burning. Call what occurs later prescribed reburns or recovery burns.

All this, too, is a suitable topic for research. Too often the failure to follow up on a wildfire has meant, particularly in montane forests, that one or two or more cycles may pass before fire returns in any form. Instead, we are presented with circumstances that could accept fire without the usual protocols of checklists and the gambler's-ruin logic of set-piece burns. We need a large cache of burning options to match the various possibilities of the postburn landscape. With some scientific guidance, modern technology could fill the cache with cool tools.

Such a strategy is not restoration as that enterprise has traditionally been understood. It does not consciously seek to return the scene to a prior conditions, nor does it shape it to a predetermined future state. It does not conform to that idealized project in which science determines and management applies. It makes do with what is at hand. It's a philosophy of resilience, not restoration. It's pragmatism with a small "p."

From its origins federal fire programs have channeled their research energies into science and policy. With a few exceptions this managerial ambition has failed to produce commensurate results on the ground, and at some point we need to admit that, taken by itself, the agenda has faltered if not failed. Besides science we need poetry, and in addition to policy we need politics. The poetry is what gives meaning to the enterprise. The politics is what gives it legitimacy and institutional leverage.

These topics are amenable to scholarship. They aren't science—they aren't topics that the scientific method can address. But they are subjects open to history, literature, art, and philosophy. Just as we use science to stiffen practices in the field, so we can look to other venues of scholarship to add rebars to the concrete of our ideas. Adaptive management, for example, is an avatar of Pragmatism, America's contribution to the philosophical tradition of Western civilization. That small "p" pragmatism we are witnessing in the field might well be improved with some upgraded logic from big "P" Pragmatism. Why not?

• • •

As the folk saying goes, we can't choose what happens to us, but we can choose our response, and in the case of fire management, we have various practical and intellectual tools to give heft and rigor to the choices we might make. We can control very little of what is driving fire in our western wildlands. In reality, the whole preoccupation with drivers is misguided, because the better analogy to wildfire is a driverless car that barrels down the road integrating everything around it. At times one factor or another may dominate, but they are all present. Their very abundance means there are many points of intervention possible.

A mashup needn't mean a mêlée. We can rationalize much of how we respond. Yet to surrender the illusion that we are in control will come awkwardly to an enterprise that has always assumed it is an exercise in objective science and disinterested politics. The fact is, we were never in control, many of our mistakes had the imprimatur of the science of their day behind them, and because the project has involved public lands and public safety it has always been inherently political. Yes, science self-corrects, as does the environment, but the legacies of errors can last for decades and the scene will never return to the future it might have had otherwise. What is different today is that the veneer of control is being stripped away by record burns, one after another, year by year. We can change only a little of what is happening. We can, however, control how we respond.

There is an adage about writing that reads, Follow your heart but use your head. Logic and craft are what transform experience and intuition into art. A similar epigram will likely apply as we confront the fires to come. We will have to follow the flames—we'll have no choice—but we will need to use our head. We'll need everything we have if we wish to turn the chaotic mashups of fire's presence into an artful management of our public wildlands.

This time I won't be patrolling Chews Ridge, pulaski in hand, looking for places where sparks and errant flames have breached the line. I'll be looking for signs that something has crossed the fuelbreaks of history. We had 50 years of unremitting effort to resist fires. We have then undergone 50 years trying to restore flames, either to desired ages past or future. We now seem to be entering a more modest era of resilience. How we handle those spot fires and slopovers will tell whether the future will mean more force and counterforce, or whether we are finally learning to live with our best friend and worst enemy.

"Science Supplies the Solution"

I T WAS AN auspicious entry to an august setting amid desperate times.

The National Academy of Sciences (NAS) had been established in 1863, amid the American Civil War, to bring scientific advice to the federal government. Its imposing building sits on the Constitution Avenue, facing the Mall. Its logo features a torch. Its Great Hall spills into the auditorium beneath a mural that depicts Prometheus carrying flame heavenward. For fire folk tasked with discussing the future of wildland fire science at a time when fires were hollowing out the U.S. Forest Service, it couldn't get more distinguished.

The Workshop on a Century of Wildland Fire Research, hosted by the Academy on August 27, 2017, gathered many of the best minds in the field. The session took place against a new political order championed by a president famous for not reading, who seemed to govern by TV and tweets, whose spokespersons preferred "alternative facts" to vetted empiricism, whose temperament seemed more likely to wield a flamethrower than a hose. The country was not openly at war with itself but was more implacably divided into warring tribes than at any time since the Civil War. Even the National Cohesive Strategy segregated the country into three realms, which, revealingly, was also a map of Civil War America.

• • •

Chief Forester Tom Tidwell announced to the assemblage both its charge and his challenge. Wildland fire was becoming an existential crisis for the USFS— burning through budgets, disrupting landscapes, shredding plans, mocking the premise behind the agency's century-old charter to apply rational, science-based solutions to the national forests. These were dire times. The 2016 fires had absorbed over 50 percent of the Forest Service's entire budget, and there was no fiscal suture in sight. Deputy Chief Carlos Rodriguez-Franco noted that a science of wildland fire had begun, largely under the Forest Service, a century ago. Now was the time to inventory and data-mine that past. Now was the desperate moment to project the knowledge gained and find an answer that would quell the flames and stanch the hemorrhaging of monies that was bleeding the agency white.

Yet like the torch on the NAS logo, the founding faith of the agency still burned bright. The chief declared, as an axiom, that "science supplies the solution." What science are we missing? he asked. What gap in research, if filled, will allow us to complete our understanding? What elusive gadget of modern technology, found or devised, will permit us to translate that revealed knowledge into practice? What was the missing link of scientific expertise that would ignite the National Cohesive Strategy into a rational program to address the nation's escalating need to halt bad fire, promote good fire, and end the fiscal immolation of its agencies?

Implicit in the chief's plea was the recognition that the wildland-urban interface had unbalanced the national fire scene. Its metastasizing spread over rural America had pulled the federal agencies away from land management and was hammering them into a national fire service that would, by endlessly suppressing fire, ultimately only worsen the scene. My mind drifted. The WUI was a true problem but a tired trope.

My thoughts finally alighted on what might well be regarded as the inter-face's founding fire—call it WUI One. I recalled images of burned houses and columns of fire engines. On the screen of my memory I watched snatches of *Design for Disaster*, the documentary produced by the Los Angeles Fire Department, and replayed its sonorous narration by William Conrad. Through the smoke of memory I saw two canonical images that 56 years later still spoke truth to the power of fire.

•　•　•

One image was specific to California—call it California Iconic. The other was a testimony in twisted irony that could have occurred anywhere but happened to fall out this time in California. Call it California Ironic.

Iconic California is the template image, endlessly recycled by photojournalists, of some doofus on a wood-shingle roof with a garden hose. Except that this time the doofus was Richard Nixon in white shirt and tie, vacantly staring up and away from the background flames while a piddling stream of water collected at his feet.

The usual interpretation is that the photo shows the indifference to hazard by Southern Californians, who then expect firefighting to plug the gap, or that politicians are ever eager for staged photo ops. But what I read in this version is that politics is not an afterthought when it comes to fire management. Wildland fire, particularly the WUI, is a matter of public safety and public assets that properly belongs in political discourse. It has its distortions and comic perversions—Nixon with a garden hose is a great illustration—but politics is a fundamental reality of the scene. The belief that fire can be managed by disinterested agents who apply the tested conclusions of science is an illusion. Politics has been there, right from WUI One.

California Ironic features another member of the great and famous. Willard Libby had, the year previously, been awarded a Nobel Prize in physics for his 1949 work on ^{14}carbon dating. In the intervening years he had joined the Atomic Energy Commission and become a public shill for fallout shelters. Two weeks before the Bel Air-Brentwood fire, he was photographed, in formal wear, complete with bow tie, in what he called a "poor-man's fallout shelter," which he had built out of sand bags and railroad ties (along with a little plastic sheeting) for $30.

The fire incinerated his shelter along with his home. His wife fled the flames with her mink coat and his Nobel medallion. A repeat photo of the shelter taken after the burn, sans Libby, failed to garner the same interest. Learning of the episode, Leo Szilard declared it proved two things. One, God exists. And two, He has a sense of humor. It would also prove that fire was less predictable than the other effects of nukes.

Or more broadly, it hinted at the limitations of science to resolve the problems that would plague the WUI. Unlike blast and much of fallout, fire was a biological construct and a creature of context. It propagated. Its spread depended on wind, fuel moisture, the arrangement of combustibles, most of which grew according to ecological imperatives or, where built by people, by cultural considerations, all of which would likely be shattered by a bomb's blast. So, too, fire's character as friend or foe, as a transient summer job or a chronic bureaucratic crisis, depended on its social and political setting. In March 2017 that context seemed dire.

• • •

The assembled scientists did their best. They talked, they traded thoughts, they proposed paths forward. But in the end, they testified, if by their deeper silence, to the limitations of science. There was no shame in that: it was the way of the world. The fire crisis was really a crisis in the American experiment itself. The wildfires burning through agencies were burning through the widening cracks in culture and politics. What the USFS needed was not a scientific breakthrough or a problem-disrupting technology but a clarity of mission, a legitimacy before the many American publics and their representatives, an intellectual civility and a cultural humility. It didn't need to upgrade its science-informed policy. It needed a poet.

The chief was proposing a world that would not have Nixon on the roof or Libby in his shelter. That was not a dishonorable ambition, just a utopian one. Until then fire officers on the ground would glance at the science as codified in decision support systems and play with new gadgets, but they would act as conditions permitted. They would try to respond in a way that was more than the sum of the parts that sent fire through the land. They couldn't hope to master fire. They could hope to listen better to what the fire was telling them, and if they were lucky as well as good, they could hope to stay ahead of the flames.

Portal to the Pyrocene

And where two raging fires meet together,
They do consume the thing that feeds their fury.
— William Shakespeare, *The Taming of the Shrew*

THE IMAGES ARE gripping. Horizons glow with satanic reds squishing through black and bluish clouds, as though the sky itself were bruised and bleeding. Foregrounds bristle with scorched neighborhoods still drifting with smoke and streams of frightened refugees, a scene more commonly associated with war zones.

But we've seen this before. Big fires are big fires, and one pyrocumulus can look pretty much like another. Communities with homes burned to concrete slabs, molten hulks of what once were cars alongside roads, surrounding forests mottled with black and green—these are becoming commonplaces.

What strikes me most about the Fort McMurray images is the mashup of foreground and background, the collision of free-burning flames with a fossil-fuel powered society. The first form of burning dates back to the Silurian era, when life originally colonized the continents. The second tracks the Anthropocene, when humanity changed its combustion habits and wrenched the Earth into a new order. At places like Fort McMurray the deep past and the recent present of fire on Earth rush together with almost Shakespearean urgency.

• • •

The plot is old, the stage setting and cast of players updated.

Monster fires are no stranger in the boreal forest. It's a fire-ravenous biota that burns in stand-replacing patches. This is not a landscape where misguided fire suppression has upset the rhythms of surface burning and catapulted flames into the canopy. They've always been in the canopy, and everything has adapted accordingly. White and black spruce and jack pine and aspen experience exactly the kind of fire they require. What fire protection has done is to encourage larger patches.

How big those patches get depends on how dry the fuel is, how brisk the winds, and how extensive the forest. In northern Alberta there is not much to break a full-throated wildfire. The Chinchaga fire started on June 1, 1950, and burned across northeastern British Columbia and most of Alberta until October 31, a total of 3 million acres. That single burn was embedded in a larger regional complex that probably summed 4.8 million acres. Comparable fires date back as far as we care to look.

Nor is a burning city a novelty. In North America the wave of settlement in the 18th and 19th centuries paralleled a wave of fire. The surrounding lands were disturbed and frequently alight with both controlled and uncontrolled fires. The towns were built of wood—basically, reconstituted forests. The same conditions that propelled fires through the landscape pushed them through towns.

Only a century ago did those urban conflagrations finally quell as urbanites turned to less combustible materials, fire codes and zoning regulations organized buildings in ways that discouraged spreading flames, fire services acquired the mechanical muscle to halt blazes early, and the wave of settlement flattened. Over the past century modern cityscapes have needed earthquakes or wars to overcome those reforms and sustain conflagrations.

Then, they began to return as a broadly rural scene morphed and polarized into an urban frontier of wildlands and cities that faced one another without intervening buffers. The middle, working landscapes, like the middle, working classes, shriveled at the expense of the favored extremes. In 1986 the term *wildland-urban interface* appeared. It was a clumsy, dumb phrase, but it referred to a dumb problem. Watching houses, and then communities, burn was like watching polio or plague return. This was a problem we had solved, then forgot to—or chose not to—continue the vaccinations and hygiene that had halted their terrors.

Initially, the problem appeared as a California pathology. But it soon broke out of quarantine and has spread across western North America. The prevailing narrative held that the problem was stupid Westerners building houses where there were fires. Most of the vulnerable communities, however, are in the southeastern United States, and if climate change modelers are correct and such outbreaks as the conflagration at Gatlinburg, Tennessee, prove prophetic, we will see the fires moving to where the houses are. That will make it a national narrative. In truth, the problem is international, each country with its own quirky combination of fire-quickening factors. Mediterranean France, Portugal, Greece, South Africa, and Australia are experiencing similar outbreaks. North America has no monopoly.

It's tempting to appeal to climate change as the common cause. Yet the burning bush and scorched town are joined not just by global climate change but also by a global economy, and, behind both, by a global commitment to fossil-fuel firepower. That makes the issue both more pervasive and, paradoxically, more amenable to treatment. It means that, while there is one grand prime mover, there are many levers and gears. Fire is a reaction that takes its character from its context. It's a driverless car barreling down the road, synthesizing everything around it.

The enduring images of the Fort Mac fire may, in fact, be its cars. Car-propelled flight, cars stranded for lack of gas, cars melted in garages, evacuation convoys halted due to 60-meter flames, relief convoys laden with gasoline. It isn't only what comes out of the tailpipe that matters but how those vehicles have organized human life in the boreal. The engagement (or not) with the surrounding bush. The kind of land use that cars encourage. The kind of industry that must develop to support those cars. The kind of city that such an industry needs to sustain it. The oil and tar sand industry that has shaped the contours of modern Fort McMurray and is in turn shaped by the internal-combusting society it feeds.

So there are really two fires burning around and into Fort McMurray. One burns living landscapes. The other burns lithic landscapes, which is to say biomass buried and turned to stone in the geologic past. The two fires compete: one or the other triumphs. At any place the transition may take years, even decades, but where the industrial world persists its closed combustion will substitute for or suppress the open flames of ecosystems. The wholesale transition from the realm of living fire to that of lithic fire may stand as a working definition of the Anthropocene. Once parted they rarely meet.

At Fort McMurray they have collided with unblinking brutality. Wildfire burned away controlled fire. The old fires have forced the power plants behind the new ones to shut down and their labor force to flee. It's like watching an open pit mine consume the town that excavates it. It's tempting to regard the incident as a one-off, a freak of a remote landscape and a historical moment. But those collisions are becoming more frequent.

That's not the deep worry, however. The deep horror is that the two fires may be moving from competition into collusion. They are creating positive feedback of a sort that makes more fire. Those images of fire on fire are the raw footage of a planetary phase change, what might end up as a geologic era we could call the Pyrocene. They will continue until, as Shakespeare put it, they "consume the thing that feeds their fury."

Disaster is not always tragedy, and Fort McMurray and the industrial complex behind it may well escape lethal consequences in the near future. So if Shakespeare seems too elevated, consider Edna St. Vincent Millay.

My candle burns at both ends;
It will not last the night;
But ah, my foes, and oh, my friends—
It gives a lovely light!

We have in truth been burning both ends of our combustion candle, and if its light increasingly seems more lurid than lovely, there are yet texts to be read in the awful splendor of its illumination.

NOTES

A TALE OF TWO LANDSCAPES

1. J. C. Ives, *Memoir to Accompany a Military Map of the Peninsula of Florida, South of Tampa Bay* (New York: Wynkoop, 1856). The quotations are from page 1.

2. This essay is possible because of the generosity of Robert Dye, who shared a day in the field with me, and then—unusually, for my tour of Florida—handed over documentation. The various publications of Paula Benshoff also proved vital in appreciating what it takes to move ideas into practice. The history of fire in the region presented in this essay is basically theirs, recast into a somewhat different frame, one that I believe they do not wholly agree with. They deserve credit for what works in the essay. For what doesn't, the fault is mine.

3. For a description of the Myakka Island concept, see P. J. Benshoff, *Myakka*, 2nd ed. (Sarasota, Fla.: Pineapple Press, 2008).

4. Quoted in David Nelson, "The Great Suppression: State Fire Policy in Florida, 1920–1970," *Gulf South Historical Review* 21 (2006): 89.

5. James A. Stevenson, "Conference Dedication of E. V. Komarek, Sr.," in *Proceedings, 17th Tall Timbers Fire Ecology Conference: High Intensity Fire in Wildlands: Management Challenges and Options* (Tallahassee, Fla.: Tall Timbers Research Station, 1991), 1.

6. The Buck Mann story is from Benshoff, *Myakka*, 191.

7. The burning pattern is from Mike Kemmerer, email message to Robert Dye, January 12, 2011.

8. For a good digest of the fire scene as the principal agents of the Myakka fire program saw it, see P. J. Benshoff, R. Dye, and B. Perry, "Fire in the Florida Landscape" (Myakka River State Park, n.d.). For an overview of the ecology, see *Proceedings of the Florida Dry Prairie Conference: Land of Fire and Water: The Florida*

Dry Prairie Ecosystem (DeLeon Springs, Fla.: Painter, 2004), http://www.ces.fau
.edu/fdpc/proceedings.php.

9. Quote from Robert Dye, email message to author, March 7, 2011.

10. For a detailed rendering of some of the experiments, see Robert Dye, "Use of
Firing Techniques to Achieve Naturalness in Florida Parks," in *Proceedings, 17th
Tall Timbers Fire Ecology Conference: High Intensity Fire in Wildlands: Management
Challenges and Options* (Tallahassee, Fla.: Tall Timbers Research Station, 1991),
353–60.

11. Numbers from Florida Park Service, *Resource Management Annual Report: July
2009–June 2010* (Tallahassee, Fla.: Bureau of Natural and Cultural Resources,
[2010]).

12. Stevenson, "Conference Dedication," 2.

ONE FOOT IN THE BLACK

1. Paula Seamon, Zach Prusak, Walt Thomson, and Ron Myers made this essay possible, and supplied documentation to back up their remembrances. I'm grateful to
them all, not least for their patient tolerance toward my efforts to mash four essays
into one. TNC, the Florida chapter, the fire initiatives, Ron—each deserves its
own panegyric. Which I'll try to do, all in good time.

2. Mark Heitlinger with the assistance from Allen Steuter and Jane Prohaska, "Fire
Management Manual," revised September 1985 (The Nature Conservancy).

3. See TNC, "Controlled Burning: Getting It Done" (TNC, 2009) and "A Decade
of Dedicated Fire: Lake Wales Ridge Prescribed Fire Team" (TNC, 2010);
"The Northeast Florida Resource Management Partnership, Draft Final Report
(April 1, 2008–December 31, 2010)"; "Central Florida Ecosystem Support Team,
Florida Fish and Wildlife Conservation Commission, Final Report (February 1,
2010–December 31, 2010)" (January 2011).

4. Two publications describe the program: Ayn Shlisky et al., *Fire, Ecosystems and
People: A Preliminary Assessment of Fire as a Global Conservation Issue* (TNC, 2004)
and Ronald L. Myers, *Living with Fire—Sustaining Ecosystems and Livelihoods
Through Integrated Fire Management* (TNC, 2006).

5. See Todd Wilkinson, "Prometheus Unbound," *Nature Conservancy*, May/June 2001,
12–20, for a survey of the national scene at the time, and in the same issue, William
Stolzenburg, "Fire in the Rain Forest," 22–27, for its work in Mesoamerica.

FIRE 101 AT STAR FLEET ACADEMY

1. Thanks to Michael Good, Fred Adrian, and Dorn Whitmore for their patient
(and occasionally colorful) explication of fire management at Merritt Island, and
to Robert Eaton for editorial comments.

2. On the fatal fire, see "Report of Board of Inquiry on Merritt Island NWR Wildfire Fatalities," Memorandum from Regional Director, FWS, Atlanta, Georgia,

to Director, FWS, Washington, D.C. (July 10, 1981), and Sebastino J. Castro and C. R. Anderson, *A Report of the Committee on Appropriations, U.S. House of Representatives, on Wildfire on Merritt Island*, Surveys and Investigations Staff, House Appropriations Committee, December 1981.

THE EVERBURNS

1. Marjory Stoneman Douglas, *The Everglades: River of Grass* (Port Salerno: Florida Classics Library, 2002), 5.
   Special thanks to Rick Anderson for a delightful day of meditations, field exhibitions, and stories. Thanks, too, to Jennifer Adams, Bonnie Ciolino, and other members of the fire management staff for help in tracking down documents and getting them copied. Anyone familiar with the Everglades fire scene will recognize my reliance on Dale L. Taylor, *Fire History and Fire Records for Everglades National Park 1948–1979*, Report T-619, South Florida Research Center, Everglades National Park (1981), which summarizes the known records. Beyond that I have referred to the 1991 and 2010 (draft) fire management plans, which summarize much of the previous history. There isn't much in park records that they miss. The best distillation of the basic fire ecology remains Dale Wade, John Ewel, and Ronald Hofstetter, *Fire in South Florida Ecosystems*, U.S. Forest Service General Technical Report SE-17 (Asheville, N.C.: Southeast Forest Experiment Station, 1980). The park filing cabinets are veritable hammocks of peaty grey literature.
2. Quotes from Daniel B. Beard, "Let 'er Burn?," *Everglades Natural History* 2, no. 1 (March 1954): 2–8; Guy J. Bender, "The Everglades Fire Control District," 1941, copy in the files of the Everglades Fire Management Office; Daniel B. Beard, *Wildlife Reconnaissance: Everglades National Park Project*, Department of the Interior, National Park Service, October 1938, 51–52.
3. Douglas, *Everglades*, 349, 376, 376.
4. Beard, "Let 'er Burn?," 8; Beard, *Wildlife Reconnaissance*, 51.
5. William B. Robertson, Jr., "A Survey of the Effects of Fire in Everglades National Park," unpublished report to the National Park Service. Submitted: February 5, 1953.

STATE OF EMERGENCY

1. Compared with most agencies, CalFire has successfully preserved a record of its past. The agency generously made available that in-house archive to me. I wish to thank Janet Upton for setting up my visit and Mary Welna for introducing me to that rich storehouse. I also wish to thank Chief Ken Pimlott for an opportunity to meet. Kim Zagaris of CalEMA went even further, not only creating a most useful interview with Richard Barrows but handing over a cache of copied documents. I regret, as so often, that an essay on a topic that the agencies themselves may regard as tangential is the outcome to what could easily constitute a book in itself (and

in CalFire's case did, published for its centennial). My task is to place the agencies within California and California within the national fire scene. My thanks to all.

2. California is exceptionally fortunate to have an extensive history of state forestry operations in the form of Raymond Clar's two-volume opus, *California Government and Forestry* (Sacramento, Calif.: Division of Forestry, Department of Natural Resources, 1959, 1967). Abridgements are available in C. Raymond Clar, *Evolution of California's Wildland Fire Protection System* (Sacramento: California Division of Forestry, 1969) and Mark V. Thornton, "History of CDF," in *California Department of Forestry and Fire Protection: 100 Years of CDF* (Paducah, Ky.: Turner, 2005), 10–20.

3. Clar, *Evolution of California's Wildland Fire Protection System*, 23.

4. Clinton B. Phillips, *California Aflame! September 22–October 4, 1970* (Sacramento: California Division of Forestry, 1971).

5. Task Force on California's Wildland Fire Problem, *Recommendations to Solve California's Wildland Fire Problem*, submitted to California Resources Agency, June 1972.

6. Many good accounts of Firescope exist. The core remains Richard A. Chase, *FIRESCOPE: A New Concept in Multiagency Fire Suppression Coordination*, U.S. Forest Service, General Technical Report PSW-40, 1980. Supplement and update with the following documents from CalEMA: *Some Highlights of the Evolution of the Incident Command System as Developed by FIRESCOPE*, unpublished timeline; *FIRESCOPE California: Past, Current and Future Directions: A Progress Report* (October 1988); and *FIRESCOPE's Future*. Useful for a more personal perspective is the treatment of FIRESCOPE in Region 5 Oral History Project, *The Unmarked Trail: Managing National Forests in a Turbulent Era: Region 5 Oral History, Volume II: 1960s to 1990s* ([Vallejo, Calif.?]: U.S. Forest Service, Pacific Southwest Region, 2009), 153–87.

7. An excellent chronicle is available in CalEMA, *California Fire Service and Rescue Emergency Mutual Aid System: History and Organization* (rev. April 2002).

8. Information from *California Department of Forestry and Fire Protection: 100 Years of CDF*, especially 21–23. The book makes a case that the change from green to blue pants was required because the old pants, which included polyester, were a fire hazard, and blue was the only market alternative. But the federal agencies had developed green Nomex pants. The choice was clearly made to align more closely with urban counterparts.

9. *California Department of Forestry and Fire Protection*, 31.

IMPERIUM IN IMPERIO

1. John Todd and J. Lopez granted several hours of their busy schedule to present the labors of their centennial celebration, to coach me in the scope of LACFD operations, and to escort me around the facility at Pacoima. It was an enlightening education. I regret my sketch can only scratch at the surface but hope those markings can hint at the essence beneath.

2. An encyclopedic survey of LACFD up through the late 1980s is available in David Boucher, *Ride the Devil Wind: A History of the Los Angeles County Forester and Fire Warden Department and Fire Protection Districts* (n.p.: Fire Publications, Inc., 1991). I have mined the text for the basics, supplemented by other documents and discussions as indicated.

MENDING FIREWALLS

1. Special thanks to Anne Fege for organizing a lively group discussion over an even more delicious dinner, and to Stephen Fillmore of the Cleveland National Forest for thoughtful commentary and maps on the regional fuelbreak network, and to Clay Howe for patiently explaining the Establishment position. I'm sure they all would wish that I would get off the fence, but the endurance of the fence itself may be the story, or so it appears to a historian of my temperament.
2. As Stephen Fillmore noted in comments to a draft, "It of course is difficult to implement landscape scale management when you don't have a landscape scale to work with."
3. "International Fuel Break," California Fire Alliance, Success Stories, handout; quote on burnout from Clay Howe.
4. John McPhee, *The Control of Nature* (New York: Farrar, Straus, and Giroux, 1990), 271.

THE BIG ONES

1. I am the beneficiary of an extraordinary effort, organized by John Swanson and the East Bay Regional Parks, to educate me in the saga of fire in the Hills. I was privileged to hear analyses by Jerry Kent, Cheryl Miller, Ken Blonski, Rosemary Cameron, Leroy Griffin, and Peter Scott, and to visit the site in the field with John and Bill Nichols. A crowded day but an exceptionally instructive one, which left me with the belief that the only contribution I might make was to shift the terms of context. My thanks to them all. I was impressed with their resolve as much as with their knowledge.
2. I rely on Philip L. Fradkin's comprehensive *The Great Earthquake and Firestorms of 1906* (Berkeley: University of California Press, 2005); for figures see 9–10, 37.
3. Grove Karl Gilbert, April 20, 1906, Field Notebook 3501, National Archives. Quoted in Stephen J. Pyne, *Grove Karl Gilbert* (Austin: University of Texas Press, 1980), 211.
4. Jack London, "The Story of an Eye-Witness," *Collier's* 37, no. 6 (May 5, 1906): 22–25.
5. Figures from David J. Nowak, "Historical Vegetation Change in Oakland and Its Implications for Urban Forest Management," *Journal of Arboriculture* 19, no. 5 (September 1993): 313–19.
6. Many sources. I found most useful the National Fire Protection Association, *The Oakland/Berkeley Hills Fire*, report for the National Wildland/Urban Interface Fire

Protection Initiative (1992); and East Bay Hills Fire Operations Review Group, Office of Emergency Services, *The East Bay Hills Fire: A Multi-Agency Review of the October 1991 Fire in the Oakland/Berkeley Hills*, report for California Office of Emergency Services (February 27, 1992).

The fire is the subject of numerous publications. I should mention in particular Margaret Sullivan, *Firestorm! The Story of the 1991 East Bay Fire in Berkeley* (City of Berkeley, 1993), and Peter Charles Hoffer, *Seven Fires: The Urban Infernos That Reshaped America* (New York: PublicAffairs, 2006), which includes several chapters on the Tunnel fire.

7.   East Bay Hills Fire Operations Review Group, *East Bay Hills Fire*, 8.
8.   East Bay Hills Fire Operations Review Group, 7.

FIRE'S CALL OF THE WILD

1.   Thanks to Bob Mutch and Dave Campbell for inviting me to join their anniversary flight, and then sharing over the course of an afternoon their deep experience with the SBW. It goes without saying that the opinions expressed in this essay are mine alone.
2.   On 1967 fires, William R. Moore, Assistant Director, WO-Fire Control, "Report on Field Travel in Region 1, August 8–19 and September 3–9, 1967," National Archives, Record Group 95, Accession 72-A-3046, Box 168, p. 6. On "practically illegal," from Linda S. Mutch and Robert W. Mutch, "Wilderness Burning: The White Cap Story," unpublished manuscript with transcriptions from recordings from the anniversary event; courtesy of Robert Mutch.

THE EMBERS WILL FIND A WAY

1.   This profile could not have been written without the help of Jack Cohen, despite his unintended efforts in person and in print to overwhelm me with a swarm of data. He left the wording, the conceptual insights, and the organizing conceit to me, which is to say any errors are mine. It was a pleasure to listen to him and to feel his radiant enthusiasm, at least some of which I hope has found a glow in my sketch.

HOW I CAME TO MANN GULCH

1.   I wish to thank Craig Kockler of Helena National Forest for sparing a few minutes at the end of a long day to pass along information regarding the Meriwether fire.

WHAT MAKES A FIRE SIGNIFICANT?

1.   Stephen W. Barrett, Stephen F. Arno, and James P. Menakis, *Fire Episodes in the Inland Northwest (1540–1940) Based on Fire History Data*, General Technical Report INT-GTR-370, U.S. Forest Service, 1997, 3.
2.   Incident reported in the *Brooklyn Daily Eagle*, August 14, 1889.

3. Reported by the *Sacramento Daily Record-Union*, Sacramento, California, August 2, 1889.
4. *New York Times*, August 24, 1889.
5. Telegraph reference from *Lawrence Daily Journal*, July 30, 1889.
6. Wallace Stegner, "A Sense of Place," in *Where the Bluebird Sings to the Lemonade Springs* (New York: Modern Library Classics, 2002), 205.

THE OTHER BIG BURN

1. I want to thank Dennis Divoky, fire ecologist, for devoting an afternoon at the end of a busy week to a tutorial on Glacier fire, and to Deirdre Shaw, archivist, for help in identifying and copying a stack of relevant historical documents. Glacier is remarkable in having records, exemplary in its ability to access them, and exceptional in the willingness of its staff to make them available. They made me feel like a historian up to his elbows in paper rather a journalist dependent only on oral interviews.
2. Superintendent memorandum, November 30, 1967, "Public Relations and Press Coverage During the Fire Emergency" and "Proceedings—Glacier Forest Fire Review: November 30–December 1, 1967," both in Glacier National Park Archives; Box 309 Wildland Fire Management (Y14); Folder 21.
3. For quote, see p. 35.
4. Thomas Zimmerman, Laurie Kurth, and Mitchell Burgard, "The Howling Prescribed Natural Fire—Long-Term Effects on the Modernization of Planning and Implementation of Wildland Fire Management," in *Proceedings of 3rd Fire Behavior and Fuels Conference, October 25–29, 2010* (Spokane, Wash.: International Association of Wildland Fire, 2011).
5. See USGS, "Retreat of Glaciers in Glacier National Park," last accessed June 11, 2019, https://www.usgs.gov/centers/norock/science/retreat-glaciers-glacier-national -park?qt-science_center_objects=0#qt-science_center_objects.

SEASONS OF BURNING

1. For a good survey of contemporary art, see Julie Courtwright, *Prairie Fire: A Great Plains History* (Lawrence: University of Kansas Press, 2011), who notes that the "beauty of the prairie fires thus became a significant piece of Plains identity" (162). The book has quickly established itself as a standard reference.
2. The saga of Bessey, Clements, Weaver, et al., is well summarized in Ronald C. Tobey, *Saving the Prairies: The Life Cycle of the Founding School of American Plant Ecology, 1895–1955* (Berkeley: University of California Press, 1981); for quotes, see 2–3. My survey seeks to place Tobey's admirable account within the context of Great Plains fire history.
3. Tobey, *Saving the Prairies*, 5, 2.
4. Tobey, *Saving the Prairies*, 3.
5. See Courtwright, *Prairie Fire*, 169–71 for an account of the South Dakota fires.

PLEISTOCENE MEETS PYROCENE

1.   My debt is deep, but begin with the catalyst, Shane del Grosso. Then go to Ken Higgins, Pete Bauman, Jim Strain, and Kyle Kelsey, and on to Jeff Dion, Dan Severson, Cami Dixon, Arnie Kruse, and Karen Smith. They began my intellectual induction into a marvelous region.

2.   For background references, I found useful Kenneth F. Higgins, "Interpretation and Compendium of Historical Fire Accounts in the Northern Great Plains," U.S. Fish and Wildlife Service Resource Publication 161 (1986); Kenneth F. Higgins, Arnold D. Kruse, and James L. Piehl, "Effects of Fire in the Northern Great Plains," Extension Circular 761 (Brookings: South Dakota State University, 1989), "Prescribed Burning Guidelines in the Northern Great Plains," Extension Circular 760 (Brookings: South Dakota State University, 1989), and "Annotated Bibliography of Fire Literature Relative to Northern Grasslands in South-Central Canada and North-Central United States," Extension Circular 762 (Brookings: South Dakota State University, 2000); Leo M. Kirsch and Arnold D. Kruse, "Prairie Fires and Wildlife," in *Proceedings, Tall Timbers Fire Ecology Conference* (Tallahassee, Fla.: Tall Timbers Research Station, 1973), 289–303; Todd Grant et al., "An Emerging Crisis Across Northern Prairie Refuges: Prevalence of Invasive Plants and a Plan for Adaptive Management," *Ecological Restoration* 27, no. 1 (March 2009), 58–65; Jerry D. Kobriger et al., "Prairie Chicken Populations of the Sheyenne Delta in North Dakota, 1961–1987," in *Prairie Chickens on the Sheyenne National Grasslands*, General Technical Report RM-159, U.S. Forest Service, 1987, 1–7; Arnold D. Kruse and Bonnie S. Bowen, "Effects of Grazing and Burning on Densities and Habitats of Breeding Ducks in North Dakota," *Journal of Wildlife Management* 60, no. 2 (1996), 233–46; Robert K. Murphy and Todd A. Grant, "Land Management History and Floristics in Mixed-Grass Prairie, North Dakota, USA," *Natural Areas Journal* 25 (2005), 351–58; and David E. Naugle, Kenneth F. Higgins, and Kristel K. Bakker, "A Synthesis of the Effects of Upland Management Practices on Waterfowl and Other Birds in the Northern Great Plains of the U.S. and Canada," Wildlife Technical Report 1 (Stevens Point, Wis.: College of Natural Resources, University of Wisconsin-Stevens Point, 2000).

3.   Arrowwood refuge is fortunate in having a cache of narrative reports, and doubly so, in having digitized most. Not all years are included, but the detailed record gives a good sense of what developed.

4.   Kirsch and Kruse, "Prairie Fires and Wildlife," 293.

5.   For a summary of the controversy, so similar to those in America, see Stephen Pyne, *Vestal Fire* (Seattle: University of Washington Press, 1997), 361–64.

KONZA

1.   I want to thank Dave Hartnett for his help in setting up my brief visit, and Tony Joern for a wonderful morning's tour and tutorial; Tom and Barb van Slyke for

smoothing my stay; and Tony Crawford and Kari Bingham-Guiterrez for rousing the Konza Collection from its lair in Kansas State University's special collections.

2. A rich literature surrounds the region. For a good introduction to the land, see O. J. Reichman, *Konza Prairie: A Tallgrass Natural History* (Lawrence: University Press of Kansas, 1987). For a sample of its folklore, see Jim Hoy, *Flint Hills Cowboys: Tales from the Tallgrass Prairie* (Lawrence: University Press of Kansas, 2006). And since the hills form its emotional core, see Courtwright, *Prairie Fire.*

3. The best summary of Hulbert's life is "Proceedings Dedication, Lloyd C. Hulbert, 1918–1986," in *Prairie Pioneers: Ecology, History and Culture: Proceedings of the Eleventh North American Prairie Conference Held 7–11 August 1988, Lincoln, Nebraska,* ed. Thomas B. Bragg and James Stubbendieck (Lincoln: University of Nebraska Printing, 1989). Online copy available at http://images.library.wisc.edu /EcoNatRes/EFacs/NAPC/NAPC11/reference/econatres.napc11.i0006.pdf. Several curricula vitae that Hulbert created at different points of his career are available in Kansas State University Special Collections, Division of Biology, Konza Prairie, Accession Number U97.5, Box 12, File 1. These gave particulars about his CPS years.

    For an overview of the CPS smokejumper program, see Mark Matthews, *Smoke Jumping on the Western Fire Line: Conscientious Objectors During World War II* (Norman: University of Oklahoma Press, 2006).

4. Report of Committee, December 1953, Konza Prairie Collection, Special Collections, KSU, Box 1, Folder 20.

5. The most thorough summary of Konza's evolution is Hulbert, "History and Use of Konza Prairie Research Natural Area," *Prairie Scout* 5 (1985), 62–93.

6. On the speculation about the origins of Hulbert's interest in fire, see "Proceedings Dedication." For a digest of his scientific conclusions, see the posthumously published Hulbert, "Causes of Fire Effects in Tallgrass Prairie," *Ecology* 69, no. 1 (February 1988): 46–58.

7. Perhaps the best summary of its research richness is the application for a seventh term under NSF's Long-Term Ecological Research network, NSF 13–588. For this round the emphasis shifted toward climate: "Long Term Research on Grassland Dynamics—Assessing Mechanisms of Sensitivity and Resilience to Global Change" (Lawrence: Kansas State University, 2014).

8. From Lloyd C. Hulbert Memorial Service, Kansas State University Special Collections, Division of Biology, Konza Prairie, Accession Number U97.5, Box 12, File 1.

PEOPLE OF THE PRAIRIE, PEOPLE OF THE FIRE

1. This piece of fire journalism is the outcome of a too-fleeting visit at the invitation of the Illinois Prescribed Fire Council to speak at their annual meeting in May 2009, spiced with field trips to Kankakee and Nachusa. I would like to thank in particular Fran Harty, Bill Kleiman, and Cody Considine for their role as hosts and tutors. Anyone familiar with the pyric geography of these sites will appreciate

that I add nothing to data or concepts. Instead, I have sought only to establish a different perspective and narrative for their understanding.

Since the prairie peninsula helps bound themes from the Great Plains, I have included the essay within my general survey of the region.

THE BLACKENED HILLS

1. Blaine Cook did yeoman work in setting up a panel discussion with the key fire staff at the forest, and then organizing a marvelous field tour. For much that is informed about my essay, he deserves credit—and has my thanks. For their particular comments I would also like to thank Amy Ham, Jason Virtue, Rochelle Plocek, Frank Carroll, Todd Pechota, and Dave Mertz. I think it fair to say that there are passages that each one—each for his or her own reasons—would disagree with. The hills deserve to move from outlier to core in the national debate.

2. On its lusty pine, see Wayne D. Shepperd and Michael A. Battaglia, *Ecology, Silviculture, and Management of Black Hills Ponderosa Pine*, General Technical Report RMRS-GTR-97, U.S. Forest Service, 2002.

3. The photos have inspired several classic rephotography projects beginning with Donald R. Progulske, *Yellow Ore, Yellow Hair, Yellow Pine*, Bulletin 616, South Dakota Agricultural Experiment Station, 1974, and most recently, Ernest Grafe and Paul Horsted, *Exploring with Custer: The 1874 Black Hills Expedition* (Custer, S.Dak.: Golden Valley Press, 2002).

4. See, for example, Peter M. Brown and Carolyn Hull Sieg, "Fire History in Interior Ponderosa Pine Communities of the Black Hills, South Dakota, USA," *International Journal of Wildland Fire* 6, no. 3 (1996): 97–105. Other data from fire management staff at Black Hills National Forest.

5. Benjamin Kleinjan, "A Brief History of the Black Hills Forest Fire Protection District," 2009, unpublished report for the South Dakota Department of Agriculture.

6. Statistics from Black Hills National Forest fire management office.

7. Environmental critiques seem sparse, but a slight introduction is available in Sven G. Froiland, *Natural History of the Black Hills and Badlands* (Sioux Falls, S.Dak.: Center for Western Studies, Augustana College, 1999).

THE JEMEZ

1. A field trip sponsored by the Joint Fire Science Program for its board on October 29, 2014, provided a marvelous setting from which to think about fire in the Jemez Mountains. I had the privilege of conversing further with Bill Armstrong of the Santa Fe National Forest and the next day with Craig Allen of the U.S. Geological Survey, who graciously shared his deep knowledge of the region.

2. The best introduction to fire in the Jemez is Craig D. Allen's "Lots of Lightning and Plenty of People: An Ecological History of Fire in the Upland Southwest," in *Fire, Native Peoples, and the Natural Landscape*, ed. Thomas R. Vale (Washington, D.C.: Island Press, 2002), 143–93.

3. Teralene S. Foxx, La Mesa Fire Symposium, LA-9236-NERP, UC-11, Los Alamos National Laboratory, February 1984; Craig D. Allen, *Fire Effects in Southwestern Forests: Proceedings of the Second La Mesa Fire Symposium*, General Technical Report RM-GTR-286, U.S. Forest Service, 1996.

TOP-DOWN ECOLOGY

1. This essay is part of an ongoing exercise in fire journalism. Much of what it contains is not available in the published literature but was acquired by a marvelous field tutorial organized by Peter A. Gordon, fire officer for the Coronado National Forest. I wish to thank Pete, Chris Stetson, Buddy Zale, Toni Strauss, and especially the deeply knowledgeable and quietly passionate Randall Smith for their time and willingness to share their experiences. Any errors of fact or interpretation are of course mine alone.

2. For an excellent summary of the events, see Paul Hirt, "Biopolitics: A Case Study of Political Influence on Forest Management Decisions, Coronado National Forest, Arizona," in *Forests Under Fire: A Century of Ecosystem Mismanagement in the Southwest*, ed. Christopher J. Huggard and Arthur R. Gomez (Tucson: University of Arizona Press, 2001). For a collection of essays on the topic, see Conrad A. Istock and Robert S. Hoffman, eds., *Storm over a Mountain Island: Conservation Biology and the Mt. Graham Affair* (Tucson: University of Arizona Press, 1995).

3. Peter Frost, "Hot Topics and Burning Issues: Fire as a Driver of System Processes—Past, Present, Future" (paper presented at a postgraduate course offered by the C. T. de Wit Graduate School for Production Ecology and Resource Conservation [PE&RC] at Wageningen University, the Global Fire Monitoring Center/Max Planck Institute for Chemistry, and the United Nations University, March 30–April 5, 2008), 2.

SQUARING THE TRIANGLE

1. Acknowledgements must begin with Bil Grauel, who helped set up a visit to San Carlos. Then, so many at San Carlos to thank: Dan Pitterle, Dee Randall, Clark Richins, Kelly Hetzler, Bob Hetzler, Seth Pilsk, and Duane Chapman, all of whom took time out of their busy workday to introduce me to how they see fire on their part of the Earth and in doing so, made Point of Pines a reference point for what the future of fire in the American West might be.

2. Useful background information is available in Dan Pitterle, ed., "San Carlos Apache Indian Reservation Wildland Fire Management Plan," Programmatic Environmental Assessment, EA No. FO-SCA-EA-02–02 (January 2003) and Dan Pitterle, ed., "San Carlos Apache Indian Reservation Wildland Fire Management Plan" (January 2003).

3. Observations on western Apache fire practices from conversation with Seth Pilsk.

4. The best summary of fire from tree rings in recent centuries is Mark Kaib, "Fire History in Mogollon Province Ponderosa Pine Forests of the San Carlos Apache

Tribe, Central Arizona" (master's thesis, University of Arizona, 2001). For contemporary developments, see Kim Kelly et al., "Restoring and Maintaining Resilient Landscapes Through Planning, Education, Support, and Cooperation on the San Carlos Apache Reservation: A Historical, Cultural, and Current View" (May 8, 2013).

5.   Sheridan, Arizona, gives the barebones chronology. For background ethnography, see Richard J. Perry, *Western Apache Heritage: People of the Mountain Corridor* (Austin: University of Texas Press, 1991). A famous inquiry into Western Apache relations with their land is Keith Basso, *Wisdom Sits in Places: Landscape and Language Among the Western Apache* (Albuquerque: University of New Mexico Press, 1996), which raises interesting questions about how these peoples built, and rebuilt, their cultural connection to the lands they were given.

6.   For a synopsis of grazing history, see Harry T. Getty, "Development of the San Carlos Apache Cattle Industry," *Kiva* 3, no. 3 (February 1958): 1–4; and Harry T. Getty, *The San Carlos Indian Cattle Industry*, Anthropological Papers of the University of Arizona, no. 7 (Tucson: University of Arizona, 1963).

7.   On the timber industry, see Jack August, Jr., Art Gomez, and Elmo Richardson, "From Horseback to Helicopter: A History of Forest Management on the San Carlos Apache Reservation" (American Indian Resource Organization, 1984).

8.   Harold Weaver, "Fire as an Ecological Factor in the Southwestern Ponderosa Pine Forests," *Journal of Forestry* (February 1951): 93–98, quote from 95.

9.   Information from Bob Gray, email to author dated August 1, 2014. The heritage of this transitional era has been ignored or forgotten by the cohort that established the modern era.

10. Harold Biswell et al., *Ponderosa Fire Management: A Task Force Evaluation of Controlled Burning in Ponderosa Pine Forests of Central Arizona* (Tallahassee, Fla.: Tall Timbers Research Station, 1973). For a brief chronicle of significant events in fire management, see also Kelly et al., "Restoring and Maintaining."

11. See Pitterle, "San Carlos."

12. Quotes and numbers from William Grauel, email to author on September 11, 2013.

13. Emil W. Haury, *Point of Pines, Arizona: A History of the University of Arizona Archaeological Field School*, Anthropological Papers of the University of Arizona, no. 50 (Tucson: University of Arizona Press, 1989), 15.

14. Haury, *Point of Pines*, 15.

15. Haury, *Point of Pines*, 58, 63.

16. J. F. Lasley, J. T. Montgomery and F. F. McKenzie, "Artificial Insemination in Range Cattle: A Preliminary Report," *Journal of Animal Science* (1940): 102–5 for the technical origins.

     On the square kiva, see Haury, *Point of Pines*, 46.

17. William Grauel, "Long Term Weather, Fire Behavior, and Risk Assessment, Skunk Fire, San Carlos Agency, June 5, 2014," gives the most useful chronology and description of conditions. The basics were supplemented by discussions with Dan Pitterle, Bob Hetzler, and Duane Chapman.

18.   Statistics on the Skunk fire from Bureau of Indian Affairs, San Carlos Agency, "Skunk Fire, Day 48, Incident Action Plan, Thurs, June 5, 2014."

A REFUSAL TO MOURN THE DEATH, BY FIRE, OF A CREW IN YARNELL

1.   In an eerie echo recall how Ahab is first described: a burned, lightning-struck snag, with a "slender rod-like mark, lividly whitish" running down a body that "looked like a man cut away from the stake, when the fire has overrunningly wasted all the limbs without consuming them."
2.   Quotes from Wildland Fire Lessons Learned Center, *Two More Chains*, vol. 3, no. 2 (Summer 2013), 2.

A WORTHY ADVERSARY

1.   Biographical material from Mike Pellant, mostly obtained by a phone interview on September 26, 2016. Other useful documents include his resume; Pellant statement before a Senate subcommittee, at https://www.blm.gov/sites/blm.gov /files/congressional_testimony_documents/congressional_20071011_Regardingthe GreatBasinRestorationInitiative.pdf; and Mike Pellant, "Reflections on 30+ Years of Tackling the Cheatgrass/Wildfire Cycle in the Great Basin," accessed June 11, 2019, http://greatbasinresearch.com/consortium/downloads/Presentations13/early -powerpoint-version-pellant-gbc-plenary-1-14-13.pdf.
2.   Mike Pellant, "History and Applications of the Intermountain Greenstripping Program," in *Proceedings—Ecology and Management of Annual Rangelands* (Boise, Idaho: Intermountain Research Station, 1994): 63–68.
3.   Mike Pellant, "Cheatgrass: The Invader That Won the West" (presentation, Interior Columbia Basin Ecosystem Management Project, 1996).
4.   Pellant, "Cheatgrass," 13–14.
5.   Statistics from *Out of Ashes, an Opportunity* (Boise, Idaho: National Office of Fire and Aviation, Bureau of Land Management, 1999), 6.

DEEP FIRE

1.   In keeping with its 10,000-year horizon, the Long Now Foundation uses a five-digit timeline. It seemed appropriate to keep that convention for this essay.
2.   Information on the Phillips Ranch fire come from Alexander Rose of the Long Now Foundation in an email dated July 25, 2016, which also included emails between Stewart Brand and Rebecca Mills, superintendent of Great Basin National Park, and several press releases.
3.   I'd like to thank Matt Martin for help in setting up a site visit, although we could not coordinate our calendars, and Matt Johnson for sending me copies of the park's fire management plan, fire atlas, and fire statistics.

4. Brand quote from Long Now Foundation website at http://longnow.org/clock /nevada/, accessed July 25, 2016.

5. Many accounts of these changes are in print, at various levels of detail. A useful thumbnail is available in Donald K. Grayson, "Great Basin Natural History," in Catherine S. Fowler and Don D. Fowler, eds., *The Great Basin: People and Place in Ancient Times* (Santa Fe, N.Mex.: School for Advanced Research Press, 2008): 7–17.

6. The standard history of the range, cave, and park is Darwin Lambert, *Great Basin Drama: The Story of a National Park* (Niwot, Colo.: Rinehart, 1991), which is especially good on the politics.

7. Stanley G. Kitchen, "Historical Fire Regime and Forest Variability on Two Eastern Great Basin Fire-Sheds (USA)," *Forest Ecology and Management* 285 (2012): 53–66, quoted line from 53.

8. Emily K. Heyerdahl, et al., "Multicentury Fire and Forest Histories at 19 Sites in Utah and Eastern Nevada," General Technical Report RMRS-GTR-261WWW, U.S. Forest Service, 2011; and George E. Gruell, "Historical and Modern Roles of Fire in Pinyon-Juniper," in Stephen Monsen and Richard Stevens, comps., *Proceedings: Ecology and Management of Pinyon-Juniper Communities Within the Interior West*, Proceedings RMRS-P-9 (Ogden, Utah: U.S. Forest Service, 1999): 24–28.

9. Gruell, "Historical and Modern Roles," 26. On the Strawberry fire: http://inciweb .nwcg.gov/incident/4947/, accessed September 2, 2016. On general park statistics, see Great Basin National Park, Fire Management Plan, n.d., 12–13.

10. A delightful popular introduction is Ronald M. Lanner, *The Bristlecone Book: A Natural History of the World's Oldest Trees* (Missoula, Mont.: Mountain Press, 2007). A meditation, particularly useful in teasing out the history of progressive understandings, is Michael P. Cohen, *A Garden of Bristlecones: Tales of Change in the Great Basin* (Reno: University of Nevada Press, 1998). For basic fire ecology, see Fire Effects Information System at https://www.feis-crs.org/feis/. Also tantalizing is a poster at a fire ecology conference by Mackenzie Kilpatrick and Franco Biondi, "Bristlecone Fire History and Stand Dynamics at Mount Washington, Nevada," which looked at the aftermath of the 2000 fire.

11. On the clock see the Long Now Foundation's wonderful website at http://longnow .org/essays/time-10000-year-clock/, accessed August 1, 2016.

12. I follow Cohen's matrix of interpretations in *Garden of Bristlecones*, 61–77. The number of conflicting accounts is remarkable. Currey quoted from "An Ancient Bristlecone Pine Stand in Eastern Nevada," *Ecology* 46 (1965): 566. Contrast Currey's bland, neutral-voice article with Lambert's in *Great Basin Drama*, 142–47, which includes photos of the tree before felling (and Currey climbing it).

STRIP TRIP

1. Clarence Dutton, *Tertiary History of the Grand Cañon District*, U.S. Geological Survey Monograph 2 (Washington, D.C.: Government Printing Office, 1882), 78–79.

2. A brief survey is available at the BLM Arizona Strip website: http://www.blm.gov /az/st/en/prog/blm_special_areas/natmon/gcp/cultural/ranching.html, accessed June 6, 2016. An unpublished history compiled by Phil Foremaster, "History, Geology, and Ranching on the Arizona Strip North of the Colorado," provides considerable details up through 1934 (see website of the Washington County Historical Society, accessed June 11, 2019, http://wchsutah.org/az-strip/history -geology-ranching.pdf).

3. Bob Davis, "The History of the Mt. Trumbull Forest (As I Know It)," unpublished report on file at BLM Arizona Strip District Office.

4. Tree density numbers from Ken Moore, Bob Davis, and Timothy Duck, "Mt. Trumbull Ponderosa Pine Ecosystem Restoration Project," in *Visitor Use Density and Wilderness Experience: Proceedings*, RMRS-P-20 (Missoula, Mont.: U.S. Forest Service, 2003): 120.

5. A master bibliography of published and unpublished reports is available on the project website, accessed June 11, 2019, https://cdm17192.contentdm.oclc.org /digital/search/searchterm/mount%20trumbull.

    For a textual synthesis see David W. Huffman et al., "Ecosystem Restoration, Assistance Agreement No. PAA 017002, Task Order No. AAW040001, Final Report, February 5, 2006." Excellent digests can be found in Moore, Davis, and Duck, "Mt. Trumbull Ponderosa Pine," and Thomas Heinlein et al., "Changes in Ponderosa Pine Forests of the Mt. Trumbull Wilderness," unpublished report to BLM (November 1999). The best summary, and by far the most readable, is Peter Friederici, "Healing the Region of Pines: Forest Restoration in Arizona's Uinkaret Mountains," in Peter Friederici, ed., *Ecological Restoration of Southwestern Ponderosa Pine Forests* (Washington, D.C.: Island Press, 2003): 197–214.

6. For a good synopsis see "Restoring the Uinkaret Mountains: Operational Lessons and Adaptive Management Practices," Working Papers in Southwestern Ponderosa Pine Forest Restoration 1, Ecological Restoration Institute (n.d.).

7. Friederici, "Healing the Region of Pines," 204–5, 213.

8. Dutton, *Tertiary History*, 92. Holmes's atlas art can be viewed at https://www .loc.gov/resource/g4332gm.gnp00002/?st=gallery. See sheet VI for Toroweap and sheets XV–XVII for Point Sublime.

9. Dutton, *Tertiary History*, 89.

MESA NEGRA

1. I want to thank Steven Underwood for helping to set up a site visit, and Keith Krause and the Mesa Verde engine crew for a delightful and informative study tour. I think it fair to say that how they see fire at Mesa Verde and how I see it differ, but then people have been seeing fire in various ways for a very long time on the cuesta.

2. Mesa Verde is unusually well endowed with fire history records. The best source is an unpublished report in the files of the Fire Management Office, Mesa Verde

Fire History, March 2002, which gives detailed accounts of fires, a bibliography of published and unpublished sources, and even archival references. A popular version was published as part of the Mesa Verde Centennial Series: Tracey L. Chavis and William R. Morris, *Fire on the Mesa* (Durango, Colo.: Durango Herald Small Press, 2006).

3. Background fire ecology and fire history research is best distilled in M. Lisa Floyd, William H. Romme, and David D. Hanna, "Fire History and Vegetation Pattern in Mesa Verde National Park, Colorado, USA," *Ecological Applications* 10, no. 6 (2000): 1666–80, and a chapter by the same authors, "Fire History," in M. Lisa Floyd, ed., *Ancient Piñon-Juniper Woodlands: A Natural History of Mesa Verde Country* (Boulder: University Press of Colorado, 2003), 261–78. For criticisms of prescribed fire and mechanical treatments, see chapters by William H. Romme, Sylvia Oliva, and M. Lisa Floyd, "Threats to the Piñon-Juniper Woodlands," 339–61, and George L. San Miguel, "Epilogue: Management Considerations for Conserving Old-Growth Piñon-Juniper Woodlands," 361–74, in the same volume.

4. Conversations with Keith Krause, assistant fire management officer, and Mesa Verde Fire Management Plan.

5. On fire impacts, see Julie Bell, "Fire and Archaeology on Mesa Verde," in David Grant Noble, ed., *The Mesa Verde World: Explorations in Ancestral Pueblo Archaeology* (Santa Fe, N.Mex.: School of American Research Press, 2006), 119–21. See also Chavis and Morris, *Fire on the Mesa*, 99–100.

6. Noble, *Mesa Verde World*, 6.

7. On the fires of abandonment, see Noble, *Mesa Verde World*, 22, 26, 131, 129, 143. The savage depopulation is well documented throughout the New World. An interesting effort to reconcile that unloading with fire history is Matt Liebmann et al., "Native American Depopulation, Reforestation, and Fire Regimes in the Southwest United States, 1492–1900 CE," *PNAS* 113, no. 6 (2016), published online at http://www.pnas.org/cgi/doi/10.1073/pnas.1521744113.

FATAL FIRES, HIDDEN HISTORIES

1. Blackwater fire quote from David P. Godwin, "The Handling of the Blackwater Fire," *Fire Control Notes* 1, no. 7 (December 1938): 381.

2. See Glenn B. Stracher et al., "The South Cañon Number 1 Coal Mine Fire: Glenwood Springs, Colorado," *GSA Field Guides*, 2004, vol. 5, 143–50, https://doi.org/10.1130/0-8137-0005-1.143; and Sallee Ann Ruibal, "Coal Seam Fire Memories Still Burning a Decade Later," *Post Independent*, June 8, 2012, http://www.post independent.com/news/coal-seam-fire-memories-still-burning-a-decade-later/, and for a later update, John Stroud, "State Works to Control Burning Coal Seam in South Canyon," *Post Independent*, June 18, 2017, http://www.postindependent .com/news/local/state-works-to-control-burning-coal-seam-in-south-canyon/.

3. The technical report (and it is technical) is available at Howard A. Tewes, *Survey of Gas Quality Results from Three Gas-Well-Stimulation Experiments by Nuclear*

Explosions, Lawrence Livermore Laboratory UCRL-52656, January 23, 1979. For a gentler summary of both fires, see Ken Lammey, "The 1976 Battlement Creek Fire—A Historical Perspective," Battlement Mesa Service Association, July 2012.

4. The fire has been the subject of a Wildland Fire Staff Ride, so a rich cache of documentation is available for it. See http://www.fireleadership.gov/toolbox/staff ride/library_staff_ride10.html.

A SONG OF ICE AND FIRE—AND ICE

1. ICE as an acronym comes from Alfred Crosby, *Children of the Sun: A History of Humanity's Unappeasable Appetite for Energy* (New York: Norton, 2006).

2. For a tidy summary, see William A. Patterson III and Kenneth E. Sassaman, "Indian Fires in the Prehistory of New England," in *Holocene Human Ecology in Northeastern North America*, ed. George P. Nicholas (New York: Plenum Press, 1988), 107–36. Also useful are Carl Sauer's historical surveys, though New England is a minor constituent: *Sixteenth Century North America: The Land and People as Seen by Europeans* (Berkeley: University of California Press, 1971) and *Seventeenth Century North America: French and Spanish Accounts* (Berkeley, Calif.: Turtle Island Foundation, 1977).

3. William Wood, *New England's Prospects* (1634; repr., Boston: Prince Society, 1865), 17.

4. Thomas Morton, *New English Canaan* (1637; repr., New York: Arno Press, 1972), 52–53.

5. Morton, 53–54; Van der Donck quoted in Gordon M. Day, "The Indian as an Ecological Factor in the Northeastern Forest," *Ecology* 34, no. 2 (1953): 337.

6. Peter Kalm, *Travels into North America*, trans. John Reinhold Forster (1770–71; repr., Barre, Mass.: Imprint Society, 1972), 210, 361.

7. Timothy Dwight, *Travels in New England and New York*, ed. Barbara Miller Solomon (Cambridge, Mass.: Harvard University Press, 1969), 4:37–39.

8. My Thoreau observations derive from David Foster's very congenial and helpful *Thoreau's Country: Journey Through a Transformed Landscape* (Cambridge, Mass.: Harvard University Press, 1999); block quote from page 10.

9. Quote from Foster, 109.

10. Quotes from Foster, 116–17.

11. Quotes from Foster, 120–21.

12. Figures from Mark B. Lapping, "Toward a Working Rural Landscape," in *New England Prospects: Critical Choices in a Time of Change*, ed. Carl H. Reidel (Lebanon, N.H.: University Press of New England, 1982), 61.

13. Charles S. Sargent, "The Protection of Forests," *North American Review* 135 (1882): 397, 392; Charles S. Sargent, "Protection of the Forests of the Commonwealth," Massachusetts Horticultural Society, February 13, 1880, published note from secretary.

14. Statistics from Lloyd C. Irland, "Massachusetts Fire History Working Paper: Revised Draft, May 24, 2012," Irland, "Vermont's Fire History 1905–2011: Initial Observations, with Analysis of Individual Fire Data, 1977–2011," and Irland, "Fire

History of New York State, 1920–2010: Working Paper," all available through the North Atlantic Fire Science Exchange website: http://www.firesciencenorth atlantic.org/search-results/?category=Technical+Report. See also Julie A. Richburg and William A. Patterson III, "Fire History of the White and Green Mountain National Forests: A Report Submitted to the White Mountain National Forest, USDA Forest Service, January 31, 2000." The site is a mother lode of regional fire information, with histories for other member states as well.

15. Lloyd C. Irland, "The Northeast's Great Sixties Drought: The Fire Outbreak," Northeastern Forest Fire Compact, draft work paper, at the North Atlantic Fire Science Exchange website. The 1947 fires have been the subject of numerous articles, three documentary films, and one popular book, Joyce Butler, *Wildfire Loose: The Week Maine Burned* (Kennebunkport, Maine: Durrell Publications, n.d.).

16. David B. Kittredge, "The Fire in the East," *Journal of Forestry*, April/May 2009, 162–63.

ALBANY PINE BUSH

1. A weird footnote to the pine bush's signature species, the Karner blue butterfly, is that it was first identified by Vladimir Nabokov, the novelist best known for *Lolita* and *Pale Fire*.

2. I was the beneficiary of a marvelous tutorial and field trip organized by Neil Gifford and Tyler Briggs and included Christopher Hawver, Zak Handley, and Steve Jackson. My thanks to all of them. They were especially helpful in tracing out the operational complexities of burning within the crazy-quilt constraints of APB.

3. I've relied on two fundamental references. One is Jeffrey K. Barnes, *Natural History of the Albany Pine Bush: Albany and Schenectady Counties, New York*, New York State Museum Bulletin 502 (New York: New York State Museum, 2003). The other, very rich for its fire theme, is the TNC report: Robert E. Zaremba, David M. Hunt, Amy N. Lester, *Albany Pine Bush Fire Management Plan: Report to the Albany Pine Bush Commission* (New York: New York Field Office, the Nature Conservancy, 1991). I thank Stephanie B. Gebauer for sending me a copy over 20 years ago.

4. Both quotes from Barnes, *Natural History*, 21–22.

5. See Barnes, *Natural History*, 22–24, and G. Motzkin, W. A. Patterson III, and D. R. Foster, "A Historical Perspective on Pitch Pine-Scrub Oak Communities in the Connecticut Valley of Massachusetts," *Ecosystems* 2, no. 3 (May–June 1999): 255–73.

6. Barnes, *Natural History*, 24.

THE WUI WITHIN

1. For their marvelous tutorial I would like to thank the entire staff of the NFPA's Wildland Fire Operations Division, especially Michele Steinberg and Molly Mowery, along with Sue Marsh and Jessica Broady of the library and archives, and James Shannon, president and CEO.

THE VIEW FROM BILL PATTERSON'S STUDY

1. Unless otherwise indicated, information comes from an interview with Bill Patterson on April 30, 2017.
2. H. Ali Crolius, "Prometheus's Gift," *UMass Magazine*, Fall 1988, 3, http://www.umass.edu/umassmag/archives/1998/fall_98/fall98_f_fire.html.
3. Email from William A. Patterson III to Stephen Pyne, April 1, 2017.
4. Email from William A. Patterson III to Stephen Pyne, April 26, 2017.
5. See Motzkin, Patterson, and Foster, "A Historical Perspective."
6. Crolius, "Prometheus's Gift," 3.

WESTWARD, THE COURSE OF THE EMPIRE

1. Ernest Hemingway, "The Big Two-Hearted River," in *The Short Stories of Ernest Hemingway* (New York, Scribner, 1953), 209.
2. Hemingway, 212, 211.
3. Hemingway, 232.
4. The best overview remains Susan Flader, ed., *The Great Lakes Forest: An Environmental and Social History* (Minneapolis: University of Minnesota Press, 1983).
5. Quotes from Stephen J. Pyne, *Fire in America* (Princeton, N.J.: Princeton University Press, 1982), 201, 210–11.
6. Among the standard accounts see Rev. Peter Pernin, "The Great Peshtigo Fire," *Wisconsin Magazine of History* 54 (1971): 246–72; Denise Gass and William Lutz, *Firestorm at Peshtigo: A Town, Its People, and the Deadliest Fire in American History* (New York: Henry Holt, 2002); Lawrence H. Larsen, *Wall of Flames: The Minnesota Forest Fire of 1894* (Fargo: North Dakota Institute for Regional Studies, North Dakota State University, 1984); Grace Stageberg Swenson, *From the Ashes: The Story of the Hinckley Fire of 1894* (Stillwater, Minn.: Croixside Press, 1979); and Francis M. Carroll and Franklin R. Raiter, *The Fires of Autumn: The Cloquet-Moose Lake Fire of 1918* (St. Paul: Minnesota Historical Society Press, 1990). Among the new contenders is Daniel James Brown, *Under a Flaming Sky: The Great Hinckley Firestorm of 1894* (Lanham, Md.: Rowman and Littlefield, 2009).
7. Jim Harrison, *True North* (New York: Grove/Atlantic: 2005), 23.
8. A classic account of how the logging was done is available in Agnes M. Larson, *The White Pine Industry in Minnesota: A History* (Minneapolis: University of Minnesota Press, 1949).
9. A good survey of state efforts is Raleigh Barlowe, "Changing Land Use and Policies: The Lake States," in *The Great Lakes Forest*, ed. Susan Flader (Minneapolis: University of Minnesota Press and Forest History Society, 1983), 156–76.
10. The sad story is nicely told in Curt Meine, *Aldo Leopold: His Life and Work* (Madison: University of Wisconsin Press, 1988), 518–20.
11. Rodney W. Sando and Donald A. Haines, *Fire Behavior of the Little Sioux Fire*, Research Paper NC-76, U.S. Forest Service, 1972, and Albert J. Simard et al., *The Mack Lake Fire*, General Technical Report NC-83, U.S. Forest Service, 1983.

12. See Seney National Wildlife Refuge, History: https://www.fws.gov/refuge/seney /what_we_do/history.html.

13. I rely on my account in *Between Two Fires: A Fire History of Contemporary America* (Tucson: University of Arizona Press, 2015), 152–53.

14. Several studies exist. A useful summary is available in Dawn S. Marsh, "The Evolution of Land and Fire Management at Seney National Wildlife Refuge: From Game to Ecosystem Management," report for Fish and Wildlife Service, December 2013.

MISSOURI COMPROMISE

1. This essay is a species of interpretive journalism that resulted from a two-day field trip to the Missouri Ozarks organized by Rich Guyette, Dan Dey, and Mike Stambaugh, as a prelude for a daylong workshop on human fire history at UM–Columbia. For some years I have followed the fascinating fire-history articles the UM Tree-Ring Lab group had published and leaped at the chance to see them and their sites in person. Others joined in: Tim Nigh, Susan Flader, Dan Drees, and Rose-Marie Muzika. To their research I have tried to provide a larger historic and philosophical context. The data is theirs. The refractive prism is mine. I also thank Mike Dubrasich for a gentle editing of a rough-pixelated manuscript.

2. Milton D. Rafferty, *Rude Pursuits and Rugged Peaks: Schoolcraft's Ozark Journal 1818–1819* (Fayetteville: University of Arkansas Press, 1996), 62–63.

3. The classic introduction remains Carl O. Sauer, *The Geography of the Ozark Highland of Missouri* (New York: Greenwood Press, 1968).

4. See, for example, Curtis Marbut, "The whole region and its vegetation was more closely allied to the western prairies than to the timber-covered Appalachians." Quoted in Tim A. Nigh, "Missouri's Forest Resources—An Ecological Perspective," in *Toward Sustainability for Missouri Forests: Proceedings of a Conference*, ed. Susan L. Flader, General Technical Report NC-239, U.S. Forest Service, 1999.

5. See Michael J. O'Brien and W. Raymond Wood, *The Prehistory of Missouri* (Columbia: University of Missouri Press, 1998), 295–96, 331–33.

6. Sauer, *Geography of the Ozark Highland*, 52–54. Quotes on Indian burning from Marbut come from Nigh, "Missouri's Forest Resources," 11.

7. Leopold quote from Susan L. Flader, "History of Missouri Forests and Forest Conservation," in Flader, *Toward Sustainability for Missouri Forests*, 20.

8. Sauer, *Geography of the Ozark Highland*, 207, 230–33, 237.

9. See E. R. McMurry et al., "Initial Effects of Prescribed Burning and Thinning on Plant Communities in the Southeast Missouri Ozarks," *Proceedings of the 15th Central Hardwood Forest Conference*, U.S. Forest Service e-GTR-SRS-101 (2006): 241. The most comprehensive summary of contemporary fire statistics is Steve Westin, "Wildfire in Missouri" (Jefferson City: Missouri Department of Conservation, 1992).

10. Details of conservation history from Flader, "History of Missouri Forests and Forest Conservation."

11. See Susan Flader, "Missouri's Pioneer in Sustainable Forestry," *Forest History Today*, Spring/Fall 2004, 2–15.

12. The UM-Columbia group under Richard Guyette has produced an ever-lengthening literature on these topics. Perhaps the central paper is R. P. Guyette, R. M. Muzika, and D. C. Dey, "Dynamics of an Anthropogenic Fire Regime," *Ecosystems* 5, 2000, 472–86. I take considerable liberties in extrapolating their concepts into a more general critique of fire scholarship.

FIRE AND AXE

1. Gifford Pinchot, *The Fight for Conservation* (New York: Doubleday, 1910), 15.

2. On the early alliance between industry and the state, see George T. Morgan, "The Fight Against Fire: The Development of Cooperative Forestry in the Pacific Northwest, 1900–1950" (PhD diss., University of Oregon, 1964).

3. For a summary, see Pyne, *Fire in America*, 338.

4. Logan A. Norris, "An Overview and Synthesis of Knowledge Concerning Natural and Prescribed Fire in Pacific Northwest Forests," in *Natural and Prescribed Fire in Pacific Northwest Forests*, John D. Walstad, Steven R. Radosevich, and David V. Sandberg, eds. (Corvallis: Oregon State University Press, 1990): 7. Data taken from J. K. Agee, "The Historical Role of Fire in Pacific Northwest Forests," same volume, 37.

5. A nice summary of the fire is available in *The Oregon Encyclopedia*, "Biscuit Fire of 2002," at https://oregonencyclopedia.org/articles/biscuit_fire_of_2002/. The best review of the controversies is the GAO, *Biscuit Fire: Analysis of Fire Response, Resource Availability, and Personnel Certification Standards*, GAO-04–426, April 2004.

6. As of this writing, an assessment of the Chetco Bar fire is still underway. For the basics see Inciweb (https://inciweb.nwcg.gov/incident/5385/) and the Chetco Bar timeline published by the Forest Service (https://usfs.maps.arcgis.com/apps /Cascade/index.html?appid=809cc1882e8d45169b9baf2669f95c5a).

AN ECOLOGICAL AND SILVICULTURAL TOOL

1. Harold Weaver, "Fire and Its Relationship to Ponderosa Pine," in *Proceedings: 7th Tall Timbers Fire Ecology Conference* (Tallahassee, Fla.: Tall Timbers Research Station, 1967), 127. I wish to thank Sonja Pyne for her help in locating newspaper references to Weaver in Oregon and Arizona.

2. Weaver, 127–28.

3. Weaver, 128.

4. Harold Weaver, "Ecological Changes in the Ponderosa Pine Forest of the Warm Springs Indian Reservation in Oregon," *Journal of Forestry* 57 (1959): 20.

5. Weaver, 7n.

6.   Harold Weaver, "Fire as an Ecological and Silvicultural Factor in the Ponderosa Pine Region of the Pacific Slope," *Journal of Forestry* 41 (1943): 14–15.

7.   Weaver, 15.

8.   Weaver, 7.

9.   Ben Avery, "Areas Burned to Cut Hazard, Help Growth," *Arizona Republic*, June 18, 1950, 13; "Forester Gets New Position in Washington," *Arizona Republic*, May 30, 1951, 11.

10. Harold H. Biswell, et al., *Ponderosa Pine Management: A Task Force Evaluation of Controlled Burning in Ponderosa Pine Forests of Central Arizona*, Miscellaneous Publication 2 (Tallahassee, Fla.: Tall Timbers Research Station, 1973), 1–2.

KENAI

1.   I want to thank Kristi Bulock for setting a wonderful primer on Kenai fire. Others who contributed to the session, and beyond, are Andy Loranger, Doug Newbould, John Morton, Ed Berg, and Mike Hill. Randi Jandt also contributed some helpful observations. For a bibliography of scientific publications about the refuge, see https://www.fws.gov/refuge/Kenai/what_we_do/science/bibliography.html.

2.   Fire history from E. E. Berg and R. S. Anderson, "Fire History of White and Lutz Spruce Forests on the Kenai Peninsula, Alaska, Over the Last Two Millennia as Determined from Soil Charcoal," *Forest Ecology and Management* 227, no. 3 (June 2006): 275–83; Starker Leopold and F. Fraser Darling, *Wildlife in Alaska: An Ecological Reconnaissance* (New York: Ronald Press, 1953), 58. For a useful summary, see "Fire Ecology and Regime Shift Due to Climate Change," Kenai National Wildlife Refuge, U.S. Fish and Wildlife Service, last updated September 26, 2012, https://www.fws.gov/refuge/Kenai/what_we_do/science/fire_ecology.html. On background fire information generally, see *Alaska Interagency Fire Management Plan: Kenai Peninsula Planning Area* (April 1984).

3.   Leopold and Darling, *Wildlife in Alaska*, 58. The ubiquitous H. J. Lutz also investigated the question of moose history; see *History of the Early Occurrence of Moose on the Kenai Peninsula and in Other Sections of Alaska*, Alaska Forest Research Center, Miscellaneous Publication No. 1, June 1960.

4.   See C. W. Slaughter, Richard J. Barney, and G. M. Hansen, eds., *Fire in the Northern Environment—A Symposium* (Portland, Ore.: Pacific Northwest Forest and Range Experiment Station, U.S. Forest Service, 1971).

5.   The bark beetle infestation has claimed the most national attention; for a journalistic account, see Andrew Nikiforuk, *Empire of the Beetle: How Human Folly and a Tiny Bug Are Killing North America's Great Forests* (Vancouver: Greystone Books, 2011), 4–30. A succinct summary of Project Skyfire is in Kristine C. Harper, *Make It Rain: State Control of the Atmosphere in Twentieth-Century America* (Chicago: University of Chicago Press, 2018), 168–75.

     The drying peat has received special attention; see E. E. Berg et al., "Recent Woody Invasion of Wetland on the Kenai Peninsula Lowlands, South-Central

Alaska: A Major Regime Shift After 18,000 Years of Wet Sphagnum-Sedge Peat Recruitment," *Canadian Journal of Forest Research* 39 (2009): 2033–46.

6. Benjamin M. Jones, et al., "Presence of Rapidly Degrading Permafrost Plateaus in South-Central Alaska," *Crysophere* 10, no. 6 (November 2016): 2673–92.

7. Conversations with Kenai staff, May 23, 2017; *Card Street Fire Fuels Treatment Effectiveness*, publication of Kenai NWR. John Morgan provided some extra details on the Caribou Hills and Card Street fires.

NORTH TO THE FUTURE

1. Figures from Stephen Haycox, *Battleground Alaska: Fighting Federal Power in America's Last Wilderness* (Lawrence: University Press of Kansas, 2016), 36.

2. See, for example, John Imbrie and Katherine Palmer Imbrie, *Ice Ages: Solving the Mystery* (Cambridge, Mass.: Harvard University Press, 1986).

3. See William Ruddiman, *Plows, Plagues, and Petroleum* (Princeton, N.J.: Princeton University Press, 2005).

CREDITS

"Coming to a Forest near You" was first published as "Box and Burn: The New Reality of Western Fire," *Slate* (July 4, 2014).

"Portal to the Pyrocene" was first published as "Welcome to the Pyrocene," *Slate* (May 16, 2016).

"Our Coming Fire Age: A Prolegomenon" was first published as "Fire Age," *Aeon* (May 5, 2015); reproduced with permission.

"Thinking Like a Burned Mountain" was first published as "Burning Like a Mountain," *Aeon* (January 2014); reproduced with permission.

"A Refusal to Mourn the Death, by Fire, of a Crew in Yarnell" was first published in *Pacific Standard* (November 12, 2013); reproduced with permission.

"Squaring the Triangle" was first published as "Squaring the Triangle at San Carlos," Wildland Fire Lessons Learned Center (August 1, 2014).

Quotations from Gary Snyder, "Wildfire News," reproduced with permission.

Quotations from Dylan Thomas, "A Refusal to Mourn the Death, by Fire, of a Child in London," reproduced with permission.

INDEX

ABOUT THE AUTHOR

Stephen J. Pyne is an emeritus professor at Arizona State University and a former North Rim Longshot. Among his many books are *Between Two Fires: A Fire History of Contemporary America* and To the Last Smoke, a series of regional fire surveys. He lives in Queen Creek, Arizona.